TERRITORIES, BOUNDARIES
AND CONSCIOUSNESS

BELHAVEN STUDIES IN
POLITICAL GEOGRAPHY

EDITED BY

PETER J. TAYLOR, UNIVERSITY OF NEWCASTLE UPON TYNE
JOHN O'LOUGHLIN, UNIVERSITY OF COLORADO, BOULDER

PUBLISHED

JAN NIJMAN *THE GEOPOLITICS OF POWER AND CONFLICT*
TOM NIEROP *SYSTEMS AND REGIONS IN GLOBAL POLITICS*
ANSSI PAASI *TERRITORIES, BOUNDARIES AND CONSCIOUSNESS*

TERRITORIES, BOUNDARIES AND CONSCIOUSNESS

The Changing Geographies of the
Finnish–Russian Border

ANSSI PAASI
University of Oulu, Finland

Foreword by W.R. Mead
Editor's Introduction by P.J. Taylor

JOHN WILEY & SONS
Chichester • New York • Brisbane • Toronto • Singapore

Other Wiley Editorial Offices

John Wiley & Sons, Inc., 605 Third Avenue,
New York, NY 10158-0012, USA

Jacaranda Wiley Ltd, 33 Park Road, Milton,
Queensland 4064, Australia

John Wiley & Sons (Canada) Ltd, 22 Worcester Road,
Rexdale, Ontario M9W 1L1, Canada

John Wiley & Sons (SEA) Pte Ltd, 37 Jalan Pemimpin #05-04,
Block B, Union Industrial Building, Singapore 2057

Library of Congress Cataloging-in-Publication Data

Paasi, Anssi.
    Territories, boundaries, and consciousness : the changing
geographies of the Finnish–Russian border / Anssi Paasi ;
foreword by W.R. Mead.
        p.   cm. — (Studies in political geography)
    Includes bibliographical references and index.
    ISBN 0-471-96119-1
    1. Political geography. 2. Nationalism. 3. Finland—Historical
geography. 4. Russia—Historical geography. I. Title. II. Series.
JC319.P32   1995
320.1'2—dc20   95-31384
                                                    CIP

British Library Cataloguing in Publication Data

A catalogue record for this book is available from the British Library

ISBN 0-471-96119-1

Typeset in 10/12pt Sabon from author's disk by Mayhew Typesetting, Rhayader, Powys
Printed and bound by Antony Rowe Ltd, Eastbourne
This book is printed on acid-free paper responsibly manufactured from sustainable
forestation, for which at least two trees are planted for each one used for paper production.

For

Eija, Ossi and Otso

Outside and inside form a dialectic of division, the obvious geometry of which blinds us as soon as we bring it into play in metaphorical domains. It has the sharpness of the dialectics of *yes* and *no*, which decides everything. Unless one is careful, it is made into a basis of images that govern all thoughts of positive and negative. . . The dialectics of *here* and *there* has been promoted to the rank of an absolutism according to which these unfortunate adverbs of place are endowed with unsupervised powers of ontological determination. (*Bachelard 1969, 211–12*)

. . . this universal practice of designating in one's mind a familiar space which is 'ours' and an unfamiliar space beyond 'ours' which is 'theirs' is a way of making geographical distinctions that *can be* entirely arbitrary. I use the word 'arbitrary' here because imaginative geography of the 'our land–barbarian land' variety does not require that the barbarians acknowledge the distinction. It is enough for 'us' to set up these boundaries in our own minds; 'they' become 'they' accordingly, and both their territory and their mentality are designated as different from 'ours'. (*Said 1978, 54*)

The state, however, differs from other social structures not only in its claims to sovereignty but also in the fact that its organization is territorial. Survival for such a unit means preserving political independence and retaining control over a specific territory whose limits are defined by an imaginary line called a 'boundary'. . . . The boundary is thus not only a line of demarcation between legal systems but also a point of contact of territorial power structures. (*Spykman 1942, 436–7*)

Countries are separated from each other by boundaries. Crossing a boundary always touches one's mind: the imagined outline takes the material form of a wooden barrier, which is never located in the place which it is thought to represent, but some tens or hundreds of metres to one side of it or the other. Yet it can change everything, even the landscape. The air is the same, the soil is the same but the road is no longer entirely the same, the graphics on the signposts have changed, the bakeries are not the same as those of a moment ago, the loaves are a different shape, the cigarette packets that have been thrown on the ground are different. . . Boundaries are lines, lines for which millions of people have died. Thousands of people have died because they did not succeed in crossing a boundary. Survival depended merely on passing some ordinary river, small hill, peaceful woodland. (*Perec 1992, 86–7*)

CONTENTS

FOREWORD

Political boundaries remain one of the most important features in human geography. As subjects for detailed geographical research they have been curiously neglected. Generations of cartographers have marked them as red lines on the map. On the ground they consist of frontier zones of varying width, perhaps to be ultimately reduced to a few strands or coils of barbed wire. Such is much of Finland's boundary with Russia — a 1269 kilometre share of what used to be called the 'iron curtain'.

Political boundaries are usually the consequence of international compromises following armed conflict. In the Nordic countries, Sweden's boundary with Norway, passing through thinly peopled terrain, was marked out on the ground in the 1740s. It has been one of the most stable in Europe. Denmark's boundary in Slesvig (Schleswig), arbitrarily defined by Prussia in 1864, was adjusted through ethnographically mixed territory following a plebiscite after the Versailles Peace Treaty. Norway's boundary with Russia, including that of the former Finnish Petsamo, was confirmed by commissioners and surveyors after World War II. Following the cessation of Russo-Finnish hostilities in 1944, a new border had to be staked out in eastern Finland in accordance with the line drawn on the map by Stalin.

Studies of the territorial and human consequences of the decisions made by statesmen on the map are relatively few. So too are investigations into the experiences of those whose lives and livelihoods have had to be adjusted to new, arbitrarily imposed frontiers. The problems are the greater when the frontiers are closed and remain so for decades. These are some of the issues which are central to Anssi Paasi's investigations.

No country in Europe has changed its geographical outline more often than Finland and, given this situation, it is possible that none has succeeded more effectively in retaining its distinctiveness. Indeed, apart

from the relatively peaceful post World War I settlements of its eastern
boundary in Karelia and western boundary through the Åland Sea, the
entire history of its changing eastern frontier has been the product of
armed conflict.

The study of geography, not least political geography and occasionally politicized geography, has a well-established history in Finland.
Geographical records began in part with the map. Finland's legacy of
cadastral mapping and of military survey is one with that of Sweden of
which it was a part until 1809. The first formal geography of Finland
was published by the Rector of Åbo University in 1795. Zacharias
Topelius, first professor of Finnish history in the Imperial Alexander
University of Helsingfors, gave his inaugural lecture on Finland's
geographical position in 1851. Thereafter, he introduced all of his
history courses with a series of lectures on the geography of the Grand
Duchy. In the schoolroom, geography entered the syllabus early, albeit
as accessory to history. In 1888 the Finnish Geographical Society was
founded and the first volume of its research series *Fennia* appeared soon
after. A docent in geography was appointed in the University of
Helsinki in 1890 and the department of geography was established in
1893. As early as 1881, there had been Finnish representation at the
International Geographical Union meeting in Venice, but it was in
London in 1895 that a formal delegation of Finnish geographers made
its impact. By then, mapping had begun to acquire political overtones
and a statement intended to put Finland as a distinct entity on the map
took shape as the world's first national atlas. The *Atlas of Finland* was
published in 1899. The fifth edition of this, the most detailed national
atlas, was completed in 1993. The research monographs of *Fennia* will
soon number 200 and the journal *Terra* appears quarterly. In all of
these academic publications, the English language is widely employed.
Geographers such as Anssi Paasi therefore have a long and strong
geographical tradition behind them.

Finland has geographical institutes at the Universities of Helsinki,
Turku, Oulu and Joensuu as well as in a number of other institutions of
university standing. It is natural that the research of Finnish geographers should focus on their own country, but they have not ceased to
make themselves familiar with developments in the subject worldwide.
The theoretical framework of Anssi Paasi's study reflects the international aspect of his work. At the national level he has added
unfamiliar to familiar data covering the frontier itself. At the local level,
his fieldwork along the frontier zone is of the essence of the subject. It is
not easy for older Finns to view the 1944 boundary objectively. It
reduced the country's territory by a tenth and led to the displacement of
a full tenth of the population. Furthermore, with the migration of all
Finns from the ceded territory and its occupation by Russians, a new

ethnographic boundary was created. Three generations of Finns now live together alongside the frontier zone. Their experiences of it and the perceptions that they have of it differ with the passage of time. For half a century it has been a closed boundary, with only one main crossing point. Relaxation along the boundary with the new political situation in Russia presages new problems. Meanwhile, Anssi Paasi has availed himself of a cross-border epilogue to his work as well as his intensive surveys along the border. His analysis of this unusual political divide, unique in its treatment, offers a model for others to consider in the context of frontier situations elsewhere.

W.R. Mead
University College London
March 1995

EDITOR'S INTRODUCTION

Nationalism has been badly misunderstood by the social sciences. No doubt other phenomena of our modern world have been equally misinterpreted, but nationalism is the most important which has been so ill treated. One result has been a chronic underestimation of the power of 'nation'. A quarter of a century ago social scientists were surprised by the rise of 'regional nationalisms' in the old states of Western Europe; since the end of the Cold War social scientists have been shocked by the vehemence of many 'new nationalisms' which have made a mockery of their hopes for a 'new world order'. Social scientists just do not seem to be able to get a reliable handle on the practices of nationalisms and their meanings.

One relatively obscure corner of the social sciences has always had the potential to provide vital insights into nationalism. Political geography claims as its core the complex relations between nation, state and territory. By grounding their studies in bounded territory, political geographers have focused on a fundamental link between state and nation that is often missing elsewhere. In addition, political geography has always been concerned with relations between different geographical scales of political processes. This is important because nationalist conflict is invariably about setting the scale of political resolution (i.e. a big nationalism versus a small nationalism and the latter against an even smaller nationalism, and so on like a Russian doll). To a large extent, however, this political geography potential has rarely been realized because of a failure to appreciate adequately the complexity of the subject matter. This book is a major exception to this generalization.

Anssi Paasi has produced a theoretically sophisticated treatment of nationalism by focusing on boundaries in the construction of territories and on discourses in the representation of others. Drawing on a range of relevant social science and humanities literature, he has woven together an original argument that links traditional geographical understanding of regions with national territorial constructions through the symbolic role of boundaries in national socialization. All this is revealed in an unusual empirical context with, first, national-level study of the meaning of Finnishness and, second, a detailed study of a community dominated by the Finnish boundary with the USSR/Russia. The integration of the theoretical and empirical is one of the highlights of this book. Here we find political geography at last beginning to provide what it has always promised: a sensitive understanding of place and space allied to an informed analysis of the political and cultural. This is a book for all social scientists who are interested in nationalisms and are looking for a fresh perspective on a neglected yet so powerful phenomenon of our modern world.

Peter J. Taylor
University of Newcastle Upon Tyne
March 1995

PREFACE

Recent years have witnessed the most substantial changes in the world system of states and their boundaries since World War II. As a culmination, the deep ideological divide between East and West, created in political rhetoric and practice after the war, has disappeared. This has not necessarily reduced the roles of boundaries, however, even though it is often argued that economic, cultural and political interaction in the present-day world is increasingly overcoming the traditional boundaries, or that the 'space of places' is turning to a 'space of flows', as Castells (1989) puts it. Regionalization is perpetually taking place, however, and new territories and boundaries are emerging on various scales extending from daily life to the global domain.

This book provides a theoretically informed analysis of the social and historical construction of territories, boundaries and socio-spatial consciousness, particularly in the case of the Finnish state and the Finnish–Soviet/Russian boundary. The rise of the Finnish state, the nation-building process and the changing positions of this state in the European and world geopolitical and ideological landscape will be traced. The aim is to combine two geographical perspectives which have typically been pursued separately in geographical research — a structurally based analysis of the construction of boundaries as part of the nation-building process, and an interpretative analysis of local, personal experience of this process. Special attention will be paid to different spatial scales in the social construction of territories, the ideas of 'us' and 'them' or the 'Other'. The key objects are hence the social production and reproduction of space and territories, the changing representations of the boundary in varying social practices, and how these representations are linked in discursive practices and rhetoric with the social context in which they were produced and maintained.

The book relies on theories and empirical approaches within geography, sociology, cultural studies, political science and history. The basic intellectual context is nevertheless that of *geography*. Instead of understanding this merely as an academic institution, the account given here starts out from a more extensive perspective and understands it both as the spatial construction of the world, an academic discipline and a field of social and intellectual practice, together with categories produced and reproduced in this practice, particularly those of space, region and place. It begins from the idea that power is inextricably and continually inherent in the definition of these three geographies and that the creation and symbolization of boundaries is part and parcel of all of them.

The themes, ideas and approaches set forth in the present book have been developed and implemented over a long period. I have been working since around 1980 with the so-called 'new' or reconstructed regional geography, the social construction of territorial communities and consciousness and identities on different spatial scales, and I have developed a specific understanding of the rise of regions, i.e. a theory of their institutionalization. While working with this topic I have moved between various spatial scales in my previous studies while using different sets of empirical research material. This book aims at being a more synthetic study of the roles and interaction of social, economic, administrative and cultural processes and practices taking place on various spatial scales, extending from the experiences of everyday life to world geopolitical landscapes. The purpose is to follow a multilayered approach, i.e. different conceptualizations are exploited simultaneously as a basis for the concrete analysis, thus combining several interrelated sets of research material, such as interviews, autobiographies, photographs, maps, novels, newspapers etc.

The meanings of a boundary in daily life are analysed through an interpretative case study of Värtsilä, a Finnish commune which was divided by the new Finnish–Soviet border as a consequence of World War II. I began my investigations there in spring 1987, at a time when the Finnish–Soviet boundary was still a fitting example of an ideological boundary: closed and mystical, a typical manifestation of the geopolitical divide between East and West. When I began to fit the theoretical framework and concrete material together, the world was changing drastically: the geopolitical divide between East and West was disappearing, many Finnish historians and political scientists were eagerly searching for evidence and even a sensation of the years of 'Finlandization', the Finnish economy was sliding into a deep recession, and Finnish society as a whole was becoming polarized between those who had a job and those who did not. At the last stage Finland entered the 'official Europe', i.e. the European Union, and concomitantly became its eastern border with Russia. Surprisingly, in the course of this long project I had unintentionally been able to witness one of the most significant changes in the European territorial system in the twentieth century both locally, during the fieldwork and by participating in more extensive socio-spatial processes. One of the aims of the present book is to document this process in geographical terms.

It would have been impossible to carry out this longitudinal study lasting more than 10 years without support from a range of interested people in Finland and abroad. Many colleagues in Finland have helped to collect and analyse the various sets of material, and I am grateful to Katariina Korhonen, Kari Vuorinen and Arto Lievonen for their practical assistance at the early stages of working with the empirical materials employed here. I also wish to thank all the people looking after the various archives for their help and patience when tracing the photographs etc. Thanks are also due to Mrs Hilkka Loire for her permission to use the archives of her late husband Erkki Loire, and to all the people who participated in interviews and also kindly provided me with photographic material.

I have benefited enormously from the fact that the manuscript or parts of it have been read and commented on by several researchers in the fields of geography, history, anthropology, sociology and political science. I am indebted to my foreign colleagues (in alphabetical order) Anne Buttimer, Stan Brunn, Bill Mead, Kevin Robins, Mike Savage and Neil Smith for their comments, discussions and numerous reprints, and particular thanks are due to Peter Taylor for his suggestions and encouragement. Even though our conceptual views differ somewhat, I have been indebted to Allan Pred in my understanding of region and place ever since his first comments on the idea of the institutionalization of regions in 1986. All these people have also helped me in bridging one of the most complicated culturally constructed boundaries in human life, that of language.

One of the key arguments in this book is that regions and territories are social constructions which have to be approached historically. In this respect I am particularly grateful to the historian Antti Laine for comments, suggestions and his longstanding friendship. In the later stages I have also benefited greatly from the comments of Jouko Vahtola, a specialist in Finnish and Scandinavian history. I similarly wish to thank the Finnish political scientists and peace researchers Vilho Harle and Pertti Joenniemi for material, discussions and comments on the changing geographies of Finland. The long cooperation and critical discussions with the anthropologist Seppo Knuuttila have likewise been extremely fruitful.

My colleagues in our small research group on 'regional trans-formation' at Oulu University, Pauli Tapani Karjalainen, V-P Raatikka, Petri Raivo and Heikki Riikonen, have always been ready to discuss the changing ideas of regions, places and landscapes and varying methodological approaches to concrete research, and I am also grateful to the students and researchers who have commented on my papers in numerous seminars and conferences in Finland, Scandinavia and elsewhere during the last few years. Finally, I am grateful to Erja Koponen and Anja Kaunisoja for drawing (or redrawing) the figures and maps, and to Kerttu Kaikkonen for her practical help when working with various versions of the text. Malcolm Hicks has carefully checked my English. The first stages of this work were supported financially by grants from the University of Joensuu. After moving to Oulu University I received a grant for senior scholars from the Academy of Finland for the academic year 1992–93, which contributed greatly to the completion of this book. Despite the kind assistance and help received from all these people and organizations, the defects that remain in it are of course entirely my own responsibility.

Anssi Paasi
University of Oulu
February 1995

ACKNOWLEDGEMENTS

The author and publisher wish to thank the following who have kindly given permission for the use of copyright material:

Department of Geography, University of Helsinki for Figure 5.1; Werner Söderström Osakeyhtiö (WSOY) for Figures 5.4, 7.4 and 6.1; Finnish Defence Forces Education Development Centre, Photographic Section, for Figures 5.6, 5.8 and 7.3; Methuen Co for Figure 5.9; Mr Patrick Enckell for Figure 7.2; National Museum of Finland for Figures 6.3, 6.4, 7.1, 7.3, 9.6, 9.7, 9.9 and 10.4; Helsinki City Museum for Figure 6.4; Photoarchives of Otava for Figure 7.5; Geographical Society of Finland for Figure 7.6; Archives of the Ministry for Foreign Affairs, Map Collections, for Figure 9.12; Metra Corporation for Figures 9.15 and 9.16; Värtsilä Local Council for Figure 9.15; Mr Helge Ratilainen for Figure 9.20; Wartsila Company/INDAV for Figure 9.14. All attempts have been made to trace the copyright holders, but unfortunately in the case of Figure 9.18 this has proved impossible. We offer our apologies to any copyright holders whose rights we may have unwittingly infringed.

PART ONE

TERRITORIES AND BOUNDARIES IN REGIONAL TRANSFORMATION

1

REGIONAL TRANSFORMATION
AND THE OTHER

1.1 INTRODUCTION

The territorial system of nations, states, various administrative or cultural regions and the boundaries between them has been in continual transformation in the course of history, reflecting economic, political, military and administrative passions, evaluations and decisions made by rulers, representatives of state organs, various classes and enterprises. Territorial units are historical products — not merely in their physical materiality but also in their socio-cultural meanings. Hence territories are not eternal units but, as manifestations of various institutional practices, emerge, exist for some time and disappear in the transformation of the world state system. Along with the rise of modernity and the state, this arrangement has become more complicated together with a simultaneous increase in interaction between states and between economic, political and cultural organizations in the world system.

The present-day world political map reflects concomitantly both stability and dynamism. Stable borders exist where the representatives of states have signed and honoured treaties recognizing their sovereignty and where various parties agree on the location, delimitation and demarcation of a common boundary. And where the (geo)political situation is less stable, the boundaries and their locations are typically potential sources of territorial conflicts (Brunn 1991, 269).

Territorial transformation in modern nation-states also constantly reflects decisions and actions taking place in the local state and the civil society, i.e. within the discourses of a variety of social groupings based on differences of sex, ethnicity, generation and race. This transformation may also reflect private social relations within the family or multifarious voluntary associations, personal local experience or — in the case of regionalist movements and mobilizations — collective spatial manifestations of local or regional identities. The rise of miscellaneous social groupings and the discourse on social identities, regionalism, nationalism and self-government in diversified spatial contexts have been, according to some commentators, a reverse of the active tendency towards centralization within states, internationalism and the liberation of colonized peoples that followed World War II (cf. Alter 1989, Mach 1993). Others have tended to link identity movements and associated phenomena with the changing experiences and practices of space, and with the rise of heritage business or regionalism, and have tended to interpret them as reflecting a new 'post-modern' reaction to the hegemony of the 'modernist project'.

Recent dramatic transformations in the territorial organization and ideological landscape of Europe after the Conference on Security and Cooperation in Europe (CSCE) meetings in Vienna in 1989 and in Paris in 1990 have again rendered the perpetual regional transformation process particularly visible. The rapid collapse of the socialist states has caused the dissolution of the geopolitical debate, characterized by the ideological 'iron curtain' between the Eastern and Western worlds. This collapse has marked the end of the ideological juxtaposition and images of an enemy — a continual vision of 'otherness' — that were fostered during the Cold War, and has again raised the question of the boundaries of 'Europe'.

These issues and the consequent rise of diverging ethno-regionalistic and ethno-nationalistic movements have profoundly transformed the territorial and ideological landscape of international politics and have given rise to new boundaries and demarcations between territorially based social groupings. In several areas on the eastern side of the former invisible curtain the ideas of friends and enemies, of identities and otherness, have been transformed as a consequence of this 'social restructuring' process. Europe has seen substantially less interstate and intrastate violence than other continents since 1945 (Wallerstein 1991, 49), but now the struggle over social and political power, particularly in the former Yugoslavia and Soviet Union, has led to violence and contested aims at redefining and signifying the boundaries and contents of the social space with new cultural and political terms. It has become evident that where minorities and majorities cannot live together, the result is displacement of the border, the people or both (Hassner 1993).

Coincidentally, the complex debate on European integration that has taken place on the Western side of the former iron curtain has pushed to the forefront the question of the diverging roles of territories and boundaries in the construction of a utopian synthesis between traditional pan-European idealism and contemporary national European technocraticism. This process is part of the broader structural transformation of the economy (and culture) of advanced capitalist societies and the new articulation of spatial scales (global, national, local) associated with the increasing transnationalization of capital accumulation (Robins and Morley 1993).

The aim of European integration, to allow people, goods and capital to move freely across boundaries, is a concrete 'small-scale' illustration of the rather abstract tendencies that were eloquently scrutinized by Castells (1989), when writing about the historical emergence of the space of flows superseding the meaning of the space of places. According to Castells, 'the ultimate logic of restructuring is based on the avoidance of historically established mechanisms of social, economic and political control by power-holding organizations'. The emergence of the space of flows, Castells argues, actually expresses 'the disarticulation of place-based societies and cultures from the organizations of power and production that continue to dominate society without submitting to its control' (p. 349). This fact typically generates local and territorial mobilization where the inhabitants of localities and regions aim at preserving and strengthening their identities and heritage. Bounded places are hence turning into spaces of interaction where local identities are constructed of material and symbolic resources that may well not be local in their origin — the most significant feature of restructuring is escalating *globalization* (Robins 1989).

Regional transformation occurring in the process labelled as 'European integration' is a broader institutional context for the social, cultural, administrative and economic changes which are taking place in the territorial structures of nation-states. A fitting illustration generated by this larger-scale institutional process is the discourse on the organization of regional administration that is going on in Finland, a small nation-state which is regarded by many commentators as located on the periphery of Europe close to the former iron curtain (cf. Rokkan and Urwin 1983, Mead 1991). The central theme of the ongoing ideological discourse has been whether a new provincial, i.e. regional, level of administration should be established, since it is argued in the political rhetoric that the provinces are regions which emerge 'from below' and not from the activities of the state organs. Supporters of this idea underline increasing efficiency of administration and increasing democracy in regional decision making, which is essential, as the argument goes on, as far as membership of the EU is concerned. The opponents, for their part, lay stress on the increasing bureaucracy and costs that would follow the creation of a new level of administration, and some insist that what is needed is a reduction in the existing number of administrative provinces and a concentration on the development of administrative cooperation between the communes, on the units at a lower level in the spatial hierarchy. Coinciding with this debate, and following the dispersion of the Soviet Union, Finnish connections with Russia are becoming more active in various spheres of social action (economic life, civil society) and the old ideological boundary between East and West is losing its traditional political and cultural role.

All these individual territorial examples point to the fact that boundaries and borders no longer play the same role in distinguishing space and place as they did. The very idea of boundary, Robins and Morley (1993) point out, has been rendered problematic, although it has certainly not been erased, as many boundaries (formal nation-state/ sub-state boundaries or cultural ones) are being blurred or eroded with the rapid increase in communication in the modern world and the time–space convergence. Even if the loyalty structures of individuals and groups are becoming both spatially and socially more complicated, one essential dimension of the regional transformation still appears to be the perpetual restructuring of various demarcations and boundaries, and the struggle over the rights and means to define the new ones as part of a system of control maintained over the state or some sub-state entity (cf. Brunn 1984, 162–3).

The functional space of the organization of economic, political and cultural activities has hence partly exceeded the absolute space of nation-states, but it has by no means replaced it. As Taylor (1993a) points out, after the rapid changes in the political landscape of Europe, questions of nation, state and territory have in fact returned to the top of the world's political agenda. A struggle over the redefinition of space is always an expression of the restructuring of economic, political and administrative practices, and also of the restructuring of the contents of social consciousness, inasmuch as one obvious aim of various ethno-regional groups is to establish a territorial counterpart for social boundaries (Barth 1969, 15). Furthermore, these practices are never abstract processes which take place somewhere above the heads of human beings, but unfailingly manifest themselves on local, regional or national scales and are in fact produced, contested and reproduced in the local everyday life of human beings.

The massive debate that has continued since the 1970s on the causes, forms and manifestations of the economic, political and cultural restructuring — which originally took place in the developed capitalist world and has now spread to Eastern Europe — has been essential for a geographical and, more generally, a sociological understanding of the nature of social and spatial transformation in the system of territories. It has been emphasized that the process of restructuring is not a pure mechanism of adjustment but a politically contested and determined process enacted by governments and organizations. Within academic geography the study of these processes has partly changed the language and methodologies by which geographers have striven to understand and explain the intertwined relation between the social and spatial. One much debated result in the social sciences has been so-called locality studies, in which researchers have attempted to analyse the manifestation of various social, i.e. economic, political and cultural, processes in specific localities (Cooke 1989a, Duncan and Savage 1991). Simultaneously with the changes in the global geopolitical system, the awareness of the role of political geography and its basic categories (state, nation) has grown and the potential contribution of political geography to the study of geopolitics and new international relations has increased (Taylor 1988b, Cooke 1989b, Johnston 1989).

1.2 REGIONAL TRANSFORMATION, THE SOCIAL CONSTRUCTION OF 'WE' AND THE 'OTHER'

In order to find a common denominator for comprehending the connection between the changing geographies of various territories, boundaries, everyday life and power and in order to develop the general perspective of this book, some ideas regarding the socio-cultural construction of socio-spatial communities will now be briefly discussed. This will lead us to evaluate generally the role of language and discourse in the social construction of spatial demarcations and boundaries — and of the world. More generally, it is a question of an analysis of *signification*, i.e. how political and cultural processes become part of the social and symbolic construction and reproduction of communities, and of how, e.g. landscapes, heritage, cultural products and rhetoric, metaphors and images are exploited in the process.

Our task resembles that of Shields (1991), who studied the process which he labelled as 'social spatialization'. This points to the ongoing social construction of the spatial at the level of the social imaginary (including collective mythologies or presuppositions) and in the form of interventions in the landscape. In his view (Shields 1991, 31), social spatialization 'allows us to name an object of study which encompasses both the cultural logic of the spatial and its expressions and elaboration in language and more concrete actions, constructions and institutional arrangements'. He analyses this process through four case studies of 'places' or 'regions' varying from the seaside resort of Brighton to the North–South divide in England.

Where the sociologically informed investigation of Shields (1991) aims to trace the social construction of spatial images and identities, the aim of the present book is more geographical: to study explicitly the process of the territorialization of space, the construction and signification of demarcations and boundaries. This book also evaluates more specifically the construction of spatiality in various spatial scales: local, regional, national and global. More attention will also be given to the concepts of region and place — the categories that Shields almost takes for granted. We will argue that these categories are of crucial importance for understanding the social construction of spatiality and the particular boundaries between social collectivities. As far as the construction of social communities is concerned, the idea of social spatialization is doubtless important, but so is the idea of *spatial socialization*: the process through which individual actors and collectivities are socialized as members of specific territorially bounded spatial entities and through which they more or less actively internalize collective territorial identities and shared traditions.

This introduction provides a scheme for a more detailed analysis of the roles of specific territorial ideologies and discourses such as nationalism, which are significant in spatial socialization. Accordingly, particular emphasis will be placed on the role of *rhetoric* in the construction of territories and boundaries. Rhetoric is usually understood as referring to the forms of persuasive argument — be they good or bad — put forward by advertisers, editorialists and politicians. According to Sayer (1992, 264), to examine rhetoric is to explore the field constituted by the relationships between 'the object, the author's intentions, language and literary processes, readers' pre-understandings, moral dispositions and self-presentation'. Rhetoric is not just an individual practice, since there are also rhetorics of organizations, of social movements, of professions and of scientific schools and disciplines. It is also being accepted to an increasing degree that the discourses put forward by the human sciences can be regarded in some sense as being akin to these more conventional forms of persuasion (Simons 1989, 2–3). Researchers probably cannot avoid at least some forms of rhetoric, since they are always operating in normative social contexts. Social research in particular, Brown (1987, 89) writes, is constructed rhetorically and can be understood as a set of meanings established in response to specific problems that emerge in different political and historical contexts. Analysis of the rhetoric hidden in accounts of the social world is hence a challenge for any critical social science.

How are socio-spatial groupings constructed? Carr (1986) sets out a thought-provoking scheme for the construction of various social groupings and the involvement of individuals in this process. He attempts to trace the trans-individual subject, i.e. how *I* and *we* are constituted in social practice, and goes on from there to discuss the constitution of divergent social groupings. He does not elaborate explicitly on the power relations, ideologies or spatial relations that constitute and are constitutive of various communities, but he does put forward some ideas for approaching this theme when deliberating on the role of rhetoric, for instance.

Dalby (1990, 17–29), for his part, discusses more explicitly how discourses construct their objects of knowledge as 'Other'. The Other has become an immensely important theme in feminist and cultural-historical studies since the 1970s. As far as the role of spatiality in the construction of otherness is concerned, this theme has become significant particularly since the publication of Said's (1978) *Orientalism*, in which he shows how the idea of 'Orient' was to a great extent a European invention. Numerous scholars — including geographers (Duncan 1993) — have aimed to analyse both theoretically and empirically the construction of borders, both physical and symbolic, and the Other. Where is the divide between self (or us) and Other to be situated, where are the borderlines between human and 'something else', and so on? Mason (1990, 2) points out that the discovery of the Other has a history, and it has socially and culturally determined forms. This holds good in the case of all symbolic and physical distinctions, all social spatializations.

Dalby (1990) shows how the ideas of otherness are significant in producing certain metaphysical conceptions and ideological structures, and how the Other provides a useful way of illuminating the categories of space and time, which confer a fundamental structure on the discourses of social and political theory. An explicitly spatial dimension for this discourse is provided by Walker (1993, cf. Bachelard 1969), who discusses the meanings of the distinction between 'here' and 'there' in the case of international relations. Walker sets forth a critical discussion of the rearticulation of spatio-temporal relations and transformation in modernity and late modernity and considers the construction of the division between inside/outside, which has characterized much of the discourse in international relations and which is typically taken for granted.

According to Carr (1986, 146–68, *passim*), a social community exists wherever a narrative account exists of a 'we' which has continuous existence through its experiences and activities at any level of size or degree of complexity whatever. 'An account exists' means that it will be articulated or formulated, maybe by one or more of the group's members, in terms of 'we', and will then be accepted or subscribed to by the other members of the community. Carr points out how such accounts or stories are told by an individual or individuals on behalf of the 'we' — using the 'we' as the subject of the action and experience, and also of the narration itself. This points to the social power roles governing people and classes, the rhetoric of which unites the group and expresses what it exists for, where it has come from and where it is going. The main stories can have rival versions, or versions based on different premises — as is the case in the struggle that is going on in the former Yugoslavia. According to Carr, much of the communal rhetoric which addresses a group as 'we' is putative and persuasive rather than expressive of a genuine unity and an already accepted sense of communal activity. Nevertheless, the story must be shared by people if it is to be constitutive of a group's existence and activity. In certain cases the 'story-teller' aims at creating a community where none existed before.

As regards the roles of scientists as 'story-tellers', White (1987) — discussing the form and content of historical writing — makes clear that narratives are not merely neutral forms of discourse which can be used to represent real affairs, but that the construction of all narratives has ideological and political implications. Scientists always have the power to classify and depict phenomena, and the symbolic power to make people believe in their accounts. In fact, scientists also signify and legitimate various practices and distinctions.

Carr's (1986) basic idea is engrossing, since it provides one perspective for understanding the eternal problem of social solidarity and the construction of territorial identities and otherness. Hence this perspective is beneficial when deliberating on the production and reproduction of territorial units as well as other social units which manifest themselves on various spatial scales. His frame affords a heuristic device for tracing and understanding the main narrative accounts and their rivalries in various social units.

Carr's position leads us immediately to scrutinize the role of *language* and discourses in the social construction of space and its various forms that geographers have by tradition labelled in their discourses as 'regions', 'places' and 'localities', for instance, sometimes emphasizing the purely academic nature of these categories and sometimes referring to their societal origin (cf. Paasi 1991a). It has been pointed out that discourses are much more than merely linguistic performances (Dalby 1988, 1990). ÓTuathail and Agnew (1992) claim that discourses are best conceptualized as 'sets of capabilities' that people have, as 'sets of socio-cultural resources' exploited by people for the construction of meaning about their world and their activities. Hence discourses are also 'plays of power' which exert various codes, rules and procedures to impose a specific understanding through the construction of knowledge within these codes, rules and procedures (Dalby 1988). Furthermore, the rules that govern practices are often implicit, not clearly articulated. These rules are 'socially constructed in specific contexts' and are understood subconsciously or taken for granted by practitioners. Hence, Dalby (1990, 5–6) writes, discourses have institutional origins and commitments, and the knowledge they produce consists of political products. Therefore discourses carry implications of power.

1.2.1 Geographers and the discursive construction of space

Geographers took an interest in the role of language and discourse particularly during the 1980s, partly in association with their analyses of the traditional categories of the discipline. Accordingly, Berdoulay (1989), Pred (1989) and Pratt (1991), for instance, called attention to the role of language when deliberating over the local scale or the idea of 'place'. Even though they placed emphasis on somewhat differing points, they nevertheless seem to agree that a place explicitly comes into being in the discourse of its inhabitants and particularly in the rhetoric it promotes. Pred (1989, 212), for example, observes that 'no local worlds exist without words or other symbols, and that all the practices, all the forms of social life that constitute the world of a place are enabled and sustained by language'. These discourses are embedded in local institutional practices, reflecting practices taking place on 'larger scales', and they serve to create and delimit social reality.

As far as the role of boundaries is concerned, Berdoulay (1989, 131) maintains that a place does not have to be thought of in terms of an area but it may be more significant to view it as a network — or as a space of interactions between what is global, national and local. Similarly Pratt (1991) points out that to say that social relations are spatial does not necessarily imply a contiguous or meaningful territory. This statement encompasses an interesting parallel with the deliberations over the nature of social communities, in which it is common to presume that the social memory of a group of people or community has a 'geographical' referent, some territorial basis. However, the construction of a community cannot be reduced merely to this basis, for communities may or may not be coterminous with a specific, contiguous territory. Bender (1978, 6), for instance, defines a community as an experience rather than as a place.

While such arguments presumably hold good in the case of everyday life and local discourse, the idea of demarcation and delimitation — or spatialization — is in any case present in all social practices, particularly those through which the local discourse 'stretches' itself to other spatial scales, for instance through regionalistic and nationalistic argumentation and inherent territorial identities (Paasi 1991a). Specifying the difference on these larger scales is a linguistic act, but also crucially a political act (Dalby 1988). Pratt (1991), in fact, is pointing to this possibility when referring to B. Anderson's (1991) *Imagined Communities*. In this book Anderson demonstrates that discourses represent or signify not just boundaries but various communities as coherent socio-spatial units. This view may lead, when such discourse is reproduced, to these places being taken as 'real', particularly when they are signified and legitimized by legal means. Such is the case with nation-states, where identity construction or nation building always occurs against the background of an external other. Hence the social idea(l) of community, a deeply historical and theory-laden category, has functioned in the course of time as an ideology in various socio-spatial scales and social contexts, particularly at times of crisis (cf. Mosse 1982). All political and territorial identities are, in a sense, fictional identities which are connected with 'imagined communities' (Hassner 1993).

Clark and Dear (1984, 84) stressed the roles of language in the construction of social world and the active role of *state* in this process. They remark how language is employed to construct or reconstruct social and political reality and how this language is 'studded with signs, icons, or symbols which may carry meanings in excess of the simple word being used'. Further, language may be used in a purposeful manner in various discourses in order to maintain the identity and cohesion of a group. Hence power relations are institutionalized in language at the same time as language serves as the medium of social action and communication.

Given the ideas of Carr (1986), what is especially interesting here are the connections between these ideas and the arguments put forward by some geographers, in particular Berdoulay (1989, cf. also Entrikin 1991). Berdoulay maintains that a geographical account of place resembles a whole staging process whereby people, objects and messages are coordinated. It is like 'telling a story'. He also proposes that the study of place has a strong narrative component. Talking about places makes a difference: 'Obviously, this is not like telling any story. It has to reflect the actual interweaving of the relationships among those people, objects and messages, which produces place and which may be viewed as a discourse'. Geographical accounts of place are thus crucial for the social spatialization process.

1.2.2 We and the Other

So far we have discussed the construction of a 'we' and the role of discourses in this task, a process through which socially constituted linguistic and semiotic constructions maintain and modify meanings in the service of power. Now we will take a precursory look at how the Other is created as an external entity against which 'we' and 'our' identity is mobilized. Dalby (1990, 17) writes that the discourse of the Other is concerned with the perennial philosophical debates within the Western tradition concerning identity and difference, although other cultural traditions also embrace such dualistic structures. Harle (1990) underlines the role of dualisms, particularly in the European mind and cultural heritage. Dichotomies such as past and present, private and public, true and false, us and them, friends and enemies, good and bad have all become a crucial part of European discourse. The idea of the European identity as superior by comparison with non-European people and cultures, the Other, is hence not new (Robins and Morley 1993).

As we have seen, the idea of the Other has become pivotal to modern social theory. In order to understand the construction of the Other, it is beneficial to point to the discussion by Aho (1990, 19) of the process of reification, by which he refers to the way in which 'figments of the imagination', words and ideas, return 'to haunt their makers by coming to be experienced as objects with independent lives'. Hence reification (or objectification) refers to the way in which people come to experience their created social world in a false or nonfactual way. Referring to Berger and Luckmann (1976), Aho distinguishes five dimensions in the process by which reification occurs: naming, legitimation, myth making, sedimentation and ritual. He is discussing in particular the construction of visions of evil, but his ideas are also more generally applicable to the construction of the Other. Hence the Other has to be labelled, these labels have to be legitimated and the Other has to be mythologized. Through sedimentation, word and myth come to have lives of their own, detached from the original act of myth making and evolve into autonomous parts of the everyday stock of knowledge taken for granted by a society (Aho 1990, 22). Finally, vivid re-enactment of the essential themes of the myth — ritual — is required.

A spatial dimension is usually inherent in the definitions of the Other, in the fact that the Other typically lives somewhere else, *there*. If the Other lives here, *we* — defined in specific narratives — are in any case different from it. In the case of nations, the discourse on culture, language etc. constitutes the 'we' and distinguishes it from the Other. The constitution of the we/they dichotomy usually exploits stereotypic definitions of both us and them, and of them in particular. Stereotypes are not confined to any particular spatial scale. In the case of nation-states these definitions can be exploited in the construction of national units (e.g. Finns, Russians) or supra-national units (Europeans, Scandinavians). They can also be exploited to constitute larger-scale geopolitical distinctions (e.g. East/West) and hence actual, physical and representational divisions of space. Hence, both social, cultural and/or physical boundaries can be exploited in this categorization, whether we call it social spatialization or spatial socialization.

In geopolitical discourse stereotypes may be 'geopolitical scripts' which constitute confrontations between our territory and theirs (Dalby 1990, 39). These scripts are essential constituents of the language of geopolitics, where geographical knowledge is transformed into

	Here	*There*
We	Integration within a territory	Integration over boundaries
Other	Distinction within a territory	Distinction between us and the Other

Figure 1.1 An analytical framework for forms of socio-spatial integration and distinction

geopolitical reasoning. Geopolitics is, as ÓTuathail and Agnew (1992) define it, a discursive practice by which intellectuals versed in 'statecraft' spatialize international politics and represent it as a world characterized by places, peoples and dramas of particular types. The most influential and durable illumination of these geographical descriptions or stories since World War II has been the story of the great ideological, political and economic confrontation between West and East that prevailed up to recent years.

Concluding the foregoing discussion, it is now possible to construct an analytical framework which provides a heuristic basis for the present study by pointing out in general terms how the construction of various territorial identities takes place in relation to certain social distinctions (Figure 1.1).

It should be noted that these four fields are purely analytical and are in practice related to each other. They characterize discourses which are typical in the modern world but which have also been historically significant. The first points to internal integration within territorial units, whether nation-states or sub-state regions. This discourse is essentially based on two keywords, we/here. Secondly, the discourse on we/there refers to intentions to integrate a social group beyond boundaries. This is the case when various ethnic minorities, for instance, strive to exceed their boundaries through integration. Various expansive state ideologies have also given expression to this (e.g. the ideology of the expansionist *Lebensraum*). The third important discourse points to the distinction between various groups within a territorial unit. Typical of this in the present-day world are refugees, who constitute a minority and the Other within a territory. The fourth discourse has been, and still is, of vital importance in the construction of otherness, pointing to the socio-spatial distinction between various territorial groupings (the Other/there).

All of these distinctions have to be constructed in social action and discourse; they do not emerge *in vacuo*. All are based on a family of concepts depicting plural subjects. Plural subjects manifest themselves in our everyday concepts of social groups, collective beliefs and social conventions (Gilbert 1989).

All these categories place a special emphasis on language, on its use in the creation of social representations and images, and on the sedimentation of power relations in this process. In fact, it is reasonable to argue that the symbolic construction of space, territoriality and boundaries is based on a dialectic between two languages, the language of *difference* and the language of *integration*. The latter aims at homogenizing the contents of spatial experience, the former strives to distinguish this homogenized experience from the Other. It is clear that power is an essential dimension in the construction of this dialectic.

1.3 AIMS OF THE BOOK

The preceding discussion has made it abundantly clear that discourses on territories and boundaries involve not merely language but also social practices, positions and relations, which in the framework of the division of labour embody power. This book evaluates the historical construction and the changing meanings of territories both in theoretical terms and by means of empirical case studies. It analyses in particular the construction of boundaries between various social systems — especially states — and tries to trace how various narrative accounts have been exploited in the production, representation and reproduction of territorial units and their boundaries, how the construction of we/the Other and here/there has taken place as a historical process. In a word, the objective is to scrutinize what lies behind the 'fields' put forward in the preceding framework. It will also attempt to analyse how state boundaries — among other 'smaller-scale' socio-spatial distinctions and demarcations — manifest themselves in the stories emerging from the everyday life of people living in border areas.

Accordingly, the major tasks of the book are as follows:

1. The first aim is theoretical and conceptual. The book will thus begin by analysing the changing idea of geography in order to understand various dimensions of regional transformation. Secondly, it will discuss the role of boundaries in the tradition of geographical thought and will put forward a culturally based interpretation of boundaries which explains them as one essential dimension in the construction of territoriality. It will then put forward a framework for understanding the emergence of various territorial units in the regional transformation. After that it will analyse in theoretical terms the character and constitutive mechanisms of nationalism and the role of boundaries in the construction of nationalism. It is here that the mechanisms constitutive of symbolic communities will be discussed, in particular the role of education in national socialization (Chapters 2 and 3).

2.   The second main task is to analyse as a concrete example of nation building and boundary construction the rise of the Finnish state and nation and the role and representations of the Finnish–Russian boundary in this process. The constituents of the Finnish self-image and of the images of Russians are also scrutinized. This analysis is reasonable, since from World War II until the dispersal of the Soviet Union, Finland has been the capitalist state possessing the longest boundary with the leading socialist state. Furthermore, the relations between the two states were for a long time characterized by stereotyped enemy images, which were connected with distinctions on broader international scales (Chapters 4, 5, 6 and 7).

3.   Whereas tasks 1 and 2 aim to analyse the construction of the Finnish–Russian boundary at the national level and in an international context, the third purpose is to study the construction and representation of boundaries at the local level, in people's everyday lives. Hence it will analyse how people living in the border area of Finland experience and produce, reproduce and transform the meanings of various territorial and social boundaries in their day-to-day life. In particular, it will scrutinize the meaning of the Finnish–Russian boundary in daily lives (Chapters 8, 9 and 10).

The purpose of this book is thus to combine two approaches that geographers have often regarded as exclusive owing to their different ontological perspectives — the perspective of analysing general structural features and processes and the viewpoint that emphasizes the importance of the meanings emerging from local everyday contexts. The aim is both to theorize and to analyse these perspectives empirically in relation to each other.

Particular attention will be paid to the *geographical* dimensions of boundary construction and the geographical manifestations of territorial demarcations. At the same time, the aim is to interpret the changing boundaries of human geography itself in the field of academic disciplines: how the story integrating geographers is changing in its content, how these changes manifest themselves in the concepts and categories that geographers employ, and how the interpretations that geographers give to these categories are modified.

2

TERRITORIES, BOUNDARIES AND THE DISCOURSE ON POLITICAL GEOGRAPHY

2.1 CHANGING SPACES AND GEOGRAPHIES

It has been obvious since the 1970s that the idea of geography is broadening. Today it embraces much more than the academic discipline *per se*. If we regard 'geography' as a linguistic sign, it seems that three essential referents for the concept arise, all of which appear to be significant for understanding the nature of regional transformation and its interpretation, both in academic and non-academic circles (Paasi, forthcoming). These are:

- the spatial (and historical) construction of the world
- geography as a social (academic) institution and institutional practices
- geography as intellectual action/categories emerging from this action

On the first score, the idea of geography refers to the fact that spatiality is simultaneously a fundamental structural constituent and an outcome of the construction of social life. This can be seen in geographical categories such as territory and region. Territory, for instance, is doubtless a geographical notion, but as Foucault (1980a, 68) remarks, it is first and foremost a juridico-political one — an area controlled by a certain kind of power. Region, for its part, is originally a fiscal, administrative and military notion.

Nature, culture, daily life and social world as a whole are essentially organized in and through spatial relations and contexts, as are the power relations which are crucial constituents of their organization. The spatial and temporal structuring of the 'social' is hence the central constituent in this organization. Spatiality is nothing mythical, not a feature separate from the 'social' or having causal power as such. As a number of authors have suggested, social space is a social product. But space is never produced, as Lefebvre (1991, 85) indicates, in the sense that 'a kilogramme of sugar' or 'a yard of cloth' is produced. Nor is it an aggregate of the places or locations of such products as sugar, wheat or cloth. Rather, 'it subsumes things produced, and encompasses their interrelationships in their coexistence and simultaneity, their (relative) order and/or (relative) disorder'. Lefebvre (1991, 73) also remarks, 'itself the outcome of past actions, social space is what permits fresh actions to occur, while suggesting others and prohibiting yet others'.

Lefebvre's (1991, orig. 1974) profound discussion of the production of space renders this abstract category more comprehensible, inasmuch as he traces various forms of spatial practice, which in a way 'put life' into the abstraction. He makes a distinction between three forms of spatial practice. The first points to the physical and material flows, transfers and interactions that take place in and across space and ensure production and social reproduction. The second, representations of space, includes all the signs and significations, codes and knowledge that allow such practices to be discussed and comprehended, in terms of both day-to-day common sense and the academic discourse that deals with spatial practices. Thirdly, he distinguishes representational spaces which refer to mental 'inventions' that imagine new meanings or possibilities for spatial practices. They are thus codes, signs, utopias, imaginary landscapes and material constructs such as symbolic spaces, built environments, paintings, museums and so on. All these forms of practice are of course related and in a way constitutive of each other.

Harvey (1989) developed this scheme by employing four new dimensions which facilitate the understanding of Lefebvre's basic dimensions. These are accessibility and distanciation, appropriation and use of space, domination and control of space and the production of space. We may evaluate the 'cross-tabulation' of the dimensions recognized by Lefebvre and Harvey to obtain a much more comprehensible picture of the complicated nature of space and its various dimensions and to understand the materiality and experiental dimensions of the production of space more easily (see Harvey 1989, 220–21). The dimensions connected with the production, domination and control of space are particularly useful for the present purposes. Thus some obvious 'keywords' arise from Harvey's synthetic frame — first the production and reproduction of state and administrative divisions of space; secondly the representation of space and exploitation of various 'territorial ideologies' in this construction, e.g. regionalism and nationalism. Also, as far as the relations between representational spaces or spaces of representation and control of space are concerned, Harvey's typology brings up the question of tradition and symbolic demarcations and, beyond these, spaces of fear, which are essential for understanding the construction of otherness. Hence the frameworks of Lefebvre and Harvey contribute to an understanding of the construction of territoriality and boundaries. Similarly, Harvey's framework touches on the role of academic disciplines in the construction of spaces and the narrative accounts regarding these, and also introduces the question of *academic geography*, the second essential meaning of the word 'geography'.

2.1.1 Geography and power

Before the establishment of academic geography, explorers and other 'pre-geographers', inspired by this general geographical or spatial dimension, attempted to produce knowledge for organizing the political and strategic power relations of the world. In a way it was realized that space is fundamental to any form of communal life and to any exercise of power (cf. Foucault 1980a). During this spatialization process, as Harvey (1989) states, geographical knowledge became a valued commodity in a society which was shaped by profit seeking. The importance of geopolitics also became apparent in this connection and territories became more effectively divided, contested and ruled. The exclusion of the Other and the inclusion, incorporation and administration of the Same turned out to be of essential geopolitical importance (Dalby 1988).

The concept of spatial organization typical of the Western cultural realm has to a great extent been shaped by the idea of property, and accordingly territories have been regarded as goods or commodities which can be bought and sold in market places (Soja 1971, Knight 1982). This has put its stamp on our understanding of the broad spatial structure of political organizations. Space has been understood as being divided into pieces, the boundaries of which are defined mathematically and astronomically.

Through this process the *space of states* was created. States as bounded 'geometrical' units became understood as absolute entities. Dalby (1988, cf. Taylor 1988a) writes that the cartographical designation of states as geometric entities with precisely definable boundaries 'hypostasizes' states and in fact denies large parts of the social world. This discursive practice of reification, Dalby argues, is crucial to the operation of geopolitics. The background for this has always consisted of new social and political distinctions between 'us' and 'them' and the construction of a new iconography and new social practices to establish the above-mentioned state functions again in this changing geography.

The signs and texts, particularly maps and cartography, which have been employed to illustrate and visualize this 'geography' — the space of geopolitics — have always been social and political instruments of power in the division of space (cf. Harley 1989). They have provided possibilities for manipulation, by their users, through various stories and 'truths', through various interpretations of the mimetic/non-mimetic relation between 'reality' and (visual) representation and through interpretations of the messages of the representations themselves. Rather than being mirrors, maps are cultural texts, which construct the world rather than reproduce it (Wood 1992). In this sense geography is also drawing and visualizing the abstract and invisible boundaries of power which emerge from social practices and inherent human relations. The mechanisms that construct boundaries are one part of the process through which individual subjects are constituted and through which society is reproduced.

The question of 'truth' and representativeness therefore becomes significant in connection with new societal knowledge. Foucault (1980b) points out that truth is not located outside power or lacking in power, and this holds good particularly in the case of knowledge that can be employed for the purposes of control and integration in society. Foucault then argues on these grounds that each society has its regime of truth, its general politics of truth, the types of discourse which it accepts and causes to function as truth. This points again to the role of rhetoric as a context and a tradition, making us more aware that there are alternative ways of telling the truth and that we are therefore responsible for the ways in which we do this (Brown 1987, 118).

Foucault (1980b) argues that each society has the mechanisms which enable one to distinguish true and false statements and the means by which each is sanctioned. Societies also have specific techniques and procedures which are accorded value in the acquisition of truth. He also points out that societies have the status of those who are charged with saying what counts as the truth. These post-structuralist arguments therefore suggest that the pursuit of ultimate truth or real meaning in a text is fruitless. One text refers to a second, and so on. The shape of space, Brown (1987, 124) argues, is one key code in social texts. This holds true both in the case of codes regarding personal space and on broader spatial scales. In both cases space is typically divided into binary possibilities, mainly distinctions between inside and outside.

Likewise it is possible to argue that power is present everywhere (Foucault 1980b), *in all social practices*. Equally, geography is everywhere — both in large-scale territorial processes and in the local contexts of everyday life and inherent experiences and meanings (Cosgrove 1989). Social life, like everyday activity, is essentially a practice of demarcation, of continually making social and cultural distinctions. 'Geography' in this sense distinguishes and connects individuals, groups and states, human beings and nature. It is also obvious that when we talk about the hierarchies of power, 'the exercise of power over people necessarily involves the creation of geographies' (Johnston 1986, 364).

But power manifests itself just as well here and now in spatially constituted everyday life as in the historical process through which cosmographic description of the world became a speciality of numerous disciplines — including academic geography. Van Paassen (1957, 21) once wrote that geographers and geographical science can only exist in a society with a geographical sense. Whatever this sense consists of, it is impregnated with power. Here we come to the basic political oppositions as to who controls geographical knowledge, who uses it and how, and how it is produced and for whom (cf. N. Smith 1990, 177).

Whereas many disciplines (e.g. sociology, political science and economics) had their origins in the practical interests of the state (social control, state management, national accumulation of wealth), geography was established as an academic discipline by the modern state to construct and document an orderly picture of the world, a picture that was valid in the social spatialization from the perspective of a given state. Geography had to produce and reproduce territoriality, to create and establish boundaries on the horizontal continuum of absolute space, nature and culture that prevailed on the globe, and to establish the basis for the stories of creating a distinction between us and the Other (cf. Agnew 1987, 74). One important vehicle for these purposes was found in maps, which in the modern Western society soon became crucial for the maintenance of state power — for its boundaries, its commerce, its internal administration, its control over populations and its military strength (Harley 1989, 12).

The crucial backgrounds for the establishment of academic geography were thus *nationalism* and *colonialism* (Capel 1981). To promote these, many of the applications of academic geography have been national in their aims. School geography in particular has been crucial for the production and reproduction of socio-spatial consciousness and spatial representations. This has been significant from the perspective of social integration, the key motive behind establishing academic geography. In the course of the history of geographical traditions in various countries geographers have been engaged in the construction of territorial representations and demarcations, in order to produce time- and space-specific truths for the needs of individuals and groups possessed of power. This has been based on the exploitation of both the language of difference and the language of integration.

2.1.2 The changing language of geography

Thirdly, geography can also be understood as a set of social and linguistic practices which in principle exceed the national scientific institutions. The 'language of geography' consists of certain keywords such as space, region, place, location etc., which are reproduced and modified in the practice of geographers, practice which essentially consists of a struggle to put forward legitimate definitions for these categories. The context for this struggle emerges from the society and academic milieu in which it takes place, since the state makes its demands on geographers in the field of social spatialization, while within geography, academic tribes create their own conceptual totems by which they strive to signify and legitimate their discipline (cf. Becher 1989). Hence positivistic, humanistic and critical geographers, for instance, have set forth their own interpretations of fundamental categories such as region and space. Ultimately these spatial categories are not merely the property of geographers, however, since 'geographical imagination' is important in many other academic fields (urban planning, history, art, architecture), where these categories have been effectively exploited (cf. Harvey 1973). The discourse on these categories has also been essential as far as the boundaries between geography and other academic disciplines are concerned.

Driver (1992) has pointed out that the history of geography — like historical writing of all kinds — presents us with various choices both in its execution and in its interpretation. The distinction between previous geographies provides an explanation for why this is so in the case of geography, and hence it is beneficial for comprehending the divergent outlooks on the history of geography: whether we interpret it on the basis of the geographicality of the world (James and Martin 1981), institutionalized academic geography (Capel 1981, Taylor 1985) or specific 'geographical modes' of thought and practice (Soja 1989).

The above three meanings of geography constitute the point of departure for the analysis to be presented in this book, which will thus concern the geographical construction of territoriality, geographical interpretations of these processes and the categories that have been employed in these efforts. In practice, the concepts of region and place and the territorial/demarcational dimension they imply are crucial. These categories have always been important concepts for establishing the disciplinary identity of geography, and they still play a key role in academic geography and more generally in recent social theory. This has again become obvious within academic geography, through the recent rise of locality studies and the so-called new or reconstructed regional geography, inspired by modern social theory, which the scholars have discussed a great deal from the 1980s onwards (Thrift 1983, Pred 1984, Paasi 1986a, 1991a, Taylor 1988a). These discussions have not been 'autonomous' in the academic field, but rather they have been an expression of a more general reassertion of space in social theory.

Most discussions of place, region and the new regional geography in general have been derived from largely theoretical and conceptual perspectives and for this reason have also remained rather abstract discourses (see, however, Warf 1988, Johnston 1991). Here I shall treat the concepts of region and place as historically constructed, created and recreated categories which can be understood in connection with the construction of territories and boundaries. Regions and places are thus not merely tools of the present analysis, but theoretical and conceptual fixed points of analysis, fixed points which, I hope, provide full information for the following concrete analysis of territories and boundaries and for a search for a kind of regional geography (cf. Wallerstein 1990).

Hence, instead of comprehending 'theory' as a ready-made framework for ordering empirical observations, the term 'theoretical' in the present context refers to the process of conceptualization: to the creation of abstractions which make various social processes and mechanisms theoretically 'visible', and finally amenable to concrete analysis (cf. Sayer 1992, 49–51).

2.2   THE QUESTION OF BOUNDARIES IN GEOGRAPHICAL THOUGHT

The global restructuring of the economy and recent developments in regional transformation in Europe — e.g. the rise of new states, changes and struggles over social spatialization, and the redefining of territories and their boundaries — have emphasized the roles of all three geographies discussed in the preceding section. Pressures for change in territorial structures emerge from social, economic and political action, new representations and the struggle to create new stories, new significations for social communities. These trends are also paralleled by developments in academic human geography and the social sciences.

Many geographers in the late 1970s and the 1980s became sensitive to social theory (Gregory and Urry 1985, Peet and Thrift 1989), while the most recent influences for a transformation in the practice of academic geography have emerged from the post-structuralist theories of culture and, more broadly, the multidimensional discourse on modern vs. post-modern society and culture (Harvey 1989, Soja 1989, Doherty et al. 1992). The most recent phase particularly includes debates on the content and form of (mimetic/non-mimetic) representations or the exploration of the interfaces between politics, space and identity (Keith and Pile 1993, Duncan and Ley 1993).

A new cultural perspective puts emphasis on the multidimensional character of diverging social, spatial and cultural frontiers and boundaries. As Morin (cited in Bennington 1990, 121) writes, the frontier is both an opening and a closing: 'All frontiers, including the membrane of living beings, including the frontier of nations, are, at the same time as they are barriers, places of communication and exchange. They are the place of dissociation and association, of separation and articulation.'

Owing to the reassertion of space in social theory, these themes have moved geographers to the centre of the dispute over the changing roles of time and space, the role of changing identities in post-modern society (Best and Kellner 1991, Rosenau 1992). In the course of these processes the boundaries of academic geography are expanding and dissolving into other disciplines such as sociology, anthropology and literary criticism. The transformation of the dominant stories is thus also taking place in the realm of academic geography.

Along with the continuous regional transformation in the world system of states and recent tendencies in social theory, the idea of the 'political' has turned out to be particularly important in shaping the diverging roles of boundaries and distinctions in territorial structure, identities and consciousness. It has become clear that the political dimension is built into various larger-scale social practices as well as into the spatial practices of the everyday lives of individuals and various social groups. These themes have been of crucial importance in the recent discourse of geographers and other social/cultural scientists (see the collections edited by Keith and Pile 1993, Bird et al. 1993 and Duncan and Ley 1993).

As far as the content of political geography is concerned, Taylor (1993a, 83) writes that the trilogy of nation–state–territory continues to lie at the heart of political geography. And as far as the content of social consciousness and various discursive practices is concerned, 'all speech is political' argues Brown (1987, 80). He remarks that relations of power and authority are aspects of all communication, both explicitly and even ostensibly political. Hence, space, place and politics become inextricably fused with each other.

This increased sensitivity to the role of language in the social and political construction of the world has also revived the role of political geography as a means of interpreting and explaining these developments. But as far as the various cultural and social roles of boundaries and the inherent relations of power are concerned, the idea of the boundary cannot be comprehended in the traditional sense provided for by political geography. Instead it becomes a part of a wider context of cultural geography. To make this difference clear, the differences and common constituents of traditional definitions of boundaries and those emerging on the basis of modern cultural and social theory will now be briefly analysed.

2.2.1 Frontiers and boundaries in the tradition of political geography

Frontiers and boundaries have been notorious as being among the most popular topics in political geography in the course of its history. Minghi (1991, 15) remarks that the interest in borders and borderlands has, in one way, been the reverse of the regional model: political geographers have focused 'on the edges, not the cores of regions'. By tradition, geographers have been interested in the influence of geographical factors on the location of boundaries and the reciprocal influence of the boundary, once established, on the development of the landscape through which it has been drawn. It has been typical to analyse boundaries on an international scale, since international political boundaries provide the most obvious manifestation of the large-scale linkage between geography and politics. As a result, boundaries have been understood above all as geographical limits of (nation-)states. As Rumley and Minghi (1991a, 2, cf. also Prescott 1987, 63) put it, international boundaries are 'palpable spatial manifestations of political control displayed in some way in the landscape and also on maps'. In the tradition of political geography, boundaries have been above all empirical manifestations of political processes, which geographers have aimed at mapping.

The role of frontiers and boundaries has varied greatly during the history of politico-geographical thought, reflecting their changing roles in the system of nation-states. The situation has been approximately the same with the idea of the boundary itself. As Jones (1959) points out, ideas about boundaries are related to their geographical and historical milieu; in a word, they have been contextual. As regards the definitions and practice of political geography, even if boundaries and frontiers have been employed at times as though they were synonymous, there prevails today a thorough agreement that whereas a frontier is a zone of contact, an area, a boundary is a definite line of separation — and not merely a line demarcating legal systems but a line of contact between power structures which at least partly manifest themselves in territorial frames (Jones 1959, 253).

Giddens (1987a, 50) is ready to conclude that borders are only to be found following the emergence of nation-states. Thus the frontier — as

a zone between states — is generally considered a historical phenomenon, and one which ended with the closing in of the world system at
the beginning of this century. 'We now live in a world with one system
so that there are no longer any frontiers: they are now phenomena of
history', writes Taylor (1993b, 164). From this viewpoint frontiers are
seen as being outwardly oriented, whereas boundaries — as
manifestations of social or national integration — are understood to
be more inwardly oriented and also necessary components of the
sovereignty of territories, marks of limits of jurisdiction.

This implicit assumption of uniqueness has caused a problem as far
as the development of the theory of boundaries and border landscapes
is concerned. It has been noted that political geography is particularly
rich in morphological, empirical and generic studies on boundaries, but
their economic and psychological functioning, for instance, has not
received the attention it merits (Muir 1975, 117).

In order to overcome the problems of the traditional approaches,
which are commonly regarded as descriptive and empiricist, some
authors have recently identified a need to appreciate border landscapes
as explicit products of a set of cultural, economic and political interactions and processes occurring in space (see the articles in Rumley and
Minghi 1991b). Within the framework of the present book, this
argument can be based on a more general discourse on landscapes,
since it has been argued that landscape is the key concept to grasp in
spatial transformation (Zukin 1992, 223).

Zukin (1992) discusses the problems and contradictory private and
public dimensions of 'post-modern urban landscapes' in her essay. Here
only some general points on the changing role of the idea of landscape
will be taken up which seem to be useful in the construction of the
conceptual basis for the present study. The most significant of these
changes is that the concept of landscape has broadened in meaning to
include an appreciation of material culture, 'text' and social processes. It
can also be employed in more abstract ways to refer metaphorically to
non-visual phenomena, such as institutional fields, cognitive structures
or the existing social order. These new meanings of landscape put stress
on the importance of spatiality and social spatialization, but they also
point to the fact that history matters — even if it is argued that post-
modernism as such puts emphasis on space rather than time (Jameson
1984). Furthermore, landscapes are always socially constructed, they are
built around dominant social institutions and are ordered by their power
(Zukin 1992). Hence landscapes are coded by society, and several
co-existing codes are in a complex way linked to different spheres of
life, e.g. social, political, cultural or economic. As a result, subjects
and landscapes are constantly transforming each other (see Berdoulay
1989, 113).

2.3   BOUNDARIES IN THE PRODUCTION OF TERRITORIES

The definitions of boundaries and frontiers put forward above were aimed at characterizing the general nature and evolution of boundaries as straightforwardly, unambiguously and exhaustively as possible. This practice may be partly interpreted as the reverse of the conclusion put forward by Prescott, according to which the 'danger of subjectivity' is probably greater in political geography than in any other branch of the subject — an opinion strengthened by the notions that boundaries are unique, that generalizations about them are not especially valuable, and that attempts to produce a set of reliable theories about international boundaries have failed (Prescott 1965, 9 and 24, 1987, 8). Boundaries and frontiers have thus been regarded as concrete and, for scientific purposes, empirical manifestations of human action and markers or indicators of more general principles of territorial organization.

The definitions considered above indicate that in the tradition of geographical thought a boundary represents most typically a line (or a vertical level) of physical contact between states and ultimately affords opportunities for cooperation and discord. A boundary typically has a physical existence, which results from its demarcation and the construction of the buildings, defences and systems of communication. Prescott (1965, 91) points out that the appearance of any boundary in the landscape is a guide to the functions applicable there, and the stringency with which they are applied. Boundaries also have an influence on the attitudes of the border inhabitants, as well as on the policies of the states concerned.

The idea of boundary will not be understood here from the perspective of a static 'territorial line' but rather from a broader, socio-culturally grounded perspective. Such an approach stresses the production and reproduction of the idea of territories and boundaries and their symbolic meanings in various institutional practices. It puts emphasis both on social spatialization and spatial socialization.

Boundary maintenance appears to be a function common to all social systems and groups, small and large, and thus part of the construction of the practices and narrative accounts by which these — more or less strictly bounded — groups are constituted. We will begin from the fact that to establish and institute something, giving it a social definition or identity, means at the same time the establishment of boundaries (cf. Bourdieu 1991, 120). To carry this out, power-holding actors, groups and classes within social systems define and symbolize the social and spatial limits of membership. At the same time, the members of these collectives usually share an iconography which helps to differentiate insiders from outsiders and define the boundaries of the political community in the discourse (Bergman 1975, 21). Boundaries and their iconographies are human creations and hence manifestations of a more general principle of territoriality, which is exploited and manipulated by various social groups and classes in the control of space, i.e. social spatialization (cf. Gottmann 1973, 137–8, Leimgruber 1991). Furthermore, they are means of reifying power, rendering visible the power emerging from social practice, from social and spatial relations.

Territories are thus, as Mitchell (1992) puts it, imbued with politics and meaning. The social construction of space and territorial representations typically exploits the basic idea of one of the 'social binary oppositions', the distinction between we and they discussed above. Thus the production and reproduction of boundaries and territoriality are two sides of the same coin, two parts of the process of social spatialization. In the control of territories the dichotomy between 'us' and 'them' is typically transposed to the enveloping physical reality and devoted to constructing an antagonism to other groups (Tuan 1974, Knight 1982). Boundaries not only separate groups and social communities from each other, but also mediate contacts between them. Mach (1993, 55) points out that they provide normative patterns that regulate and direct interactions between members of social groups, the rules of how to cross boundaries, as well as the rules of exchange of people, various goods and symbolic messages.

The production of territories, boundaries and their symbolic representations are always spatially and historically contingent processes. In this sense, the study of boundaries should be contextual, but this context is not merely that located in the border area itself. It is, as far as nation-states are concerned, 'located' in the broader socio-spatial practice and consciousness of the society and state, whose territoriality and sovereignty are compressed and symbolized within the boundaries representing one specific form of state power. In fact, the proper 'context' is the continual process of nation building, i.e. the process of creating viable degrees of unity, adaptation and a sense of national identity among the people — providing a narrative account for the community.

2.3.1 Boundaries, ideologies and hegemony

As Ratzel has already pointed out, boundaries between states are an expression and measure of state power (Giddens 1987a, 49). Here it may be argued that the production of spatial distinctions and territorialities is always a struggle over the right to put forward (new) definitions. This means that the question of the production/reproduction of various social distinctions and identities is also essential in the case of spatial demarcations. In this connection the concepts of ideology, hegemony and mentality become indispensable for fathoming the constitution and architecture of socio-spatial integration. They are essential for understanding the production of the integrating stories of various communities. Nevertheless, one note is inevitable: even though these categories are exploited here, they are not adopted in an over-rationalized and reductionist manner; human beings are to be regarded as creative agents, not as puppets stuck in deterministic social structures. The action of agents is contextual, not universal: space and time make a difference. The contexts, for their part, are active networks of people, not passive environments. As Thrift (1986) puts it, action — in context — is always 'joint action'.

The concept of ideology has been employed in a myriad of ways to denote ideas and beliefs, misrepresentations, illusions, false conscious-ness etc. (see Thompson 1990). In this connection a rather 'neutral' interpretation of the concept is adopted, to refer to the process of producing meanings, signs and values in socio-spatial life, i.e. sig-nification. Nevertheless, various symbolic forms intersect with power. This points to the simple fact that most societies are divided by class, gender, age, ethnicity and or caste, and these social forms of differen-tiation are contextual (Cosgrove 1989, 124). Cosgrove points out that a different position in society obviously implies 'a different experience and consciousness', a different culture, even if the state as representative of a national culture typically strives to introduce 'at least the rudiments of a common culture into every schoolroom'. Thus power is contested between various groups, and this makes the concepts of ideology and hegemony significant. Ideologies tell people what is possible and what is right and wrong. Also, they 'structure the limits of discourse in society, and are present in all aspects of everyday life, including family, school, neighborhood and workplace' (Clark and Dear 1984, 53).

Ideology is generated above all by the people, groups and classes who hold power, who legitimate the political power. Nevertheless, ideology and its manifestations may be as necessary to subordinate groups in their spatially and temporally contextualized struggle against the social order as they are to dominant groups in their defence of the *status quo* (cf. Thompson 1990, 53, Eagleton 1991, 1–31). Thompson (1990, 56) points out that we cannot 'read' the ideological character of symbolic phenomena from the symbolic phenomena themselves, but only by placing them in the socio-historical context within which they may (or may not) serve to create and sustain relations of domination.

Symbolic forms are not, Thompson (1990, 58) writes, merely representations which 'serve to articulate or obscure social relations or interests which are constituted essentially at a pre-symbolic level'. Rather, he argues, symbolic forms are continuously and creatively implicated in the constitution of social relations and the social world.

The concept of hegemony is particularly useful for understanding this connection. In principle it refers to the power of dominant groups to persuade subordinate groups to accept their moral, political and cultural values as the natural order (Bocock 1986). Whereas domination is usually directly expressed in political forms and is explicit and compelling at times of crisis, hegemony refers to cultural institutions and hence to the structural features of society where ideologies, for instance, become an established part of consciousness and forms of experience. Hegemony is thus the lived system of constituting and constituted meanings and values (see Williams 1988, 125–32). The media, science, art, education and behavioural norms, for instance, are important vehicles for the operation of ideological processes. Similarly, various symbols and signs can be heavily loaded ideologically (e.g. national symbols).

Hegemony points to the predominant forms of practice and consciousness (Williams 1976, 144–6, 1988, 125–32). Williams writes that hegemony creates for most people a feeling of reality, a feeling of something absolute which is based on the lived reality and which is difficult to break through for most members of society in most arenas of life. Hegemony is never fully achieved, for it is always contested: the dominance of power-holding élites is always being challenged by those in subordinate positions (Jackson 1989, 53). These positions can have a regional, class, generational, racial or sex basis, which points to the fact that discursive practices are more than simple ways of producing texts, and that they do not involve only language but also practices and social positions which embody power. Some people and groups are always more active and powerful in the production of 'we-ness' and otherness, whereas most people are reproducers (cf. Dalby 1990, 6, Mach 1993).

The concept of mentality, widely employed in historical writing during the last few decades, provides a useful heuristic category for comprehending how ideologies and hegemonic structures are related to the daily life of actors and become a part of the practical orientation through which members of various groups internalize and come to share specific stories, ways of interpretation and evaluation without being clearly conscious of them. Mentalities are typically regarded as features of large collectives which change slowly. Modern daily life is being shaped continually by innumerable rules, based partly on laws and statutes and partly on invisible practices through which various groups continually establish boundaries between us and the Other, between what is permitted and what is prohibited. Nevertheless, as Pred (1989, 214) remarks, the discourses of daily life within any region are always marked by 'diversity and stratification, a wide variety of voices and accompanying gestures, a multitude of speech communities, a polyphonic plurality of perspectives'.

As far as the role of boundaries is concerned, the question of ideology, hegemony and mentalities becomes part of a more general issue of the production and reproduction of space and territoriality, and finally a question of a perpetual regional transformation. Hence it is necessary to deliberate over the role of history in this process. The basic idea of this work is that both past and present can and must be approached historically (cf. Driver 1988). In the case of territories and regions, our understanding of the present must thus be based on their 'becoming' rather than on their 'being' (cf. Pred 1984, Paasi 1986a). This presupposes a definition of the concept of history — in relation to regions and boundaries — that will be exploited in order to structure the organization of empirical illuminations. Hence the question of the production and reproduction of regions becomes a crucial one.

2.4 THE INSTITUTIONALIZATION OF REGIONS

To understand the nature of regional transformation and the emergence and disappearance of regions and localities, it is necessary to start with a brief discussion of the role of history in this process. It was common in the socio-spatial analysis of the 1980s to refer to various levels of history (as well as spatiality), i.e. not to see history merely as a collection of events that have occurred at various times in various localities, but to approach it in a manner that aims to identify various 'rhythms' through which social life and inherent power relations are historically constructed and reproduced. Pred (1990, 13) compresses this idea neatly when arguing that, 'At any given moment, the institutionally embedded power relations associated with people practicing in place vary in their geographical extent (and historical depth), in their territorial expanse (and long-term duration), ranging from the gendered, patriarchal spaces of the dwelling, farmstead, or village to the governed space of the nation-state and then scattered domains of the multi-national corporation.'

Braudel's (1980, 27) distinction between various 'histories' is probably the best known of the structurally viewed concepts of history. Braudel makes a distinction between three forms of history. The first is traditional, short time-span history which deals with events and individuals. Economic and social history puts cyclic movement in the forefront and is interested in conjectures, so that it 'lays open large sections of the past, ten, twenty, fifty, etc. years at a stretch ready for the examination', and finally the history of the *longue durée*, very long time-spans, which has to do with structures and realities which time uses and abuses over long periods.

As far as the forms of temporality entering into social life are concerned, Giddens (1987b, 144–8) provides a much more detailed socio-theoretical analysis. He distinguishes — and this distinction can be made only analytically — first the *durée* of day-to-day life, by which he refers to the ordering of social activities, their structuration, via their daily repetition. This form of temporality naturally intersects with the *durée* of the individual's life-span, which Giddens distinguishes as the second aspect of temporality. Both of these forms, for their part, interweave with the *durée* of institutions, which refers to the fact that all societies endure beyond the lives of those individuals whose activities constitute them at a given moment.

Thus both Braudel and Giddens have put forward ideas that seem to contribute to understanding the temporal forms of social life, but they do not help us very much to understand how these forms of temporality and history are related to the rise of territories and the communities producing and reproducing them. It is clear, however, that the day-to-day lives of individuals, and even their life-histories, are organized in relation to various territorial units which have to be produced. As Knight (1982, 517) reminds us, 'territory is not; it becomes, for territory itself is passive, and it is human beliefs and actions that give territory meaning'.

The construction of spatial boundaries is always a part of the construction of territorial units in space. But what are the 'regions' that constitute territories? It has been typical in geographical thought to regard regions as mental instruments of classification, naturalistic objects, communities, and so on. Recent connections with social theory have inspired geographers to develop regional concepts which put emphasis both on the material basis of region formation and the role of a region as a 'medium' of social reproduction (cf. Paasi 1991b). Few authors have emphasized the historical nature of regions (see Pred 1984, Lee 1985, Paasi 1986a, Taylor 1991a).

As far as the parameters of regional transformation are concerned, I have referred to the construction of territories in my previous publications (Paasi 1986a, 1991a) as the *institutionalization of regions,* an abstraction which aims at rendering 'visible' certain key dimensions of regional transformation. Institutionalization thus refers to the process during which specific territorial units — on various spatial scales — emerge and become established as parts of the regional system in question and the socio-spatial consciousness prevailing in society. Society itself is simultaneously being transformed as part of the larger context, as recent tendencies in Europe aptly indicate (see Figure 2.1). Regions are not 'organisms' that develop and have a life-span or evolution in the manner that some biological metaphors — so typical in Western social thought — would suggest. Rather, following Dear and Wolch (1989, 6–7), regions and localities are understood in this

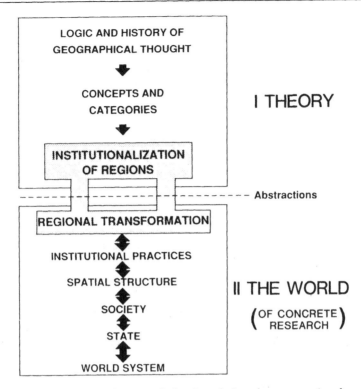

Figure 2.1 Conceptual framework for the relations between regional transformation and the institutionalization of regions

framework as being a complex synthesis or manifestation of objects, patterns, processes, social practices and inherent power relations that are derived from simultaneous interaction between different levels of social processes. Furthermore, these processes operate on varying geographical and historical scales. Through the institutionalization process and the struggles inherent in it, the territorial units in question 'receive' their boundaries and their symbols which distinguish them from other regions.

To understand the nature of the institutionalization process, it is useful to characterize abstractions that 'make visible' the stages of this usually long process in regional transformation. Four stages appear to be of importance: (1) the constitution of territorial shape, (2) symbolic shape and (3) institutions, and finally (4) the establishment of the territorial unit in the regional structure and social consciousness (Paasi 1986a, 1991a, Figure 2.2). The institutionalization of regions thus refers to the process through which various territorial units are produced and manifest themselves in various social and cultural practices,

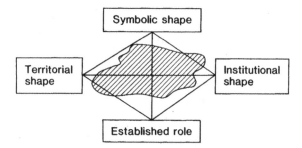

Figure 2.2 A conceptualization of the process of the institutionalization of regions

such as politics, economy and administration, which in turn will be produced and reproduced consciously or unconsciously by people. The institutionalization of a region is always a manifestation of the spatial divisions of labour and power relations that are embedded within it. Hence some individuals, groups and classes are always more active in the production of regions, while most people tend to be reproducers. The stages of the institutionalization process are usually simultaneous and can be distinguished from each other analytically. This distinction is useful, however, in order to clarify how various social practices are constituting themselves and constituted through these historically contingent processes, or 'stages' (see Paasi 1991a).

The development of territorial shape refers to the localization of social practices through which regional transformation takes place and a territorial unit achieves its boundaries and becomes identified as a distinct unit on some scale of the spatial structure. The roles of boundaries thus vary greatly. In the case of nation-states they are commonly strictly defined, symbolized and sanctioned, but in the case of smaller localities they can be much more diffuse and less influential — hardly visible in everyday life. Through the formation of the symbolic shape, specific symbols for expressing and demarcating the territory will be established. These become part of the symbolic ordering of space and time that provides a framework for experiences through which we learn who or what we are in society (cf. Harvey 1989, 214).

Territorial symbols are often abstract expressions of supposed group solidarity, embodying the actions of political, economic, administrative and cultural institutions in the continual reproduction and legitimation of the system of practices that constitute and demarcate the territorial unit concerned. Thus symbols are instrumental in the sense that they serve to evoke powerful emotions of identification with territorial groupings and can generate action. Symbols are 'keywords' in the dominating story of a territorially based community.

The most important symbol is doubtless the name of the territorial unit or region, which usually 'gathers' together its historical development, its important events, episodes and memories and joins the personal histories of its inhabitants to this collective heritage. Names of regions and other localities conform to the most classic definitions of symbolism. Naming, giving a name to someone or something, Moscovici (1981, 196) remarks, always means withdrawal from a 'troubling anonymity', a settled position in the culture's 'identity matrix'. Cohen and Kliot (1992, 655) point out that these names are part of the process of attaching meaning to one's surroundings and act as sources of information, facilitating communication, helping us to know things, and serving as repositories of values.

The emergence of institutions is the reverse of the increased use of territorial symbols and signs. Individuals and groups are socialized through institutional practices into varying territorial memberships through which the perpetual production and reproduction of social consciousness take place at various territorial levels. These memberships connect the inhabitants with the symbols of the region in various practices — and simultaneously demarcate the Other. The establishment of a region is an abstraction referring to any continuation of the institutionalization process after the region has an established, but not necessarily administrative, status and a specific regional identity in the spatial structure and social consciousness. Thus it can be argued that these territorial units have specific *structures of expectations* (Paasi 1986a, 1991a). This abstraction is employed here to refer to time–space-specific, regionally bounded, institutionally embedded schemes of perception, conception and action which — as a part of the dominating narrative account of the territorial unit in question — serve as significant structures in socio-spatial classification. The structures of expectations of regions are constitutive of the story which expresses where the territorial unit has come from and where it is going.

2.4.1 Regional identity

It is profitable to discuss one more abstraction which facilitates our understanding of the language of territorial distinctions — regional identity. During its institutionalization process, a region achieves a specific identity (the identity of the region), which cannot be reduced merely to the regional consciousness (regional identity) of the people living in it or to symbolic values. The identity of a region is typically expressed in structures of expectations, and hence it is useful to link it with institutionalization, which includes the production and reproduction of regional consciousness and images of the region in the inhabitants and other people living outside the region, and material and symbolic features of the region as parts of the ongoing process of social reproduction. This points to the fact that the formation of social identity and social reproduction are one and the same (Abrams 1982, 262).

It is important to note that 'regional identity' is a theoretical, conceptual category that characterizes the multidimensional relationship between individuals, groups and society and is manifested in different forms in social practice depending on the historical, time–space-specific manner in which language and its expressions are employed to depict this relationship. Probably all languages contain some emotionally laden expressions illuminating the relationship between human beings and ties with their environment, native localities and native country (cf. Paasi 1986a). In diverging local contexts there usually exist locally coloured vocabularies referring to local identities (cf. Hummon 1990, 4), and the adoption of these expressions is part of the 'territory-building' process on various spatial scales. In everyday speech these expressions are typically manifestations of regional identity and the reproduction of the structures of expectations of a region.

Regional identity weaves together elements which are significant in the institutionalization of a region and which are represented in its structures of expectations. Structures of expectations operate as a framework for social classification among the inhabitants and those living outside the region, i.e. in the construction of we-ness and other-ness. The distinction between the identity of the region and the regional identity of the inhabitants is crucial for understanding the (analytical) difference between the historical construction of a region and the life-histories of its inhabitants. Furthermore, this distinction makes it clear that it is essential to problematize the distinction between the production and reproduction of space and collective representations, how this takes place through various institutional practices, how material and mental spatial 'fix' will be created through these practices, how various spatial groupings are created and defined, how 'we' are distinguished from 'them', and so on (Paasi 1986a).

Also, when deliberating over the 'we' or community aspect of territories and regional identities it is crucial to make a distinction between ideal and factual communities (Paasi 1986a). The former points to the spoken or written identity of a territory, i.e. it refers exactly to the discourse through which the idea of a regional identity is maintained. These discourses are inextricably connected with the social division of labour and inherent relations of power. Some actors are specialized in the production of territorial distinctions and identities, e.g. politicians, journalists and representatives of tourism or territorial organizations, producing representations of territorial identities on various spatial scales because of their position in maintaining the hegemonic structures of society. The notion of factual community, for its part, refers to those communities, in practice associations, in which the actors really participate. The production and reproduction of territoriality can be among their explicit aims.

Accordingly, regional transformation consists of an institutionalization of regions taking place continually on various spatial scales and potentially having different forms on different scales. The production of territorial units always reflects the social division of labour and associated power relations. Since the institutionalization of regions is part of this continual regional transformation, it always means the de-institutionalization of some other territorial units, i.e. the change in the importance of various institutional practices through which the region exists (cf. Paasi 1991a). It should be noted that regional transformation takes place simultaneously on all spatial scales, e.g. at the local, regional, national and global levels. Hence regional identities are also organized hierarchically. In regional transformation some of these identities, and their material and symbolic base, can be thought of as being in danger because 'their' identities are expanding into 'our' territory. This situation can create conflicts between territorial units, whether states or sub-state entities.

Some aspects of territorial transformation become more visible in the institutionalization of territorial units and some less. The state is the central apparatus through which important social practices, e.g. legitimation, take place and in which the economic, administrative and political prerequisities for the constitution of various sub-state territories are created. Likewise, many of the integrative forms employed in continuous nation building and the creation of social order and consciousness on various spatial scales — e.g. the ideas of nationalism, regionalism or territorial identities — are created by state-engendered practices through the political socialization system. Similarly, the status of symbols varies on various scales. The name of the territorial unit in question, for example, is different in the case of nation-states from that in the case of sub-state regions. In the first case it is an intrinsic component of the political landscape of the nation and an expression of the sovereignty of the state, whereas on the local scale names can provide more immediate expressions of the connections between personal histories and the continuity of the locality.

On the grounds of the above discussion it is now easy to discern that the construction of territoriality is essentially a process of production and reproduction of mental representations which arises on the basis of diverging social and material practices and interests. The production and reproduction of territoriality and identities are part of the continual struggle over social classifications, a struggle 'over the monopoly of

power to make people see and believe legitimate stories', to make them
'know and recognize', to create 'the legitimate definition of the divisions
of the social world', and thus to 'make and unmake social groupings'
(Bourdieu 1991, 221). The production and reproduction of space are at
the same time the construction of social distinctions and more or less
marked, visible and invisible boundaries between social groupings and
classes. These distinctions can then have a more or less explicit
territorial basis.

To employ the terms of the recent 'culturalist turn' in geography and
other social sciences, collective representations of territories and
boundaries can be regarded as cultural *texts*. They are not merely
neutral or mirror-like reflections of the shapes and physical-cultural
landscapes of territories and their boundaries, but fabrications that
people have engendered to reflect past and present socio-cultural relations which for their part evince diverging forms of ideologies and aim
at signifying and legitimating the distinctions between 'us' and 'them',
'our' culture and way of life and 'theirs'. The expanded concept of text
that geographers have adopted from cultural researchers and literary
theoreticians embraces various cultural productions such as paintings,
maps and landscapes, but also social, economic and political institutions
that together constitute the story of the community. Hence this viewpoint comprehends texts as 'constitutive of reality rather than mimicking
it' or 'as cultural practices of signification rather than as referential
duplications' (Barnes and Duncan 1992, 5–8). This point of departure
implies that the construction of territories and boundaries cannot be
understood in terms of empiricism by limiting the analysis merely to
visible phenomena. Thus an essential task in any study of territory and
boundary construction is the conceptualization of diverging territorial
ideologies and their construction.

3

TIME, SPACE AND CONSCIOUSNESS: CONSTRUCTING NATIONALISM AND COMMUNICATING BOUNDARIES

3.1 THE NATIONAL IDEAL

Social spatialization, the construction of territoriality and the rise of regions were traced in the previous chapter in rather general terms. The various territorial units and scales nevertheless make a difference in the institutionalization process. The most important institutionalized territorial frame, which cannot be 'escaped' in the modern world, is undoubtedly the (nation-)state, the most prominent of the social forms that modernity has produced. The state is a political organization covering a particular territory. It has specific forms of territoriality, surveillance capacities and a monopoly over the means of violence (cf. Giddens 1991, 15). A nation, for its part, is a community of people with a common identity, which is typically based on shared cultural values and attachment to a particular territory. The nation-state is the most powerful combination of nation and state. By tradition, a nation-state has meant a sovereign state inhabited by a group of people who see themselves as one. By the 1950s the term turned out to refer in the practice of the social sciences to any 'sovereign state', no matter how ethnically or nationally heterogeneous that state might be, and the nation and state citizenry became the same thing (Murphy, forthcoming). Hence in the current world there exist nations without states as well as states with more than one nation ('plural' states), where the state can adopt various strategies to control the situation (Smith 1991, 15, Short 1993, 91–9).

In the tradition of political geography it has been typical to concentrate on individual states, and analysis has correspondingly been directed towards the separate treatment of each state. Also, it should be noted that both 'state' and 'nation' as such are abstractions. It seems to be typical for the (nation)-state under discussion to be above all a developed Western capitalist state. Even though it is possible to find some key mechanisms of control and security in almost all states that are significant for the production and reproduction of the state, the state should in the 'real world' always be comprehended contextually. Whereas geographers commonly argue that space makes a difference, it can be argued that 'state makes a difference' too (cf. Paasi 1994).

Concentration on the state also underestimates the existence of an inter-state system and concentrates the explanation at the state level (Taylor 1993b). The existence, limitations and symbols of boundaries between states, as well as the sanctions of acting against them, are usually confirmed in laws and statutes to express the sovereignty of the state and exercise territorial power in the interacting system of states. Hence the state determines the specific rules by which society will be organized and individuals will operate in their daily lives (Johnston 1991, 195). In fact, spatial units become territories only when boundaries are used to affect behaviour by controlling access (Sack 1986, 19).

The state has for a long time been a key category in the tradition of political geography, and it has been understood in various ways. Nevertheless, a typical criticism since the 1980s has been that by tradition the roles and nature of the state have not been adequately treated theoretically in geographical thought (cf. Taylor 1983, 1993b).

As far as diverging perspectives on the need for a theory of states are concerned, one distinction which can be made is between *status quo*, reforming and radical attitudes (Macpherson, cited in Taylor 1983). The first of these, which accepts and upholds the main features of the existing democratic-liberal state, has dominated much of modern political science, particularly in the USA. Reformers, including non-marxist socialists and social democrats, accept the normative values of the late nineteenth-century liberal theorists but believe that the present-day state does not put those values into practice, while radicals do not accept the existing state and social order and reject the theories that uphold or attempt to reform it, preferring to replace both with Marxist theory and practice (see Taylor 1983, 10).

The role of the state will be treated in the present context on two levels. First, the state is empirically a specific region or territory with relatively well-defined, internationally recognized boundaries, the political identity of the people living in which is constituted through specific forms of nationalism (cf. Dear 1986). Theoretically, the state is recognized through its basic functions — hence it is a set of institutions which aim at production and reproduction of the society at the level of the economy, administration, politics, culture etc.

Accordingly, it has been argued that the state must have at least three main functions (see Johnston 1989). The first is to secure social consensus, whereby all the people in its territory accept specific rules for action in society, the second is to ensure adequate conditions for production, which points to the provision of an infrastructure within which production and exchange can take place, and the third is to guarantee social integration by ensuring the basic welfare of all citizens — particularly the poor and exploited groups within society (Johnston 1989, 295).

Every state is a part of the international system of states, however, and hence another specific dimension has to be added to the preceding frame, i.e. 'the repressive sub-apparatus', as Dear (1986) labels it. This consists of the mechanisms of internal (intranational) and external (international) enforcement of state power, and includes the civilian police and the armed forces. In most countries the state is very much a power container, but it is typically its citizens rather than external foes who have to confront its effects (Taylor 1994). As far as the basic military medium for maintaining territoriality is concerned, this incorporates the existence of armed forces and/or a specialized guarding system for the control of boundaries. Security policy is displayed in foreign and defence policy doctrines and hence contains an internal and foreign policy perspective. Internal and external policy, for their part, are dependent on factors such as the history and geography of the state and the interstate system, international political developments, prevailing images of threat, economic development, technical development, the struggle for political power between various social groups and classes, regional policy etc. (Visuri 1989, 15–21). These elements constitute the hegemony of the state. Hence we see that paradoxically it is *internationality* that creates a world in which the maintenance of armies and weapons appears to be natural and where the arts of violence are regarded as the summit of human solidarity (Rée 1992, 10).

All these state functions are of course historically contingent, which is most strikingly illuminated by the fact that the number of independent states has continually increased, now exceeding 190. Since the basic functions of the state are to a great extent directed towards the reproduction of itself, it is particularly concerned with answering people's needs for identity, stimulation and security — sometimes regarded as the basic human needs (Johnston 1989, 296). It is by no means self-evident why this specific territorial context, the state, has become the key institution for satisfying such needs.

There exists one key difference between the state and other territorial contexts, however: as Bloom (1990, 53) points out, the state does possess one overwhelming advantage over other territorial entities which might compete with it for the loyalty of its citizens. This is the monopoly of force, and power — often accompanied by control of the media and the education system. The state is centred on a specific, delimited territory, over which it has authoritative power, and *in extremis* it can claim the life of its citizens. Mann (1984) talks about infrastructural power, which points to the ability of the state to penetrate daily life within civil society and to implement political decisions. Territoriality is essential to this process, and hence the greater the infrastructural power of the state, the greater usually is the territorializing of divergent forms of social life.

The crucial role of the modern state provides one cornerstone for the argument which follows. It should nevertheless be noted that day-to-day life always takes place in some time–space-specific territorial context within states. While there may also be other territorial units, there are two necessarily territorial contexts in the modern world which constitute the basis for the production and reproduction of human life — the state and the locality (cf. Agnew 1987). Even though these contexts are settings for action, they are not static frameworks but are themselves in a process of transformation, a fact which holds good in the case of both states and other territorial units.

3.2   SOCIAL INTEGRATION AND SPATIAL SCALES

The process of nation building aims at binding the state and its inhabitants — a nation or nations — together. Nation building is a metaphorical expression which has the connotation of an 'architectural' or somewhat mechanical model for the construction of nation-states, in accordance with which a nation can be built following different plans, using different materials, rapidly or gradually, by different sequences of steps, even independent of its environments (Deutsch 1963, 3). It is by no means necessary, however, to regard this metaphor in a mechanistic way as a construction process which pursues a 'straightforward logic' and plans conceived by 'superior beings of heroic stature' (Friedrich 1963, 28). Rather, nation building can be regarded as a theoretical abstraction referring to a process which has both internal and external interdependence in space and time. It can also comprise the contradictory motives and actions of individuals, groups and classes in the struggle over the right to define social practices in various spatial contexts. Nation building is hence also a time–space-specific, context-dependent process.

The idea of nation building paves the way to two fundamental problems of social theory: the question of social integration and the relation between structure and agency, collectives and individuals.

Individuals become members of groups — and citizens of nation-states — by adopting collective interpretations and purposes in various social practices.

Collective interpretations are representations that are created by human beings and are effective only through human action, but they also seem to have a relative independence of various social practices. Even if we do not interpret them in the manner of the sociologist Durkheim (1966) as 'social facts' located in the society outside individuals but which have power over them, the roles of these elements of a common world view are without doubt unifying and normative (cf. Manninen 1977, 16). They provide individuals with continuity and hence join them with common stories, contain a specific perspective on nature, society and human beings, and also guide the individual and collective search for knowledge and creative action. Hence these representations are constituents of the practice of life, where action, knowledge and emotions are fused together. In this connection consciousness and material practice are inextricably bound to each other.

Since Durkheim (1964), one of the key concepts of sociology has been *solidarity*, referring to the fact that the 'glue' of any social system is to some degree based on individuals' tendency to internalize their society's values, norms and accepted patterns of behaviour. In the case of a nation-state it is reasonable to comprehend social integration, and particularly 'solidarity', on the basis of legitimacy, so that if the members of the society share solidarity with the state, they approve the basic rules of the political system as legitimate and acceptable and are ready to act on behalf of them.

Durkheim's basic conception is that social life emerges from two sources: from the similarity of the consciousness of individuals and from the social division of labour. In a modern society the division of labour becomes the integrating mechanism. Hence, where individuals often specialize in the tasks that most naturally correspond to their abilities, they become more dependent on others and on the whole of society, which is the result of their coalescence.

Identification is a private psychological as much as social act, even if it has been typical in the social sciences — as was the case with Durkheim — to explain 'social' by 'social'. Bloom (1990, 52) suggests that political ideologies and ideas of nationalism cannot of themselves evoke identification. Identification comes only if 'it interprets and provides an appropriate attitude for an *experienced* reality'. This experience, Bloom remarks, may be politically manipulated, but any symbol or an ideology without a relevant experience is 'meaningless and impotent only in terms of evoking identification'. It is here that the communal rhetoric becomes crucial in the construction of the collective *Besinnung* (cf. Carr 1986, 157). Contrastingly, ideologies symbolize reality in specific ways, so as to motivate commitment and action — they are rhetorically persuasive and simultaneously enable people to appreciate what kinds of 'communities' and symbolic landscapes (villages, provinces, nation-states) exist in general. Community ideologies thus also provide accounts, and through them a shared sense of integration and a moral landscape (cf. Hummon 1990, 37–40). This is the ultimate basis for the production and reproduction of the languages of difference and integration.

There have by tradition been two extreme views in geographical thought — and several others between them — regarding the problem of national identity and integration. These parallel the visions of the nature of the state. According to the critical interpretation, national integration can best be understood in relation to the ideologies generated by the state, while such ideologies for their part represent the interests of the hegemonic groups of the state, e.g. in the control of education or the mass media, and ultimately contribute to the maintenance of the capitalist social formation, the accumulation of capital and the exploit-ation of the working class. This brings us back to the vision of state as 'a committee for organizing the affairs of the bourgeoisie' (Taylor 1993b, 179–90). At the other extreme are probably the deliberations of some humanistic geographers, e.g. Tuan (1975), in which the nation (and national identification) is approached in a relatively non-problematical, taken-for-granted way as one specific level in the multilayered idea of place which is understood as the centre of meaning and territorial identification.

The question of levels of identification leads us to the theme that has been pivotal in the history of geographical thought — the question of scale. A number of geographers have discussed the role and production of scales as part of the territorial organization of the world, sometimes putting forward distinctions in an *ad hoc* manner and sometimes aiming at providing more theoretically grounded divisions. One example of the former is the division between imposed, overtly political hierarchies, e.g. district, city, county, state, and those which arise from everyday life, e.g. household, street, neighbourhood, community (Webb 1976, cited in Bird 1989, 37).

Societies not only produce space but they also produce scales, which are produced and contested in social practice (N. Smith 1990, 1993). They do not arise *in vacuo,* and hence cannot be reduced merely to the categories that academics generate in order to classify the phenomena of the social and physical world. Examples of the more theoretically grounded discussions of scale are the frameworks of Taylor (1993b) and Agnew (1987), which aim at conceptualizing the social content and dynamics of diverging spatial scales.

A useful heuristic device for approaching the abstract distinction between spatial scales is the framework put forward by Taylor (1982, 1993b), who singles out three underlying socio-spatial scales, the scale of experience, the scale of ideology and the scale of reality. He calls the scale of state 'ideology' since it aims at separating 'reality' (global reality of the world economy) from 'experience' (the local everyday life environment in which reproduction takes place). Taylor's materialistic framework begins with the organizing principles of the capitalist world economy, and it is useful for the present purposes since it indicates the role of the state in the construction of national consciousness and identities — whatever the ultimate purpose of the exploitation of the latter may be. Taylor (1993b) does not stress the autonomy of the state, however, nor the internal dynamics of nation-states in economic, political and cultural development (cf. Cooke 1989b).

Taylor's framework thus identifies the role of the sphere of experience, even if it does not develop it. The scale of experience includes the scale of home, and also the scale of body, which is now beginning to be considered within geographical discourse (N. Smith 1992, 76). The framework proposed by Taylor (1982, 1993b) helps to appreciate the connection between processes taking place on various spatial scales and in various social spheres, and illustrates the essential functions of the scales. The question of the relations between structure and agency needs to be evaluated in a more detailed manner, however.

Agnew (1987) discusses scales more precisely when deliberating over the relations between structure and agency. Structures that originate from a more extensive order (national, regional, global) are at the same time imposed on, or reproduced and modified in, smaller territorial units within the spatial realm concerned (cities, countries, regions). Territorial states are made up of localities in which the socio-cultural and economic relations on which the social and political order depend are constituted, and which are hence interrelated by state-building and social practices connected with this process. Agnew (1987, 1) points out that the social bases of response and resistance to state-bounded institutions are best viewed in terms of the histories of places or localities, rather than in terms of universal or national structures (such as class as a category) or in terms of an individual agency.

Places, according to Agnew (1987, 28), contain three major elements. The first is 'locale', which refers to the settings in which social relations are constituted, the second is 'location', which points to the geographical area encompassing the settings for social interaction as defined by social and economic structures operating on a wider scale, and the third is 'sense of place', or the local structure of feeling. The key point for Agnew (1987, 28) is that local social worlds of place or locale cannot be comprehended apart from the 'objective macro-order of location' and the 'subjective territorial identity of sense of place'.

3.2.1 Nation building and everyday life

The fundamental constituent of every nation is its population, joined together by certain common characteristics or sentiments. The most significant of these is probably the sense of belonging, identification with a particular social group and portion of land, the size of which and the manner in which its boundaries are drawn will vary from case to case. In these divergent contexts human beings differentiate themselves from others who do not live in 'their' area. People are also classified as a result of political and administrative practice as inhabitants of various ready-made territories (Weilenmann 1963), possessing various institutional practices by which they produce and reproduce social spatialization and become members of villages, communes, provinces and states. Although for most of their everyday lives people are not conscious of their 'community' perspective, the 'matrix' or 'order of things' provided by the community constitutes the self-evident framework which shapes their experience of the community and influences their action towards other communities (Hummon 1990, 11).

Most individuals thus operate in their everyday lives in various overlapping categories of roles in the (spatial) division of labour. They identify themselves with a number of groups distinguished by more or less visible social, cultural and spatial boundaries. Tester (1993, 6) writes, when discussing the principles of Simmel's sociology, that without boundaries the 'space of our worlds' would indeed be infinite. At various stages of socialization individuals associate themselves with new roles, and at certain rank(s) within one or more social hierarchies (cf. Meyrowitz 1985, 52). In the course of their lives people are consequently socialized into sundry territorial and other communities. Furthermore they participate in the production and reproduction of various territories in the institutionalization process. People learn values and morality from their territorial community, and absorb its common sense or mentality in divergent discourses. Language serves as a building block for other forms of collective consciousness and identities — whether they are locally confined or extend over a wider area (cf. Pred 1989, Billig 1991).

Group identity always involves the notion of otherness attached to those not in the group and those who do not share the same situations and stories with a specific group. Herzfeld (1986, 82) provides the basis for this: if all ethnic and national terms are moral terms in the sense that they imply a qualitative differentiation between 'insiders' and 'outsiders', moral value terms are also to some extent 'negotiable markers' for the lines of social or cultural inclusion and exclusion.

Recent events in the former Eastern bloc have proved distressingly that the territorializing of group identities does not inevitably occur in

peace and harmony. Pred (1989, 221) remarks that, both metaphorically and physically, each community and place has its own sites
of confrontation, spaces of struggle and arenas of contention, even if the
matters contested are embedded in geographically extensive processes of
non-local origin. Many of the moral conflicts and dilemmas facing
individuals have their origins in the fact that people identify themselves
with several territorial communities, stories and discourses at the same
time.

As Rokkan and Urwin (1983, 67–9) write, identity can usually be
broken down into at least four component parts: myth, symbol, history
and institution. First they define the mythical aspects of identity as 'a set
of beliefs that creates an instrumental pattern for behaviour in the sense
that these beliefs afford aims for their followers'. Several sets of beliefs
may of course feed into the cultural myth, but Rokkan and Urwin put
particular stress on religion. They emphasize the role of nationalism as
a 'civil religion', which in fact defines the nation as holy. Secondly, the
symbolic element, for its part, represents the enduring expressive aspect
of culture. This transmits the values of a culture from individual to
individual, from generation to generation. Language is the supreme
expressive component of identity.

Thirdly, the history of a given territorial unit is important for its
population independent of any scholarly or academic historical
standards. The object of interest in this 'lore' can vary from memories
of significant past events and decisions through 'the celebration of ritual
acts' to micro-level socialization into an ideology which is congruent
with the aims of nation and state builders. Rokkan and Urwin (1983)
point out that what actually took place in the past is usually of minor
importance in this lore or story. Fourthly, they stress the role of various
institutions, e.g. laws, privileges, education and taxation systems, in the
construction of identities.

3.3   THE CHANGING ROLES OF NATIONALISM

It has been argued that the state is nowadays a natural, almost
unquestionable part of our world and everyday life (Mach 1993, 95,
Taylor 1993a). This also holds good in the case of two key processes of
nation building: national integration and nationalism, which are also
typically so 'naturalized' that they are rarely questioned today. Through
these processes the physical and political space of a nation will be
transformed into cultural spaces, which are then typically represented as
being internally homogeneous — and this homogenization typically
takes place in relation to the Other.

Löfgren (1989), analysing the nation building process in the case of

Sweden, distinguishes between two cultural 'paradigms' in nation building: patriotism and nationalism. The former is based on the love of God, King and country by *subjects* of the state, and thus emerges from the internalization of the values represented by these fundamental symbols, while the latter is based on broader ideas of a *Volksgemeinschaft*, a shared history and culture, an idea of 'our' destiny, as well as an idea of equality and fellowship. Hence, Löfgren points out, nationalism contains 'political dynamite' and can be exploited both to mask class interests or to fight them.

An influental theoretician of nationalism, Anthony D. Smith (1979, 1), argues that of all the visions and faiths that compete for loyalties in the present world, the most widespread and persistent is the national ideal: none of the numerous ideologies has been so successful in penetrating every part of the globe, no other vision has set its stamp so thoroughly on the map of the world. Alter (1989) writes that nationalism is a political force which has been more important in shaping the history of Europe and the world over the last 200 years than the ideas of freedom and parliamentary democracy or of communism. Since the eighteenth century nationalism has spread from Western Europe all over the world with European colonialism, and the liberation of colonies after World War II rapidly increased the number of nation-states — which became an inescapably universal phenomenon. All in all, nationalism is ultimately a thoroughly international ideology which, paradoxically, is imported for national ends (Löfgren 1989).

Nationalism has become an essential constituent of our sense of identity, and it treats national identity, a specific form of collective identity, typically as a system of absolute values (cf. Herzfeld 1986). The inseparable connection between human beings and nations becomes apparent in the following statement by Smith:

We are identified first and foremost with our 'nation'. Our lives are regulated, for the most part, by the national state in which we are born. War and peace, trade and travel, education and welfare, are determined for each one of us by the nation-state in which we reside. From childhood, we are inculcated with a love of country and taught the peculiar virtues of our nation. And though in later life some may dissent from the patriotic ideal, and a few turn 'traitor', the vast majority of citizens will retain a quiet loyalty to their nation, which in a moment of crisis can swell into a fervent devotion and passionate obedience to the call of duty. (A. Smith 1979, 1)

There has been much discussion in recent years — both theoretical and practical — about the future of nationalism, e.g. about the changing roles of various territorial and ethnic identities and the 'rationalization' and '(post-)modernization' of world views. It has been argued, for example, that social consciousness has changed in the course of time from traditional to modern, and modern consciousness is permeated by representations from science, so that concepts originating from science fill our minds and our conversations, our mass media, our popular books and our political discourses (Billig 1991, 68).

The renewal of the nationalist enterprise is viewed in recent textbooks of political geography as part of a broader shift from modernism to post-modernism (Short 1993, 169), and some authors in the current literature on post-modernism argue that post-modern individuals are characterized by the absence of any strong singular identity, that the world would perhaps be better without high levels of political participation and that there would be a greater chance of tolerance during an era in which national and cultural boundaries were more easily crossed and redrawn. For other post-modern writers, however, nationalism is among the range of new political movements they in fact seem to support (see the examples in Featherstone 1991, 146–7 and Rosenau 1992, 141–4).

It can be argued that nationalism has to some extent given way to decentralization and the acknowledgement of local, regional and subcultural differences in the Western world (Featherstone 1991, 142). Also, the connections between national and capitalistic interests in Western Europe, which were originally fundamental to the rise of nation-states, seem to have become weaker along with the development of transnational production and the globalization of monetary markets (Robins and Morley 1993). But at least some forms of nationalism appear to be resilient in spite of the transformation of modern societies and in spite of the increased international cultural, economic and political networks of cooperation and communication and the rise of supranational mass entertainment — not to mention the rise of local or regional cultures.

For instance, recent tendencies in the former Eastern Europe, as well as the rise of extreme nationalist movements in the West, are clear indicators of the continued power of nationalism. Thus it seems that the idea of nationalism co-exists powerfully with new elements of consciousness and that its power is probably becoming spatially more diversified. This may be characterized by the conflict between ethno-nationalism and state-based bureaucratic nationalism. Anthony Smith (1990, 24) concludes his evaluation by noting that 'the nation and nationalism, accommodated or not, separately or together, will, it seems, continue to provide humanity with its basic cultural and political identities and political organizations well into the next century'. This is obvious in spite of continual territorial transformation, since the construction of new territories is based on new forms of nationalism and civil religion.

3.3.1 Aims and strategies of nationalism in the territorialization of space

What, then, is nationalism? According to Anthony Smith (1978), the term has been employed in three main ways. First it refers to the entire process by which nations and nation-states have come into existence. This is typically designated as the nation building process in the literature — even if it is also concerned with state building as well as national unification. Secondly, nationalism concerns specifically one of this broad set of processes: the formation of national consciousness and solidarity. In this sense, Smith writes, nationalism is equated with national sentiment, and this definition is typically favoured by historians and social psychologists. Thirdly, nationalism is regarded as an ideological movement. Here the analyst aims to trace the rise, courses and effects of ideological movements which aim to reach an autonomy and identity among certain units of population.

Nationalism is always a product of a particular set of political, economic and cultural circumstances — it is contingent (Johnston et al. 1988, 11). In principle, nationalistic thought is not interested in local or regional variation and particularities in landscape or in the social composition of the nation, unless these are a 'barrier' to nation building. As far as the history of nation building is concerned, nationalism has typically witnessed the absorption of smaller regions into larger units of territorial, political and economic organization. The aim has been the redefinition of boundaries and the reorganization of internal social and territorial divisions so that loosely federated societies and territories could be transformed — with varying success — into functioning socio-geographical units that would possess as high a degree of cultural homogeneity as possible. This redefinition has taken place as much 'through force as through negotiation' (Mac Laughlin 1986, 300).

Nationalism has been treated in a variety of ways in the tradition of the social sciences. A. Smith (1983) distinguishes between three main approaches. The first is the 'development perspective', in which a distinction can be made between modernization theories and uneven development theories. In the former nationalism becomes a part of the process by which traditional societies disintegrate and transform themselves into modern societies, while in the latter the uneven development of capitalism over the globe is used to explain the rise and incidence of ethnic nationalism. 'Communitarian approaches' constitute the second perspective, with stress placed on the role of community, community spirit and a sense of solidarity based on shared beliefs and sentiments. Thirdly, 'conflict approaches' emphasize social and political conflicts. Here enmity creates and sharpens group boundaries and serves to mobilize the members of a group so that they become aware of their ethnic or national unity.

Typologies of the various versions of nationalism have usually focused on the presence or absence of a state and whether the nation still has to be constructed (Johnston et al. 1988, 2–3). Here the question of the existence and representation of boundaries becomes essential. In the nineteenth century the convergence of analogies between nature and nationalism provided a vision according to which society, rather than remaining merely an ideal type or abstraction, became coterminous with the boundaries of the national state (Agnew 1987, 74). Nationalism consequently holds to the political principle that political and national units should be congruent (Gellner 1983, 1). This makes the role of boundaries particularly important in the construction of the narrative account that constitutes the kernel of territorial identification.

A. Smith (1979, 3) argues that the national ideal necessarily leads to nationalism, which can be characterized as a programme of action to achieve and reproduce the national idea. Nationalism is a complicated and even contradictory phenomenon which manifests itself on various spatial scales, but all states, large and small, exploit their own versions of it to legitimate their existence — and to create a vision of their historical continuity (Taylor 1991b).

The solidarity that is a prerequisite for nationalism is based on the possession of land, on the existence of territory. Territory is hence an intrinsic element in instilling people with a sense of kinship and attaching them to a place that they feel is 'theirs'. It is impossible to achieve a feeling of identity and culture without inscribing personal life experiences into a discourse defining the story of a recognized homeland (cf. A. Smith 1979, 3). The aim of nationalism is typically to create a national identity which is not based on the boundaries of language, race, tribe or religion but on state boundaries.

Nevertheless, many theories and empirical studies of nationalism commonly neglect the spatial dimension — the fact that nationalism is a specific form of territoriality. Nationalism is two-faced with respect to space, as Anderson (1986) writes. It looks inwards in order to unify the nation and its constituent territory, and it looks outwards to divide one nation and territory from another. Williams and Smith (1983) claim that whatever else nationalism may be, it is always concerned with a struggle over the control of land. It is equally a form of constructing and interpreting social space. At the root of the national ideal there is often a specific view of the world and a certain type of culture (A. Smith 1979, 2). In this vision boundaries are an essential constituent: the world is really and naturally divided into specific historical and cultural units labelled as nations, all of which are regarded as distinct and unique. Nations are not unchanging social entities, however, for as

Hobsbawm (1990, 9–10) points out, they matter merely in so far as they relate to a certain kind of modern territorial entity, the nation-state. Furthermore, nations belong to a particular, historically recent period and are subject to continual change.

It has been argued that the state has become a 'natural' part of the world which we take for granted (Taylor 1993a). This fact leads us to an essential ideological viewpoint exploited by nationalism, that is *naturalization*, as discussed by Thompson, for example (1990, cf. Herzfeld 1986). Naturalization can be regarded as an ideological strategy by which socially and historically created states of affairs are treated as natural events or as the inevitable outcome of natural characteristics. Naturalization typically eclipses the historical and social character of entities. In the case of nationalism it is also possible to talk about *eternalization*, i.e. socio-historical phenomena are deprived of their historical character by being portrayed as permanent, unchanging and ever recurring. In this framework, Thompson (1990, 65–6) argues, customs, traditions and institutions which appear to stretch almost indefinitely into the past — so that all traces of their origin will be lost and no question of their end is imaginable — take on a position which cannot easily be disrupted. Furthermore, 'they become embedded in social life, their apparently ahistorical character is re-affirmed by symbolic forms which, in their construction as well as their sheer repetition, eternalize the contingent'. These strategies facilitate the understanding of why territorial transformation, the emergence and disappearance of various territorial entities, is so difficult to understand both in practice and in the intellectual history of geography, for instance.

Territory as such is not sufficient as the basis of identification, but a complex group of other elements are needed which strive to connect the actors of the territory culturally. Also, specific mechanisms for creating solidarity are required, as well as symbols to express the physical, social and psychological integration of people. Territoriality manifests itself indirectly in the symbolism that is used. In many cultures symbols are more directly connected with more or less abstract expressions of power, group solidarity and authority rather than with territoriality. Symbols are the means whereby this abstract idea is expressed (cf. Beattie 1964, 71).

The process of nation building also partly reflects the emergence of the division of labour in modern society. As Durkheim (1964) argued, in a society which is characterized by a division of labour, regional divisions become 'artificial'. This is apparent from the perspective of the traditional local social life. As far as the functioning of the division of labour and its exploitation for the administrative control of the society are concerned, the role of territorial demarcations may in fact increase.

In any case, all national governments try to make persuasive use of the idea of a common territory. They try to construe a legimate story about it and thus aim at creating an illusion of commonality for geographically diverse areas and various population groups in order to forge a sense of national identity (Knight 1984, 169–70). As Knight points out, there are always various groups within states — sometimes hundreds — which do not have any attachment to their supposed national territories or to the nation, some of which may have regionalist aims of creating new nation-states, while others may have loyalties connected with other nation-states, and still others may have no attachments to states or nations.

All in all, nationalism is primarily a territorial form of ideology and one part of the hierarchical structure of regional consciousness. It aims at 'circumscribing' and signifying territories in space, at creating feelings of belonging and of producing and reproducing social order. One part of the principle of territoriality is the marking and communication of boundaries. This points to the second important dimension of territoriality, classification by area (Sack 1986, 32). Nevertheless, territoriality is not merely a means of creating and maintaining social order, but is, as Sack (1986, 219) writes, 'a device to create and maintain much of the geographic context through which we experience the world and give it meaning'. Since it is a fact that the territorial system of states is in continual transformation, the boundaries of a territory and the means by which they are communicated are similarly not unalterable.

Even though nationalism is by definition taken for granted, Hobsbawm (1990, 11) puts forward three comments to indicate its complicated nature. First, he argues that official ideologies of the state 'are not guides to what it is in the minds of even the most loyal citizens or supporters'. The state has the sovereign power, however, to control the actions of citizens through law, statutes and its legitimate right to force. Hobsbawm's second argument is that we cannot assume that for most people national identification excludes or is always or ever superior to the remaining set of identifications which constitute social being — these identifications are in fact various sides of the same phenomenon (cf. A. Smith 1991). This harks back to the fact that territorially based identifications are constituted hierarchically. Thirdly, national identification and what it is believed to imply may vary and shift in time, even in the course of brief periods. Such a comment points to the connections between national identifications and activities of the state — how people comprehend the role of the state in various conflicts, etc. Here we enter into the question of national socialization — how people are instilled with the idea of citizenship in their everyday lives and in life-history.

3.4 NATIONAL SOCIALIZATION

While the social construction of the spatial has been labelled social spatialization (Shields 1991), now we will analyse the process which may in turn be labelled *spatial socialization*: the discourse in which inhabitants become members of specific products of social spatialization — that is territorially bounded spatial units — and adopt specific modes of thought and action. Socialization is generally defined as the process of 'becoming', the transition from one role to the next, which ultimately aims at acquiring the 'shared but special' information of the reference group, and of other groups (cf. Meyrowitz 1985, 57).

The socialization process occurs in various socio-spatial contexts, including the family and its friends, various social groups, social institutions, formal establishments and so on. Power is an essential dimension for adopting collective representations, opinions, attitudes and practices. Olsson (1994, 118) argues that the powerful techniques of socialization and naming are everywhere: from the myths of the Old Testament and the communion of the Christian Church to the laws of parliaments and the outbursts of family tyrants.

Thrift (1987) speaks about 'subjectification' when referring to the process by which people are formed by a number of interacting institutions that shape them 'from the cradle to the grave'. He puts particular emphasis on the role of space in this process, and contends that spatial arrangement is a key strategy in the systems of discipline that various social institutions exploit to produce particular kinds of subjects. Furthermore, subjects are always produced in specific local contexts, even if the aims of the subjectification and the institutions involved in this process are not themselves local.

As far as the adoption of the territoriality of nation-states is concerned, we can talk about national socialization. The general national ideology, the dominant meaning system concerning the socio-spatial construction of a state, is the basic frame for national socialization. It takes place in the context provided by the language(s), a context which itself may be a source of conflict and resistance. This is owing to the fact that very few nation-states are culturally homogeneous, so that divergent interests exist as regards the content of the dominating culture (cf. Mikesell 1983). Thus national socialization forms a part of a broader political socialization which refers to the ways society transmits its political culture from generation to generation and produces and reproduces its hegemony in the process. People are not passively socialized into different social institutions, but rather they perpetually construct both institutions, themselves and others (see Thrift 1986, 92).

Historical continuity is essential for understanding the nature of the territoriality of nationalism at both the local and the national level.

Anderson (1986, 1988) points out, for example, that nationalism is simultaneously 'forward-looking' and 'backward-looking'. Hence associations with the past are central to its territoriality, for territoriality is the receptacle of the past in the present. During the nineteenth century national history as a whole became an important part of the syllabus in schools and legislators were willing to spend public funds to train and employ teachers of national history and to support academic research into it (Shils 1981, 58–9).

It appears to be typical of nation-states, as well as other territorial units, that they create a continuity for their existence through stories, which also provide a common past which does not diverge historically or symbolically from the regions or nation-states of the present. This is an integral part of the process which Hobsbawm (1983) labels 'the invention of tradition'. This points to practices that usually automatically imply continuity with the past. The invention of traditions includes both 'traditions' that are actually invented, constructed and formally instituted and those which emerge rapidly but whose origin is difficult to date.

The problem of historical continuity reflects the fact that the established territorial unit is preferably realized and apprehended as a continual *subject* of the narrative account in question, a representation of *we*, not an object of *them* (cf. Paasi 1986a, 125–6). Smart (1983, 81) provides a fitting illustration of this persuasive construction of territorial continuity:

Since the French Revolution, nationalism has helped to promote the writing of history as each national group tries to create, so to speak, its own past. It is now common to think of history in national terms — we talk of Italian history, American, Canadian, French, Indian, Cambodian, and so on — using modern political groupings in order to define the past. There was no consciousness as such of being Italian three hundred years ago: there were different regions of the peninsula under different rules. The German statesman Bismarck was not altogether wrong in referring to Italy as merely a geographical expression. But Italy emerged as a self-conscious nation and it is from this standpoint that we look backwards to the 'Italian' past. Anyway, there is a strong trend in modern times to see history as a grouping of national histories, each of which illuminates the nature of the nation in question. In brief, the story of Italy or of the United States becomes a means of creating a consciousness of being Italian or being American.

In the context of this book, nationalism may be defined as a social process by which certain historically contingent forms of territorial identities, symbols and ideologies are instilled into the social and individual consciousness. Through this process individual experiences are colonized by collective ones to join them in the communal story. This process takes place in the various social practices of daily life, in particular through education and the media. The prerequisites are a literate culture and a centralized education system supervised and run by the state which tends to monopolize legitimate culture and take care of the historical and geographical socialization of children. Through various media the ideas of imagined political communities — inherently limited and sovereign — will be constructed and reproduced (B. Anderson 1991, 6). According to Anderson, nations are imagined as limited since even the largest of them has finite, elastic boundaries beyond which lie other nations.

It is obvious that the singularity of locality is nowadays disappearing owing to the expansion of the activity of the media, so that there exist fewer distinctions between localities. Or even if various localities are still different from each other, they are probably not as different as they were previously. Where the diffusion of literacy and printed materials supported distinctions between groups and created new experiential worlds, the rise of the electronic media in particular has created a blurring of many previously distinct social roles and has given rise to new groups which were formerly isolated and distinct (Meyrowitz 1985, Robins and Morley 1993). These developments have not diminished the strength of nationalism. One reason for this is doubtless national education, which aims at creating national consciousness and creating or maintaining the necessary linguistic discourse.

Although no one denies the role of the media in shaping world views today, it is common to stress the *education system* as the principal force. Rousseau, regarded as the first to outline the nationalist theory, wrote that 'it is education that must give souls a national formation, and direct their opinions and tastes in such a way that they will be patriotic by inclination, by passion, by necessity' (Rousseau, cited in Birch 1989, 15). Regardless of the prevailing social system, education is the main institution that builds up social integration in the modern state, and it is connected with the production and reproduction of the dominant values and ideologies in society.

Education is an essential factor in *secondary socialization*, in which people internalize institutional or institution-based 'sub-worlds' (cf. Berger and Luckmann 1976, 158). The social relations manifesting themselves in secondary socialization are generally anonymous and formal. Teachers are always representatives of institutionally specific meanings and ideologies, beliefs and belief systems. Teachers, civil servants, politicians and business leaders are key intellectuals who are important as implementers of specific cultural programmes and hegemonic structures (cf. Johnson 1992).

Education builds up central ideologies and symbolic forms in the integration of social reality. As Berger and Luckmann (1976, 141) point out, ideologies generate solidarity. Education always has a specific civilizing function, but an education system is always also an expression of *symbolic violence,* as Bourdieu and Passeron (1990) label it, since it aims to create a view of a common culture prevailing in a society, usually hiding the connections between intellectual culture and economic and politico-administrative use of power. A feeling of a common culture is usually based on the employment of '*Gemeinschaft* language' (Paasi 1986a, 134), consisting of such keywords and dichotomies as we/them, ours/theirs etc., which integrates individuals to form a plural subject. From the perspective of social integration, education thus stresses the *moral* dimension of a world view. This moral basis forms the ultimate background for the argument put forward by a number of authors that the belief in one's own nation and national identity in modern nation-states is built into the human personality and is generated by mass education, mass literacy and the homogenization of culture (cf. Cooke 1989b, 282).

The textbooks employed in the education system are one important medium through which the state apparatus, with its socializing goals, reaches individuals and groups. Efforts towards social integration can manifest themselves in positive, constructive identity building, but equally well in the form of negative terms or attitudes which build boundaries and distinctions between us and the Other (cf. Knight 1982). The teaching of regional geography is crucial in both respects, and for this reason geographical textbooks are important indicators of the prevailing, dominant socio-spatial consciousness — and beyond, of the *Zeitgeist.* Forms of social integration are historically contingent, since the content of education changes together with the content of world views: the facts and 'truths' of today can become no more than *stereotypic* opinions in the future. This has been the situation in the history of the teaching of regional geography. In any case, there are doubtless profound differences between states as regards the relations between the personal and the collective identity, and between culture and the representation of these relations (cf. Dumont 1986).

In the tradition of political geography, an education system is usually regarded as the most significant socializing mechanism of the state, inculcating children with society's values, traditions and political and social culture (Bergman 1975, 270). Birch (1989, 9) writes that control over the education system is a matter that no modern state can neglect, as the central elements of a national identity are mediated though it. National identity is naturally only one of the purposes of national socialization, however, as other important aims are connected with the reproduction of the economic and political structures of the society. It is useful to point out here that the framework of 'state–ideology–social reproduction' implicit in the foregoing discussion of the construction of collectively shared values and beliefs should not be understood in a simple, deterministic (or class-reductionist) way (for a more extensive discussion on this theme, see the critical analysis of the roles of ideology in modern societies by Thompson 1990).

It is through national socialization that the nature and symbols of the nation-state and the boundaries of the state are signified and legitimated, and its relations to other nations and states are canonized at the same time, codes of behaviour in relation to boundaries usually being strictly defined. Socialization, including national socialization, is a process which continues throughout life. In the process of producing meanings, signs and values, the ideological forces creating collective consciousness aim at homogenizing it. A typical illustration of this is the ideology of nationalism, by which nation-states ultimately operate and are reproduced both materially and intellectually.

National symbols are an essential part of national culture and integration. In spite of the fact that they join people together, they also usually have an 'indexical' normative power for the people for whom they have been made, to make them behave in specific ways (Tarasti 1990). In principle other symbols, e.g. those of the international labour movement, have operated in this way; expressing and directing the identification of groups and aiming to create a new hegemonic consciousness which manifests itself in various spatial frames (cf. Klinge 1981). The situation is the same with national symbols as with those employed in concrete local border landscapes. This indexical power does not always emerge from the symbols themselves, but they can also be explicit icons of the hegemonic power prescribed in laws and statutes.

Nevertheless, it should be noted that individuals do not merely react passively to these social forces and processes, they also actively participate in and shape the processes during which they are socialized. Thus there is a sentimental attachment to the symbols of the nation, while through certain spectacles, such as parades, flagdays, poems, plays, etc., individuals perceive themselves as being members of a particular national group. As Bloom (1990, 59) remarks, for the individual to internalize the symbols of the nation, the nation — in one representational or symbolic form or another, direct or indirect — must become part of the actual experience of the individual. Furthermore, the experience of this contact must be such that it actually benefits the individual, in terms of psychological security, to identify with the nation. This points to the ability of a state apparatus to signify and legitimate the prevailing forms of social, political and economic life.

As far as the role of national spectacles is concerned, Tarasti (1990, 192) points out that history which has been transformed into a spectacle is always in part the reality of the culture and values. Conversely, a critique of spectacles can be interpreted as an attack on cultural values themselves. This leads us back to the concept of hegemony. Williams (1988, 136) writes that the real condition for hegemony is the self-identification which is directed at the hegemonic forms themselves: a diversified, internalized socialization, which can be approved but which can also be based on resigned identification of what is inevitable and necessary. Spectacles as symbolic events (e.g. coronations, weddings, etc.) can, however, articulate both consensus and conflict in the form of ritual and rhetoric media events (cf. Dayan and Katz 1988).

3.5 NATIONAL STEREOTYPES AND THE OTHER

It has become obvious in the previous chapters that the social identity of a human being is constructed in relation to all the numerous groups, acting on various spatial scales and in various places, that he or she belongs to. These groups vary in their personal importance, and some of them may also be sources of conflict. Most of these groups have specific symbols, limits and boundaries, even though not always fixed or material ones. But as a whole it is easy to agree with Bourdieu (1991, 120) who points out that to institute or assign a social definition and identity is also to impose boundaries.

The meaning of boundaries and the instrumental nature of national symbols are probably most explicit in the case of nation-states. Even though these 'social' boundaries are in close connection with physical boundaries demarcating nation-states from each other, they are above all vehicles of socio-spatial distinction, or social spatialization, which manifest themselves in what may be labelled collective representations. These are in some instances based on personal experience, but much more significant is the collective dimension of the representations. This dimension is connected with the collective memory of the society, and thus its ideological and hegemonic structures — fundamental national demarcations between 'us' and 'them'. Through these 'structures' each nation becomes defined not only in terms of its own, but in purely relational terms as well. Each nation is defined by its relations to other nations, i.e. *difference* is constitutive of national identity (Morley and Robins 1989, Mach 1993). It is also by no means a necessity for the boundaries of a state and those of a recognized national group to coincide.

By tradition it has been common to construct collective representations of national identities, internal identities of groups, by 'depersonalizing' the members, that is by assuming stereotyped collective features that are common to all members of a group. The fusion of the individual and the collective can also be seen in the language of nationalism. As Rée (1992) points out, the word 'national' has entered into regular partnership with words usually reserved for discourses on individual personality: national *spirit*, national *life*, national *character*, national *identity* etc. Deeper analysis also reveals marked differences between these concepts, as evident in Perry Anderson's (1991) discussion of the relations between national character and national identity. National identity is hence an eminently normative force, a discourse that arose after World War II and finds its natural habitat on the symbolic political plane. The increase in international migration, the shaping of world markets and the establishment of international organizations has doubtless detracted from the role of the concept of national character, and broader 'discursive identity spaces' have also been produced during recent decades. Illustrative of this trend is Schlesinger's (1991) discussion of the construction of Europe as a cultural entity.

What are stereotypes? One the one hand, collective stereotypes can be interpreted as one form of behaviourism, having as its theoretical base a crude neo-Darwinism which regards nations as being repositories of a pool of genes giving rise to particular national characteristics (Bloom 1990, 9–11). On the other hand, collective stereotypes can also be regarded as arising from a more culturally based context, e.g. by looking on them as manifestations of a common heritage or ideological construction.

Stereotypes are hence simplified beliefs. National stereotypes, e.g. assumptions regarding national character, are fitting illuminations of public stereotypes associated with states, nations or cultural groups. They can be employed in propagandistic manipulation to represent certain states of affairs, since propaganda is known to appeal not to rationality but to feelings, myths, prejudices and finally action. Stereotypes are time and space specific and can change drastically during periods of change in economic and political conditions: friends turn into enemies and enemies become allies (cf. Klineberg 1950, 1951). Thus stereotypes can be part of the prevailing ideological structure, but can also act against it. In the course of time, stereotypes have been used to support and legitimate nationalistic aims, the power of authoritarian governments or the military arms race. The basic idea in the exploitation of stereotypes is that it is the friends who define the enemies; it is the friends who control the classification and assignment.

Bauman (1990) argues that all supra-individual groupings are primarily processes of collectivization of friends and enemies. National states also collectivize friends and enemies, even if they are, Bauman goes on, 'designated' primarily to deal with the problem of strangers, not enemies. Nationalism in particular is primarily an inward-looking ideology.

Nevertheless, the nation(-state) demarcating itself from the outside world and the political self-assertion of a new sovereign nation are typically manifested in the stereotypical 'enemy pictures' that have lined the paths of all national movements (Habermas 1988). At times of crisis these stereotypes can be created and manipulated. One example from the years of World War II would be the Dutch press, which was controlled by the Nazis in an attempt to create a favourable image of German activities (Duijker and Frijda 1960). The role of national stereotypes and the idea of national character have also been noted within political geography (Muir 1975, 77–8, 93–5). This is easy to understand since one of the basic assumptions in nationalist theories is that human beings are 'social animals' who inherit their character and aims from a community which shares the same purposes (Birch 1989, 14). The interesting question is where these purposes originally emerge from. This harks back to the concept of ideology.

Whatever interpretations exist as to their origin, national stereotypes are a part of the world view of people, and since they are abstract, simplified representations of different cultures and peoples, the ideological dimensions are inevitably important in their constitution and manifestation. When we think of the contents of geography teaching in schools, for example, one interesting question is to what extent the aim of education has been to describe or represent stereotypes or to create them in the minds of children. The problem is then not whether 'national characters' or other stereotypes exist or not but whether people believe in their existence, making their observations and distinctions on this basis and acting accordingly (Paasi 1984a, 24–6). This is the key to comprehending the power of stereotypes and the main reason for analysing their role in the processes of both social spatialization and spatial socialization.

4

METHODOLOGICAL CONTEXTS

The historical production and reproduction of territoriality and boundaries will be investigated explicitly in the concrete analysis to follow, together with their manifestation in the context of a small nation-state, Finland. The most important territorial form of ideology to be discussed is nationalism, which will be analysed as part of nation building, understood as a process which aims at the social, regional, political and institutional integration and harmonization of the divided sections of a people. Hence it points to the process during which the inhabitants of a state's territory come to be socialized as loyal citizens of that state, adopt its legends and identify themselves with its symbols (cf. Carr 1986, 156, Bloom 1990, 55).

Because of the complicated and contextual nature of nationalism, its territoriality and the construction of boundaries, these will be interpreted empirically and theoretically with reference to how one specific boundary — that between *Finland and the Soviet Union/Russia* — has been construed historically as part of nation building in Finland. An analysis will be made of how this boundary has been symbolized and represented in divergent social practices at various times and of how inherent narrative accounts of its role in Finnish society have been constructed. Beyond this, there will be an appraisal of the kinds of ideological processes and power relations — emerging on various territorial scales — that have lain behind this process. The general aim is thus to analyse the ideological and cultural construction of the territory and one of its boundaries.

The boundary between Finland and the Soviet Union/Russia will not be regarded as merely a border line between two states or social systems, but will be approached in a theoretically informed way, regarding it as an object of a continuing cultural struggle and ideological signification through which different social forces — manifesting themselves on various territorial scales — strive to distance and join actors with respect to these territorial units. In Finland this boundary has served in the course of its history as one of the symbolic signs of the ideological differences between two states. It has been regarded as the 'outpost of the West' and a 'boundary of peace'. During the Cold War it was a taboo, but it is currently becoming to an increasing degree a phenomenon which is exploited economically. Hence, the boundary has constituted historically a landscape feature around which have emerged myths of war and peace.

It can be argued that the boundary with Russia/the Soviet Union has been one essential ideological constituent in the historical construction of the grand narrative of Finnish nation building and the cultural 'semiosphere' of the state, i.e. the continuum of cultural signs that compounds what is usually comprehended as being 'national culture'. The context of a national culture is always a spatial unit which can be defined temporally and territorially on the basis of its human landscapes, language, history etc. These constitute the basis for the texts and signs which are represented as forming this culture. Furthermore, these texts are powerful in their relation to their background, the semiosphere, against which they exist. They are not powerful in themselves or when divorced from the semiosphere (Tarasti 1990, 199). Even though texts, symbols and signs can contain certain common, universal dimensions, their meanings can be understood best in the context of the semiosphere itself. Nevertheless, from a semiotic point of view these signs are always arbitrary, artificial, constructed and covenanted. This fact accentuates the role of power relations in their construction, and hence there is always a political dimension as to who can create, represent and signify the distinctions between 'us' and 'them' (cf. Morley and Robins 1989).

The representations of state boundaries understood as texts expressing territoriality are laden with strong visible/non-visible, local/non-local, ideological and metaphorical dimensions, and with continually changing meanings. All these dimensions have been constitutive of the divergent representations of the Finnish–Russian boundary — how it has been communicated, produced and reproduced in documents and various forms of rhetoric in the course of time and how the power relations of Finnish society manifest themselves in the production of meaning. Sensitivity to these perspectives forms an essential strategy for the empirical analysis below.

An exceptionally wide-ranging debate on the role of space and time in the constitution of agents, everyday life and the social world in general has emerged within geography and sociology during the 1980s and early 1990s (Eyles 1989, Giddens 1984, Gregory and Urry 1985). As far as the relationship between everyday life and the spatial constitution of the social world is concerned, the discussion has advanced in relatively abstract terms. There has been deliberation over the constitution of daily time–space routines or biographies in specific local communities and localities on the one hand, and over the manifestion of 'general' social processes in these localities on the other.

Arguments within general social theory have been set forth regarding both the time–space constitution of individual action and more general social practice, and similarly regarding the various historical scales on which social action takes place (Giddens 1987b). Nevertheless, scholars have not paid much attention to the *concrete* process through which individual life-histories become involved with more general socio-spatial processes in various social practices in concrete time–space-specific contexts, or how various social and spatial identities are constructed.

Since the early 1980s geographers have been inspired by the work of structuration theorists and theoreticians of praxis in their efforts to overcome the oppositions between structure and agency, objectivism and subjectivism, etc. (Bourdieu 1977, Giddens 1984). In geography, Thrift (1986) in particular has discussed the role and social construction of 'individuals', their personality and identity, while in recent debates within social and cultural theory the language of identity, the construction of subjects and the language of subjectivity in various spatial contexts are emerging among the relevant themes to be consulted in the social sciences, as indicated in the recent collections by Lash and Friedman (1992) or Keith and Pile (1993). Divergent definitions of identity persist within recent debates.

It is possible to talk in more general terms about a rise in cultural studies that has also taken place in human geography following the economism that dominated much of human geography during the 1970s and 1980s. Social scientists have become sensitive to the fact that a hermeneutic or ethnographic moment is part and parcel of all research analysing social action (Giddens 1987b, 66). This can be seen in geography in efforts to emphasize the connections between the critical and humanistic approaches and their potential contributions to each other (Kobayashi and Mackenzie 1989), and in the stress laid on ethnographic perspectives in geographical research (Sayer 1989, Gregory 1989). As an outcome of these trends, geographers have urged a more sensitive ethnographic and interpretative approach since the end of the 1980s, one which would do justice to the complexity and openness of the daily lives of particular people in particular places; with a view to revealing how places and landscapes are interpreted and how these interpretations are simultaneously expressions of the social life of individuals (Eyles and Smith 1988, Duncan and Ley 1993).

Some steps are also taken in this direction in the present book, by first conceptualizing the production of socio-spatial consciousness on various spatial scales and its reproduction in everyday life, and then analysing the Finnish social reality, which aims at interpreting and thus understanding both the structural, i.e. political and ideological, frameworks of the production of spatial representations of boundaries and what they are supposed to demarcate, as well as the meanings emerging from the everyday lives of people living in concrete border area contexts.

The basic intention of this work is thus to 'read', that is to reflect and interpret contextually, the various dimensions and changing cultural significations and metaphors sedimented into the texts that the boundary is understood and represented as constituting. The starting point is thus an understanding of this signification process and the material basis that forms the background to it. This approach can also be labelled as interpretative geography. As Eyles (1988, 2) writes, interpretative geography is concerned with the understanding and analysis of meanings in specific contexts. Nevertheless, the concern of this work is not merely how people act or give meaning to their own lives in the specific context of a border area, but rather to broaden this perspective both in time and space, i.e. to interpret the role of the boundary historically on various spatial scales, particularly in local and national contexts. This calls for the adoption of a strategy in which particular research materials and methods have to be chosen which can reveal the contextual nature of the research object.

4.1 SOCIO-SPATIAL CONSCIOUSNESS

It was observed in the brief discussion above of the ideas of collective representations put forward by Durkheim that a more active view of actors is needed in relation to the representations: it is not enough to explain the social merely by the social. In any case the Durkheimian idea is ultimately rather static.

An attempt will be made in this chapter to develop abstractions to approach the role of representations and forms of consciousness historically, both on the scale of a specific society and on the scale of everyday life, analysing in particular the role of spatiality in these representations. The abstraction connected with the scale of society may be labelled *socio-spatial consciousness*. The scale of everyday life will be approached by employing the category of *social representation*.

Socio-spatial consciousness is an abstraction which aims to illuminate the social construction of spatial (and social) demarcations. Beyond this, it points to socialization into various territorial demarcations, themselves 'living' simultaneously in regional transformation. This concept is used to explain the relations between 'structure' and actors in the production and reproduction of socio-spatial reality. In this connection the idea of structure is employed in a similar manner to that of Giddens (1984, 17), who associates it with rules and resources which are recursively implicated in the reproduction of social systems. Giddens points out that structure exists as *memory traces* — the organic basis of human knowledgeability — and *instantiated in action.*

Of particular significance for developing the idea of socio-spatial consciousness is Bhaskar's (1983, 84) comment which places more emphasis on the continuity of institutions than does Giddens. Bhaskar remarks that social structure is always 'given' from the perspective of intentional human agency, which nevertheless reproduces and transforms it. The past, mediated by collective representations, is always an already processed past; it has been lived, interpreted and partly forgotten (Knuuttila 1992, 14).

Identity, as Morley and Robins (1990) point out, is also a question of memory, and particularly memories of 'home'. Nevertheless, memory is not simply an individual phenomenon but instead has a base, a link with the community, and is fundamentally a community experience (Ferrarotti 1990, 30). 'Individuals' in this position are not regarded as abstract entities but as active, experiencing, feeling persons, who have a transformative capacity and whose personal histories are the medium for and the outcome of the realization of institutional practices and the reproduction and shaping of the division of labour. Memories do not simply float in the minds of human actors but can also be 'concretized' in a materialized form, i.e. in the form of books, memorials, carnivals, rituals, habits, results of scientific research etc. On this basis the ideas of tradition and the past are constantly being reworked and represented within historical experience, e.g. that of a particular nation-state (Morley and Robins 1990).

Thus socio-spatial consciousness is treated as a form of 'collective consciousness' that can for analytic purposes be *abstracted* from the social life of individuals — thus it is neither a sum of consciousnesses or individual 'mental maps', nor even a collective mentality of actors living at some specific time. This consciousness is understood as being metaphorically 'located' partly 'above' or 'behind' — in the sense of past time — individuals living in a specific time and place. In practice, of course, it 'stretches' an individual actor through various institutional practices, a part of the continuity that the community and society constitute and are represented as being. It is nevertheless simultaneously and perpetually present in various social practices, such as education, culture, politics, economy, administration, communication etc., through which individuals reproduce themselves and the structure, and through which the power relations of the society operate. 'Since the history of the individual is never anything other than a certain specification of the collective history of his group or class', Bourdieu (1977, 86) writes, 'each individual system of dispositions may be seen as a structural variant of all the other group or class habitus, expressing the difference between trajectories and positions inside or outside the class'.

However, even though individuals reproduce various social institutions in their everyday lives, they have not initially produced most of them, and thus the reasons for their existence may remain hidden in the everyday world, which is from the outset an inter-subjective world of culture. The everyday world is a world of culture, Schutz (1978, 135) writes, because individuals are always conscious of its historicity, which they encounter in tradition and habit. In this world, Schutz argues, the 'already given' refers back to one's own activity or the activity of others, of which it is the sediment. To illuminate this continuity, Schutz (1970, 119) employs the expressions 'the world of our predecessors' (which acts on us while itself being beyond the reach of our action) and 'the world of our successors' (on which we can act but which, for its part, cannot act on us). People do not always simply adopt this world as given but may also resist it and introduce conflicting interpretations.

The existence of socio-spatial consciousness is therefore based on the fact that most of our institutions usually originate from the action of previous generations and tradition, and are then reproduced and modified, contested and partly abandoned in the actions of the following generations. Tradition includes material objects, beliefs, images, practices and institutions, for instance (Shils 1981). The role of tradition becomes clear in the much cited statement by Marx (1963, 15): 'Men make their own history, but they do not make it just as they please; they do not make it under circumstances chosen by themselves, but under circumstances directly encountered, given and transmitted from the past'. As far as the conceptualization of structure employed in this book is concerned (i.e. structure understood as memory traces instantiated in action), the next, less often cited sentence in Marx's book is at least as important: 'The tradition of all the dead generations weighs like a nightmare on the brain of the living' (1963, 15). This argument makes it clear that material circumstances are not merely located somewhere above actors, as mystical causal powers, but instead become 'flesh' in discursive and non-discursive practice, in action.

Inasmuch as tradition is always manifested in a specifically spatial context, people simultaneously 'make their geography' too (Giddens 1984, 363). 'Geography' is thus not merely a passive spatial setting in which social life occurs, but something that is actively produced, reproduced, maybe maintained or transformed, in the struggles of individuals, in their local, day-to-day practices of life and in the collective forms of practice on larger spatial scales.

Thus, even if we adopt an active concept of the actor, past practices, arrangements and beliefs undoubtedly have normative power in the constitution of social life. Social consciousness constantly represents tradition and collective forms of memory to some extent. Nevertheless, 'the past does not control the present to the point of insisting upon its own exact reproduction', Billig (1991, 22) writes, since 'each echo is itself a distortion'. Consequently the collective memory of society always shows partial continuity together with a new interpretation of the past in terms of the present (Coser 1992).

It is important to remember that institutions cannot have minds of their own, even though they may have social power over individuals and groups (Douglas 1987). But even though memory is sustained by institutional structures, it is individuals who act in various social groupings and who produce, reproduce and modify social institutions in practice.

Socio-spatial consciousness thus refers to various *social* forms of consciousness and ideologies that partly constitute spatiality and spatial identity but are also themselves constituted by spatiality and collective identity. Here we can make the analytical distinction between regions and places which helps us to understand the construction of spatial identities. As we have seen, regions can be treated as structures whose institutionalization consists of their territorial, symbolic and institutional shaping. Places, for their part, refer to the dialectics of spatial experiences which individuals acquire during their personal life-histories, which can take place in various localities. Marcus (1992, 315) aptly remarks that the identity of anyone or any group is produced simultaneously in many different locales of activity by many different agents for many different purposes.

The production and reproduction of socio-spatial consciousness does not take place apart from material interests and practices, but since the constitution of consciousness is a historical process, consciousness cannot be simply reduced mechanistically to this material basis. One aim of this investigation is to illustrate the *reciprocal* connection by analysing the rise of the Finnish state and its principal territorial ideology, nationalism. Whether we imply by nationalism popular sentiments, organized movements or state policies, these are all expressions of its territorial character (Anderson 1988, 18). Territoriality is treated here in the same manner as by Sack (1986, 19), i.e. as an attempt by an individual or group to affect, influence, or control people, phenomena and relationships by delimiting a geographical area and asserting control over it. Territories require a constant effort to establish and maintain. In the case of nation-states, nationalism provides the means for this.

How can one analyse social consciousness, and how can one approach the historical construction of socio-spatial consciousness? It is reasonable to suggest first that social consciousness displays itself in the actions and discursive and practical consciousness of the members of social collectives. They doubtless also have 'material manifestations' in newspapers and books, maps and drawings, paintings and various memorials which reveal and strengthen the element of historical continuity in consciousness. In a way, the historical sedimentation of action and thought 'reveals' itself in these concrete manifestations. But it should be noted immediately that, although it is possible to analyse and interpret the contents of socio-spatial consciousness through various concrete manifestations, we should not simply believe that their content can be completely reduced to these manifestations. There also exist continually 'non-material' forms of social consciousness in social practice, e.g. political rhetoric, as well as various forms of social action which do not inevitably have any explicit materialized form (Paasi 1992).

The materials employed here to deconstruct Finnish socio-spatial consciousness consist of various documents and reports illuminating the nation building process in Finland, the construction of the Finnish–Soviet boundary as part of this process, and its appearance in the socio-spatial consciousness of the Finns. Furthermore, materials illustrating Finnish–Russian relations can be employed in a historical context. More particularly, as indicators of the (geo)political socialization of the Finns, an analysis is made of ideas and representations concerning the nature of boundaries, and especially that between Finland and the Soviet Union, contained in Finnish school geography textbooks. The ideological and political connotations of the boundary will first be debated in a historical context and its position in the socio-spatial consciousness of the Finnish people will be discussed.

Particular attention will be paid in the analysis to the rise of stereotypic images in Finnish geographical education and the stereotypic images of Russians that have been constitutive of the Finnish nation building process. The aim is to trace how these stereotypes have developed and altered in content, and to contextualize the stereotypes within the changing field of international relations. This is an important task when attempting to comprehend the construction of national identity. Löfgren (1989) points out that in order to create a symbolic community, 'identity markers' have to be created within the national area in order to achieve a sense of belonging and loyalty to the 'national project'. Furthermore, this identity also has to be marketed to the outside world as a 'national otherness'. The analysis of stereotypes is hence beneficial for the deconstruction of the ideological constituents of national identity.

Geographical textbooks are thus employed here as indicators of *Zeitgeist*, i.e. to expose traits of the 'official' opinion expected of national socialization by the state. It is doubtful whether there exists one specific, clearly distinguishable *Zeitgeist* in a society. Rather,

Zeitgeist is regarded here as a heuristic abstraction referring to a particular societal context of ideas, beliefs and thoughts. It is extremely difficult to get at what Williams (1961, 64) describes as the 'felt sense of the quality of life at a particular place and time', or the *'structure of feeling'*. This structure of feeling refers to the culture of a period, the particular living result of all the elements in the general organization. It is something that people can hardly express in their daily lives.

Geography textbooks are particularly significant in this respect, because no other documents employed in education have such a power for the creation of socio-spatial consciousness or representations and images of foreign peoples and cultures. They are probably the best 'official' material available for studying the construction of the languages of difference and integration. The situation is the same with textbooks in history: no other documents can have such a systematic power for the construction of images of the past. The importance of education in history for the construction of the national past and the struggle over legitimate methods and approaches can be seen, for instance, in the recent debate between the supporters of 'national history' and 'social history' that has been taking place in Britain (see various articles in *History Workshop* 1990). Textbooks are also indicators of the time–space-specific politics and control mechanisms of the society in which they are published — and for this reason their texts and contexts have to be treated together.

The point of departure when employing the above-mentioned materials, documentary and fictious, when analysing the 'landscapes' of Finnish social consciousness, is the idea that Daniels and Cosgrove (1988, 1) put forward explicitly in their discussion, that in order to understand a landscape it is necessary to understand written and verbal representations of it as constituent images of its meaning or meanings. A landscape is hence an 'intertext', an organised 'ensemble of fragments of various texts' which have been produced according to different logics, and it is in this 'intertext' that new meaning structures are built (see Berdoulay 1989, 135). Similarly, the changing 'landscapes' of social consciousness must be approached by analysing and contextualizing the changing representations.

4.2 TOWARDS LOCAL LIFE

Socio-spatial consciousness hence provides one specific abstraction for use when discussing the nature of collective representations, particularly at the level of the society. Attention will now be directed to the nature of representations on the scale of everyday life, which is where socio-spatial consciousness is ultimately reproduced. It is clear, however, that people come to terms with a relatively small number of these representations in their everyday lives, and hence — even if we might agree with Geertz (1973) that human thought is consummately social in its origins and social in its functions, forms and applications — nevertheless, the daily perceptions and experiences of ordinary people can hardly involve any immediate contact with most of the phenomena contained in these representations. Instead, most of them intervene in everyday life through the mediation of various institutional practices (education, media, etc.) and this takes place through a socialization process which continues throughout the life of the individual.

A particularly useful category on this scale is *social representation* — one of the recent key categories in European social-psychological discourse (Moscovici 1981), and one which has also been employed in the case of representations of social space, particularly urban space. Social representations have then been regarded as 'mental maps' of a certain kind, in the manner of traditional behavioural geographers (Milgram 1984, Pailhous 1984). When conceptualizing social representations, these simple spatial metaphors will be replaced by an attempt to find a more theoretically grounded basis.

Moscovici (1981, 181) defines social representations as a 'set of concepts, statements and explanations originating in daily life in the course of inter-individual communications', and points to the close connection between this category and the Durkheimian idea of collective representations. Social representations for Durkheim encompassed a vast class of intellectual forms (including science, religion, myths and the ideas of time and space). Moscovici also states that social representations are equivalent to the myths and belief systems of traditional societies, i.e. they are a contemporary version of common sense. But the study of social representations focuses on the ways in which human beings aim at grasping and comprehending things around them, and on solving 'commonplace puzzles' affecting their daily lives, puzzles that have concerned them since their childhood and that continue to be of concern to them (1981, 182).

Thus the theory of social representations aims at outlining how people realize, explain and express the complexity of stimuli and experiences emerging from the environment, both social and physical, in which they are living. It points out that importance needs to be attached to the realities of everyday life and of practical consciousness as one part of it (Halfacree 1993). Hence it provides a mediating category when moving from more general socio-spatial consciousness (and the role of boundaries in this) towards the interpretations and representations of spatiality emerging from everyday life. Social representations — as well as socio-spatial consciousness — are constitutive of both practical and discursive actions: they both guide and constrain action.

But socio-spatial consciousness represents a broader form of consciousness, which reflects the ideological and hegemonic structures of society, and hence it also reflects the power relations which emerge from the social division of labour.

The whole concept of 'social representation' is a complicated one and its definitions can be labelled as universal and particular. The universal definition assumes that all societies and social groups possess common sense and practical knowledge in some form or other, while the particular definition assumes for its part that social representations are only to be found in certain societies (Billig 1991).

Social representations can be characterized by the following definitions, gathered together by Billig (1991, 61). Regarded as universal concepts, they point first of all to 'all the knowledge and understanding that a society, or subgroup of a society, has about a given object'. They can also be understood either as a system of values, ideas and practices or as an elaboration of a social object by the community for the purpose of behaving and communicating. In this practical and partly taken-for-granted sense, social representations recall the idea of mentality discussed above in connection with ideology.

Moscovici (1981, 192) points to two social and psychological processes involved in the creation and maintenance of social representations — 'anchoring' and 'objectification'. These processes serve to 'familiarize us with the unfamiliar'. Anchoring permits cognitive integration of the represented object in the pre-existing system of thought: it 'draws' the individual into the cultural tradition of the group while at the same time developing that tradition. To anchor means 'to classify and label or name something'. Moscovici (1981, 189) writes that the angle from which a group will try to cope with what is unfamiliar will be determined by 'the images, concepts and languages shared by that group', and what matters is 'not the group's judgment but its conventions and memories'. Thus representations are rooted in the life of the group. Billig (1991, 65) states that all groups and societies have specific systems of naming and categorizing, and thus anchoring, as a constituent process of thinking and perceiving, refers to a social psychological universal. The territorialization of social practices on various scales is one illustration of this 'categorizing process'. Representation, as a whole, is basically a classifying and naming process, 'a method of establishing relations between categories and labels' (Moscovici 1981, 193).

Objectification refers to the way in which social reproduction transforms unfamiliar abstract concepts into familiar, concrete experiences, and thus it describes the materializations of an abstraction. Objectification transforms an abstraction into 'something almost physical, translating something that exists in our thoughts into something that exists in "nature"' (Moscovici 1981, 192). Billig (1991, 65) writes that, whereas any belief can be anchored, regardless of its content, only certain kinds of beliefs could be said to be objectified. According to Billig, these beliefs are essentially non-religious, and have originated from abstract, but non-religious concepts. The passage from esoteric scientific theory to day-to-day discourse is a prototype objectification (1991, 65).

The relation between social representations and socio-spatial consciousness is based on the fact that one category of essential 'objects' with which people have to deal concerns representations of we and the Other — stories which express the self-consciousness of various social communities occurring on divergent spatial scales. Social representations involve the amalgamation of personal experiences with sociospatial consciousness, a tradition produced and reproduced in various institutional practices. At times these may be in conflict, and people may try actively to change this general consciousness.

Similarly, the representations of space — state boundaries in particular — and their manifestations in daily life doubtless make a difference in terms of the present discussion. There are particularly good reasons for studying the representations of people who live in border areas, as it is obvious that many of the representations and meanings that people living elsewhere attach to the boundary reflect a more general social consciousness and not so much their own experiences. This must be particularly true in instances where the boundary is located far away from the context of daily life and experience. Those who live in a border area have a personal contact with the immediate border landscape and its elements and possible restrictions and experience of these. In addition, they usually have experience of the reactions of those living elsewhere towards the boundary.

Whereas the first key 'corpus' of theoretical insights and empirical research materials will be employed to reveal the content of socio-spatial consciousness and representations of the boundary at the national level of institutions, the second aim is to explore the local forms of culture, the construction of socio-spatial identities and the social representations of space among the people living in Finnish border areas. An attempt will be made to discover how individual actors become parts of an institutional and social history, how through various discourses and struggles they make sense of their lives in various overlapping spatial contexts. Neighbourhoods, villages, provinces, nations — all of these constitute overlapping and intertwined entities, systems of values, ideas and practices, which are themselves by tradition categorized and constitute a basis for further categorization and naming of other entities.

The present discussion of the local meanings of the Finnish–Russian boundary is based on a case study of the commune of *Värtsilä* in eastern Finland, the institutionalization of which has taken place mainly since the nineteenth century in the wake of the rising iron industry. What makes this commune a particularly exciting object of research is the fact that it was divided in two by the new Finnish–Soviet state boundary as a consequence of World War II. An analysis is thus made of the institutionalization of Värtsilä, of the rise of the factory community and of the territorial identities experienced by the inhabitants of the old commune and of the new, smaller one. The role of the state boundary among the spatial representations and symbolic boundaries that the inhabitants use to make sense of their everyday life will be analysed, recalling the contributions of several authors in the volume edited by Cohen (1986a), in which they focus on the 'symbolic' element in boundary marking in various British contexts. The national boundary examined at Värtsilä is nevertheless much more explicit than any of the 'boundaries' discussed in the chapters in Cohen's book (cf. Cohen 1986b).

To illustrate the role of Värtsilä in the regional transformation process, the analysis will be pursued by exploiting various documentary materials, personally written life-stories and observations and interviews conducted with people born in the commune before and after the war. Since these people belong to different generations, they have differing personal histories in relation to the construction of the boundary and its representations. The meanings attached to the border in their territorial setting and its role in their everyday lives will be interpreted. The personal experiences of the inhabitants are mirrored through interview material and various documents, collected and partly recorded during 1987–92. On the basis of these materials an attempt will also be made to understand how people have experienced recent changes that have taken place in the relations between the Eastern and Western blocs on the local scale and how these changes have affected their discourses.

On these grounds, the purpose is to deconstruct the ways in which people in their everyday lives produce local and 'non-local' meanings regarding the boundary and how they reproduce them. The aim is thus to analyse local forms of thought in relation to larger social and ideological structures. The methodological point of departure for the following discussion is the fact that space and time are fundamental, organizational categories of human life and experience. Also, as Sack (1986, 216) points out, these are not merely naïvely given facets of geographical reality but are transformed by, and affect, people and their relationships to one another.

4.3   A METHODOLOGICAL COMMENT

Thrift (1991) has sought grounds for a new regional geography and has pointed to the neglected role of the 'subject' in the construction of regional geographies (cf. Paasi 1986a, 1991a). He points to five tasks that geographers have to take into consideration when developing new regional geographies. What is needed, he argues, is an understanding of (1) the constitution of self and identity, (2) the constitution of autobiography and memory, (3) the emotional repertoires available to be drawn on, (4) the different forms of knowledge that discourses make available and the modes of structuring that knowledge (metaphors, rhetoric) and, finally, (5) how context is used as a way of establishing identity, cueing memory, releasing emotions, and activating certain discourses, along with the knowledge peculiar to them.

Some of the claims put forward by Thrift represent new challenges, and some are already part of the practice of geographical research. One point is absolutely clear on the basis of Thrift's arguments, however: what is needed is a continual effort to reconceptualize the concepts of geography in order to avoid essentialism, which easily places limits on geographical thought and practice. This also applies to the conceptualizations of subjects themselves.

The object of this book is to integrate two methodological frameworks, which can be labelled structural and interpretative, or macro and micro perspectives — perspectives which have been common to geographers, anthropologists and historians during recent years (Roseberry 1989, Comaroff and Comaroff 1992). These perspectives are not mutually exclusive but are necessarily complementary. Rüsen (1992, 279) remarks that a profound micro-historical representation has to refer to its macro-historical prerequisites if it wants to avoid rendering its object non-historical. As researchers we have to postulate some kind of modernization process as a background to daily life, even though people have not necessarily observed or accepted it. By contrast, Rüsen avers that a historical process must be filled with narratives that tell of human action and sufferings, and that provide a human view complementary to macro perspectives. Gregory (1989, 88) is doubtless right when he suggests that if we cannot provide descriptions of the relations between people and places 'so vivid that they move our emotions', then we will radically reduce the quality of our geographies. Description thus requires theoretical, empirical and stylistic depth in analysis.

An attempt is made to integrate both 'big stories' regarding nation building and the establishment of the boundary and 'small narratives' that illustrate the daily experiences of ordinary people, who are the subjects of regional transformation. Hence the basic idea is that these national and local perspectives should be both constituted and constitutive for each other. For a regional geographer this means that two essential problems have to be faced. First, how to make what in traditional regional geography has been somewhat mystically called a 'synthesis' and, secondly, how to *represent* it, i.e. how to solve the eternal problem of form and content.

These are essential problems, since it has been pointed out that one central dilemma in the new regional geography is the question of how to write up such research (Thrift 1990). The new regional geography — which could be called interpretative geohistory — inevitably contains a historical dimension, which renders it possible to trace the social processes, actions and experiences that are constitutive for each other even if they are spread across a number of temporal scales. The historical dimension was lacking from traditional regional geography in the spirit of exceptionalism, but an attempt is made in this work to create a regional geography of the construction of the Finnish nation by employing and critically synthesizing both theoretical and empirical perspectives. Briefly, the aim is to write a synthetic narrative of various 'stories', both big and small, and their contexts. Since the theme is a complicated and multidimensional one, both conceptually and empirically, the task can be metaphorically likened to peeling an onion: the object is to analyse both conceptually and empirically various time–space-specific 'layers' which ultimately constitute a whole, and not separate pieces of knowledge.

Hence the approach may be labelled as historical, although this label is not treated as an *a priori* assumption regarding the logic and continuity of the research topic. Rather, 'historical' is understood as a sedimentation of various social and cultural practices, at times conflicting, manifesting them on different spatial scales. The concept of *genealogy*, to employ the expression of Foucault, perhaps characterizes the aim best, since the past is not the object of the research but material which is interpreted conceptually and critically in order to make the historical construction of spatiality visible and understandable. It is hoped that this approach will open up a perspective on the disjunctive forms of representation that signify a people, a nation, or a national culture (cf. Bhabha 1990, 292). The aim in employing this method is to interpret the roles of the three geographies in the social and historical construction of a particular social reality — an entity called Finland. The purpose is to illustrate how the state makes a difference in this specific case.

PART TWO

THE INSTITUTIONALIZATION OF THE FINNISH TERRITORY

5

NATIONALISM, GEOPOLITICS AND CHANGING TERRITORIES: THE CASE OF FINLAND

5.1 THE RISE OF THE FINNISH STATE AND NATION

Whatever definitions of nationalism are accepted, A. Smith (1983) argues, any convincing theory of it must possess at least three characteristics. First, it should combine a clear-cut statement of the objectives of the theory, i.e. which aspects of the broad field of nationalist phenomena are to be studied, with a clear definition of the terms which differentiate the phenomena under scrutiny from other related phenomena. Secondly, it should combine an internal approach which would search for the sources of cultural identity and community and draw on an empathic understanding of the meanings with which the participants endow their attachments and activities. Thirdly, it should take into account an external approach which begins with a broad comparative perspective and historical framework within which it locates and classifies the various local and individual manifestations of nationality and nationalism, with a view to explaining their diverging roles and various features.

The aim of this chapter is to combine divergent theoretical approaches by outlining the process of institutionalization of the Finnish territory, i.e. its territorial, symbolic and institutional shaping and its establishment in the course of the transformation of the European system of states. This analysis will be based on documents through which the building of the Finnish nation has been analysed by historians, sociologists and political scientists. This will provide the foundation for a more detailed analysis of the changes in the Finnish–Russian boundary and of its representations, a problem that has been a completely neglected part of the 'non-spatial' analysis of the nation-building process.

A number of historians have analysed the early formation of the Finnish–Russian boundary in detail (Korhonen 1938, Mikkola 1942, Julku 1987). But the above argument concerning the non-spatial character of boundary analyses also holds good in these cases. Hence they do not regard boundary formation as an expression of territoriality or of the institutionalization of Finland, inasmuch as before the rise of the nation and state, boundary construction — or rather the marking of frontiers — was not an expression of the sovereignty of a state or of nationalist ideologies. In any case, such documents as do exist provide a general historical basis for an analysis of local forms of territoriality and local experience of boundary changes.

The institutionalization of nation-states is a complicated socio-spatial process in which the rise of a state is typically a result of power politics and the rise of the nation a result of nation building, i.e. political and cultural integration, which in turn is based on the exploitation of nationalistic ideologies. Bloom (1990, 55) writes that inherent in nation building is the fact that a state has already been created and that the nation, or community of solidarity, is to be built within it. As Bauman (1990, 154) points out regarding the historical process of nation building, 'nationalism was a programme for social engineering, and the national state was to be its factory'.

It is common to regard nationalism as the child of the French revolution. Nationalistic ideology, for its part, was preceded by the idea of the specific passions or characters of nations, as propagated by Voltaire and Montesquieu, for instance. These ideas were especially common during the eighteenth century (Kemiläinen 1964, Kristeva 1992, 181). It was then that the idea of national character in the construction of nationalism became significant for the construction of we-ness. What is one's 'own' and what is 'alien' had to be instilled into the minds of citizens in the political socialization process.

Nation building emerged along with the rise of capitalism and became integrated into the transforming global economy (cf. Agnew 1987, 37). Both capitalism and industrialism have decisively influenced the rise of nation-states, but as Giddens (1987a) points out, the nation-state system cannot be reductively explained in terms of their existence: the modern world has been shaped through the complicated interaction of capitalism, industrialism and the nation-state. Nationalism and nation building were inextricably linked with this process.

Poulantzas (1978, 99–107) employs the expression 'spatial matrix' to depict the spatial organization and 'temporal matrix' to illustrate the creation of the historical continuity connected with various modes of production. He is ready to conclude that the first fruits of territory, considered as constitutive elements of the modern nation, are written into the capitalist spatial matrix. The modern state aims at monopolizing the procedures of organizing space, and its spatial matrix is materialized in various apparatuses — such as schools, a centralized bureaucracy, a prison system or an army — in turn patterning the subjects over whom it exercises power (1978, 104).

Simultaneously, the temporal matrix became segmented, serial and divided. A consciousness of a specific history was created and imparted, which rendered possible the representation of the historical orientation of the nation-state. The intimate relation between the spatial and temporal matrix is fundamental to national unity, which becomes, Poulantzas (1978, 114) writes, 'historicity of a territory and territorialization of a history — in short, a territorial national tradition concretized in the nation-State; the markings of a territory become indicators of history that are written into the State'.

The nation-states of Western Europe and the system of states they formed strengthened during the sixteenth and seventeenth centuries. The treaty of Westfalen (1648) was the fundamental basis for the formation of the European system (or 'geography') of states, since it established the existence of equality among territorial units, and states became sovereign (Knight 1984, 171). England, France, Brandenburg-Prussia, Sweden and the Russian, Austrian and Ottoman Empires were those of particular importance. The rise of the Westphalian state system brought about a change in the political organization of territories, but also a change in people's understandings of place and space. It modified the ideas of 'our' and 'their' territories, the relations of political territories to other phenomena and, finally, thinking about frontiers, boundaries and the socio-spatial organization of society itself (Murphy, forthcoming).

When capitalism developed and large numbers of people received political rights, nationalism became the central ideology for identifying the people of Western Europe with modern states (B. Anderson 1991). This also created the prerequisites for the development of education systems, in which the national and state organizations set forth their own education standards and principles. These developments were time–space-specific, so that the education system in each state had its own characteristics.

Nationalism — as it has developed in Europe since the end of the eighteenth century — is a specifically modern manifestation of collective identity (see Habermas 1988). Nationalism satisfied the need for new identifications in a situation in which the traditional social structure was eroding. Whereas the traditional social system was based above all on descent, individuals now became emancipated within a framework of abstract civil liberties. The basic task of nationalism was to move the group loyalty based on kinship onto a local and regional level. The history of the construction of nation-states has been based more on the aim of establishing the ideological means to weld together a common identity for people than on the exploitation of power and violence (cf. Soja 1971). The central bases for this identity were usually geographical connections, language, history, the homogeneity of attitudes and values and so on (Duchacek 1975, Sugar 1969).

Habermas (1988) points out that nationalism differed from previous acts of identity formation in several respects: (1) the idea of establishing identity was drawn from the secular heritage prepared and mediated by the emerging *Geisteswissenschaften*, (2) nationalism brought the shared cultural inheritance of language, literature and history into coincidence with the organizational form of the state, and (3) there prevailed a tension in national consciousness: (a) between the universal value orientations of democracy and the rule of law; and (b) in the form of particularism of a nation marking itself off from the outside world.

The development of nation-states consists of several levels of organization, the most important of which are internal organization and external relations. The rise and development of a nation-state and nationalism can be fully understood only in a framework which takes into account its relations with other states. New states usually emerge through conflicts in the original system of states.

The emergence of the Finnish state is a typical example of the role of a political unit created by *external political forces*. This for its part was an expression of more general political and social processes. The Napoleonic wars and the subsequent peace settlement of 1815 created vast upheavals throughout the continent. Rokkan and Urwin (1983, 38) mention as important events the granting to Prussia of important possessions on the Rhine, the transference of Norway from Denmark to Sweden, the handing over of Belgium to the Netherlands and the transference of Finland from Sweden to Russia.

It has been argued that, whereas the Napoleonic Wars gave rise to Finland as an autonomous unit, World War I formed the general background to its independence (Alapuro 1988, 19). Finland is in many senses located between East and West, having been controlled successively by Sweden and Russia up to 1917. The Swedish domination lasted until 1809, when Finland became an autonomous part of Russia, a buffer state with its own system of government and economy. This was the beginning of the *political* history of Finland: the geographical entity of Finland was now laden with a new content and became a national unit. Klinge (1980, 1982) writes, however, that the inhabitants of the area that later became Finland did not form a unified whole. He argues strongly that Finland was *made* and not born and that this creation was not at all self-evident, let alone inevitable. A new social space had to be created. Despite its autonomy, Finland was above all a territorial expansion of Russian military power and from the Russian strategic point of view a safety zone for St. Petersburg. This may be seen in the fact that Russia had armed forces in Finland throughout the autonomy period, in the fortification of the coastal areas of Southern Finland and in the fact that Finland had no foreign policy of its own. As a manifestation of its autonomy and separate national economy, Finland had a customs boundary with Russia and the Russian citizens did not have civil rights in Finland.

It is typically argued, particularly in the culturally based interpretations of nation building, that it is the educated classes, not the popular masses, that constitute the nation at an early stage of development, inasmuch as the intelligentsia usually have the means to make and remake histories, codify a language, create spatially extended systems of education and so on (Cooke 1989b). In Finland only those with political rights, i.e. the groups in the assembly of the representatives of the estates, constituted the nation. Jussila (1979, 17) notes that even at the beginning of the twentieth century this group accounted for only one tenth of the adult population. The 'intelligentsia proper' constituted merely 2% of the population, and the 'people' 98% (Kemiläinen 1984). Under the Parliamentary Act of 1906 the parliament became unicameral and every Finnish citizen aged 24 years or older was entitled to vote. The change from a four-chamber system was the most radical in Europe at that time. For the first time in Europe, women were able to participate in legislation.

When the Russians occupied Finland in 1808–9, the Finns themselves did not feel that the nation had changed, only the monarch, and even at the end of the nineteenth century Russia was still regarded as the new fatherland on the basis of the Emperor and the ruling family (Jussila 1979, 17). This association with the monarchy was weak, however, compared with the class, industrial and territorial bonds. Jussila remarks that in a corporative class society people were first and foremost members of their class, and inasmuch as only those belonging to the upper classes usually travelled outside the parish and the country, the local community — village, parish or town — was the first unit of belonging for the majority of people, and the nation took second place. The Finnish nation, then, was a concept employed by the Fennoman, 'pro-Finnish', intelligentsia: the nation was a necessary subject of history, and a body which at the same time was quiet and humble enough to be governed (Alapuro 1993).

The Finnish territory was shaped as a specific socio-spatial unit during the period of Swedish domination, and it was also shaped to constitute a specific territorial entity — which nevertheless did not have independent state organs. When the autonomous state of Finland was established, the Finnish people were labelled as a 'nation', but there was no debate as to whether they had any feeling of 'togetherness'.

Kemiläinen (1986) writes that even if there were no nationalistic tones, people already had an idea of which country they were living in, i.e. Finland, during the late Swedish period, even though the area still belonged to Sweden.

The concept 'Suomi' (Finland) which came to symbolize the state had by tradition several meanings depending on the context and sources (maps, travelogues etc.) and referred originally to the southwestern part of the country, and expanded later to include a larger area. Historians tend to emphasize the year 1581, when the Swedish King John III declared Finland a grand duchy and the name gradually began to be used for the whole area with a Finnish-speaking population (Vehvilä and Castren 1972, 41). It is interesting that during the sixteenth and seventeenth centuries a number of learned persons, such as Olaus Magnus, regarded Finland as a 'state' and represented this idea in maps (Kemiläinen 1986, *Vanhoja Suomen karttoja* 1967). In fact, Sweden delivered up to Russia under the peace treaty of 1809 a group of administrative provinces, not an entity called 'Finland' (Tommila 1989, 51). The country was represented as a clear spatial entity on some Swedish maps published during the last decade of Swedish rule, however (see Figure 5.1), and it should also be noted that the idea of Finland as a free state between Sweden and Russia had already been put forward in the mid-eighteenth century in a propagandistic manifesto issued by the Russian Empress Elisabeth during the war between Sweden and Russia (Manninen 1993). According to Tanner (1936, 23) the word 'Finland' (Findlandi) appeared on a map for the first time in 1427. The first map of Finland with Finnish placenames was published only in 1846, after which several versions were published during the second half of the nineteenth century (Paasi 1986b, see Figure 5.2).

In spite of these facts, the rhetoric of national socialization provided by Finnish teaching of history insists that Finland has existed for almost 9000 years. In a high school history book, for instance, Vehvilä and Castren (1972, 5) begin by telling readers how the Ice Age came to an end around 7000 years BC and when the lithosphere rose, the features of Finland began to take shape! Here the authors follow the lines set forth in 'official' Finnish histories where these arguments have at times been strengthened by using maps representing the boundaries of present-day Finland, even when the phenomena discussed go back to ancient times (see Luho 1964). This is a typical illustration of national history writing, and particularly of the popular ideas of national history which still reflect the nationalistic themes of nineteenth-century history books. According to this way of thinking we tend to emphasize the present boundaries to provide a territorial context for the names of states and ideas of their past (cf. Klinge 1975, 25). The question of the nature of Finland's history was debated during the nineteenth century,

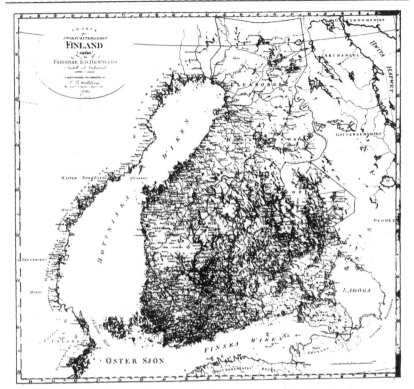

Figure 5.1 Storfurstendömet Finland (the Grand Duchy of Finland) on Hermelin's map (1799). Source: Map collections of the Department of Geography, University of Helsinki. Reproduced by permission

and the theme has raised polemic discussions during the twentieth century. The key question has undoubtedly been whether Finland had any history or national consciousness before 1809. In any case, the Finnish state was ahistorical at that time since it was only just coming into existence (Kemiläinen 1986).

When Finland became an autonomous part of Russia, the area of Lapland was attached to it to conform to the 'natural boundaries' of the region (see Puntila 1971, 23). This argument reflected the more general European discussion and beliefs about the nature of boundaries, which will be considered in detail later. 'Old Finland', the present southeastern part of the country, and Karelia were incorporated into Finland in 1812, and some new areas from Karelia in 1833. Hence the territorial shape of Finland was established, to become a basis for the construction of national symbols, institutions and consciousness.

The construction of national identity and its territorial expressions were partly manifestations of interaction between two states, Finland

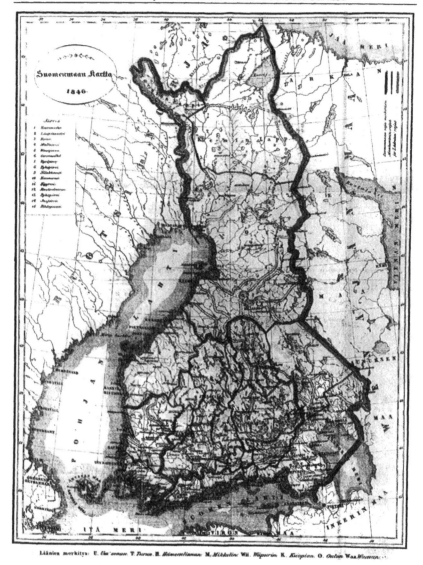

Figure 5.2 The first map of Finland to be published in Finnish (1846). Source:
Appendix in *Lukemisia Suomen kansan hyödyksi*, 1846

and Russia. The construction of a national identity became possible and
even desirable not merely because of the interests of the élites in
Finland, but also because of the strategic interests of Russia, for it was
realized that the loyalty of the people was best guaranteed by means of
local élites and traditions (see Klinge 1980, Kemiläinen 1989).

A significant part of the history of a number of countries concerns the struggle for territory (Prescott 1987, 1). This also holds good in the case of Finland. The changes in Finland's eastern boundary have been manifestations of changes in the power relations between Sweden, Finland and Russia (Puntila 1971, 19). After being under Swedish rule for several hundred years, and after several wars against Russia, a range of major changes in the boundary between the two entities since 1323 and a long ideological struggle over the role of Finland between Sweden and Russia, Finland finally became an autonomous part of Russia in 1809. This situation lasted over a hundred years, until 1917, when it became an independent state (see Figure 5.3).

The boundary established between Finland and Russia (and later the Soviet Union), which remained in broad outline the same for the period 1833–1940 — the main exception being the Petsamo area — was regarded as stable and sanctioned by history, the only real boundary, while earlier stages were experienced as though they had been merely temporary stages (for a summary of the history of the boundary, see Sihvo 1989). Originally the boundary was above all that between the Eastern and Western worlds: there was no 'Finland' in the present sense of the word (Klinge 1982, 12), not to mention Finland as an institutionalized territory.

The opposition between the Eastern and Western heritage, which was later to become a major ideological argument in Finnish–Russian relations, is clearly illustrated in the iconography of the provincial coat of arms of Karelia, originating from the sixteenth century (see Figure 5.4). Kenneth Boulding (cited in Minghi 1991) once wrote that 'Good fences only make good neighbors. When they are not made out of sabres.' In the case of the province of Karelia the relations between East and West have been symbolized by a Western sword and Russian scimitar since the sixteenth century (see Rancken and Pirinen 1949, 60-1). These symbols are indicative of the historical representation of Russia in Finland.

Several definitions for types of nationalism have been put forward in the course of time, e.g. state nationalism, unification nationalism, liberation nationalism, renewal nationalism and separation nationalism. The last of these, typically connected with the disintegration of existing sovereign states, was particularly significant during the nineteenth and early twentieth centuries, and was also the primary basis for nation building in the case of Finland (Taylor 1993b, 204). The question of nationality emerged in Finland after its separation from Sweden, but Finland was for a long time a *state-nation* or political nation based on bureaucratic practices — and not a cultural nation based on a sense of belonging to a common heritage and history (cf. Alter 1989, 14). Not even the upper classes had a clear national consciousness.

Some theorists of nationalism place particular emphasis on modern

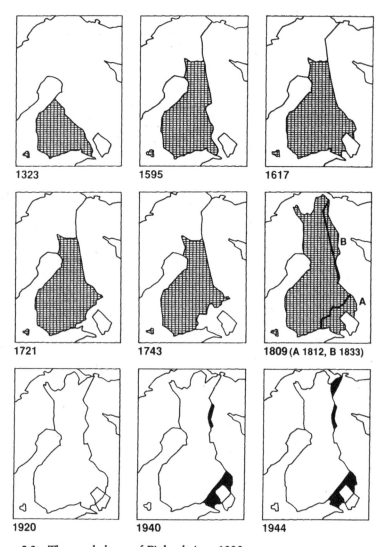

Figure 5.3 The areal shape of Finland since 1323

bureaucracy when shaping its constituents (see Smith 1979). Modern bureaucracy is often understood as being interventionist and efficient; it requires boundaries and territory as a context within which to operate. It also needs a modern type of person as the agent and, finally, the modern intelligentsia which has a clear understanding of the forces — whether historical, cultural etc. — of communal change. This is seen to constitute the basis for cultural nationalism. Cooke (1989b) is critical of this model, arguing that reality is often different, and that national

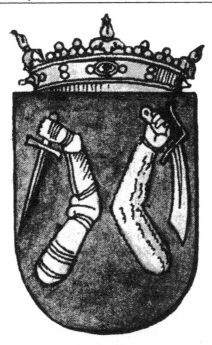

Figure 5.4 A symbolic representation of the confrontation between West and East: the coat of arms of the province of Karelia, dating from 1562. Source: Rancken and Pirinen 1949, 61. Reproduced by permission of Werner Söderström Osakeyhtiö, Helsinki

movements are typically led by 'culturally informed, economically sensitive political parties which struggle to create spaces into which national claims are channelled'. Cooke points out that most national movements 'do not consist of bureaucrats but workers, the unemployed, small professionals and the petit bourgeoisie'.

The emerging nationalism in Finland originally had two sources (cf. Kemiläinen 1989). First, the foundation of the state provided motivation for the rise of political consciousness. Secondly, the emerging romantic movement raised the question of culture, above all the role of the Finnish language, and hence promoted the rise of a cultural nation. There were about 1 million inhabitants in Finland in 1812, of whom 87% were Finnish speaking (Tommila 1989, 51). The élites who were afraid of Russian influence had been successful in keeping Swedish as the official language of administration even after Finland became an autonomous part of Russia.

Finland had a class society, with a small upper class consisting of the Swedish-speaking élite and a lower class, including the peasantry, which was isolated from it. The representatives of the peasantry were not usually able to move into the upper class, a situation which was maintained by state policy. The power of the upper class was thus to a great extent based on its position in the state apparatus and not on land ownership, for example, as in many other countries (see Halila 1980, Alapuro and Stenius 1987).

5.2 ECONOMIC AND CULTURAL INTEGRATION

The early European nation-states developed almost inseparably from the rise of capitalism. Since the Finnish state was created by a political decision, it was strengthened economically only after that, and the role of the state was essential in this process. Until the 1840s Finland maintained close connections with its former mother country, Sweden, and its own economic integration took place only during the 1850–60s (Alapuro 1988, 29). The strengthening of the nation-state and nationalism were organically connected with the rise of industrial capitalism. After losing the Crimean War, Russia had to choose a new political and social direction, and Emperor Alexander II began a more liberal policy which rapidly reached Finland as well (Klinge 1980).

The changes did not occur merely in the spirit of liberalism, since the state was also a strong 'regional policy' director aiming at establishing and developing the capitalistic system. To promote more liberal development, the state increased the rights of individuals to move both spatially and socially. The construction of canals and a railway network was significant for economic integration and more effective communication. Industrialization was supported by new privileges, and in this way the impediments to the capitalist transformation were demolished. The state no longer merely controlled the economy and collected and redistributed the surpluses of production, but instead built up activities aimed at reproducing the capitalistic mode of production and at promoting the accumulation of capital. In this way, Alapuro (1988, 32) writes, the political unit formed by external forces was also gradually integrated economically, and this transformation also accelerated the connections with the international economic system. As far as significant institutions and symbols in the institutionalization of the Finnish territory are concerned, Finland adopted its own currency, the Finnish Mark, in 1860, and began to create its own monetary policy. Finnish stamps began to be issued even earlier, in 1856, and the subsequent expansion of the postal network together with the land and water transport network was also significant for the integration of the whole territory.

Soja (1971, 39) points out that 'cultural territoriality' becomes significant in societies along with the rise of states. Finnish society during the second half of the nineteenth century was characterized by state control, and also by *cultural movements* and the struggle over language and culture. A new civil society began to develop from 1855 onwards (Konttinen 1991, 181). The straightforward bureaucratic system weakened and new organizations of citizens gradually emerged to influence the development of the state. A number of changes, partly reflecting broader international developments, took place during the second half of the nineteenth century that affected the shaping of the territorial structure of Finland as far as its inhabitants were concerned.

We can briefly conclude that the 'discovery' of history and folk tradition and the shaping of *political publicity* as a general condition affected the development of the national consciousness; the central points of the emerging bourgeois public sphere came to be the press and public opinion (Pulkkinen 1987, Smeds 1987). The 'cultural' theories of nationalism put emphasis on the roles of communication and the media in the development of solidarity with a specific culture and territory. Such attachment takes place through the use of a dominant language to convey membership of the national community (Cooke 1989b, 280, Schlesinger 1991). Several authors have stressed the importance of printing (newspapers, books etc.) both as a capitalist enterprise and/or as a medium in time–space transformation, e.g. in the creation of 'imagined communities' such as nations and the ideologies behind them (B. Anderson 1991, Calhoun 1991). The nationalization of culture has been typically linked with the creation of this public sphere (Löfgren 1989). A central medium for (territorial) ideologies is the control of publicity. Thus publicity is inextricably bound up with power.

As we have seen above, the fundamental 'medium' of various (territorial) ideologies, including nationalism, is *language*. As Rokkan and Urwin (1983, 68) remark, language is only one of several expressions of identity, but it is the most pervasive and obvious sign of distinctiveness. They point out how the use of language is a 'collective act' which all people in a territory must share, and hence it becomes politicized when élite groups establish the standards of written communication and may put forward claims for its recognition in public life. In a word, language is the medium through which discursive stories about us (and them) are produced and reproduced. But whereas language is a medium for the discourse of integration, it is also a medium for *difference*.

Ideologies also operate through language, which is the basis for socialization. Ideologies are not forces operating somewhere above people, but creative, constitutive elements in social life. As Johnson (1992) indicates, the education system was a cornerstone in the language policy of Ireland, and the same observation has been made in many other countries. As far as the construction of political publicity is concerned, the question of language becomes important. The controversial element in this respect is usually the written language or the language spoken for public purposes. The language or languages spoken within private communication usually raise no major problems even when co-existing with public languages, for each occupies its own space (Hobsbawm 1990, 113).

Pred (1989) writes that a local language is frequently the scene of ideological contestation, because it is frequently a site of struggle and resistance. In certain respects this also holds good in the case of social practices extending over larger spatial scales, e.g. in the nation-building process. In Finland, for instance, the creation of cultural consciousness was to a great extent a struggle over the status of the Finnish language, since Swedish was still the major language among the dominant élite of the state. The question was at the same time one of a struggle for *social power* between the classes. The leading and most influential class, the nobility, aimed at defending its class status and was unfavourably disposed towards the education of the peasantry up to the 1850s and even later. It was only in 1863 that a statute was issued to raise the Finnish language to an equal status with Swedish in all matters that concerned the Finnish-speaking population, and it was only in the late 1880s that the number of students in Finnish-speaking secondary schools became comparable to that of Swedish-speaking students. According to the Romantic movement, which spread to Finland from Sweden and Germany, all nations have their own character, which is bound up with their language, history, poetry etc.

On these grounds, activists started soon after 1809 to trace a *national-ideological* Finland, to follow the roots of the Finnish language, to collect and search for poetry, to write Finnish history and so on (cf. B. Anderson 1991, 74–5). It has been said that there is probably no other country in which the marriage of folklore studies and nationalism produced such dramatic results as it did in Finland (Wuorinen 1935, Wilson 1976). The aim was to construct a heroic national past and to analyse it to find models on which the future could be shaped. The cornerstone for this became the *Kalevala* (1835), an epic based on folk poetry, which was compiled by Elias Lönnrot from poems discovered in Finnish and Russian Karelia. As far as the relationship between the history of geographical thought and the geography of constructing 'Finnishness' is concerned, it is interesting to note that Alexander von Humboldt refers to the *Kalevala* in his *Kosmos* and lays emphasis on its descriptions of nature (Leiviskä 1949, 135). The publication of the *Kalevala* formed a background for the rise of scientific studies of folk poetry and gave a new direction to Finnish intellectual life (Haavio-Mannila 1973).

Another important episode was the founding of the Finnish Literary Society in Helsinki in 1831, with the purpose of spreading knowledge of the motherland and its history and furthering the use of its language.

The aim was also to develop literature in the Finnish language for both educated 'compatriots' and the 'lower class' (Klinge 1984, 125). The development of a sense of history among leading politicians did not proceed rapidly, however. In 1875 Professor Yrjö Koskinen, speaking in the first annual meeting of the Finnish Historical Association, pointed out that 'The Government of our country, leaning on the statement given by the Consistorium of our University, has not regarded this effort — the study of the past of our nation — worthy of help and support'. He severely criticized the view that the Finnish nation lacked a national history and proclaimed the axiom that 'a nation without history, also lacks a national character' (Koskinen 1876, 2). Koskinen's own solution was that nationality 'lives' not in language but in 'national spirit, which unites the members of a nation with common ideologies, feelings and interests'. Ideologies, feelings and interests, for their part, manifest themselves in 'deeds', just as in the language and literature they manifest themselves in 'statements'.

While the first national awakening was kindled at the beginning of the Russian autonomy period, the second was set in motion in the mid-1840s by J. W. Snellman, inspired by Hegel's philosophy. Snellman regarded nationality as a product of the historical process and the national spirit as a cultural form typical of the nation in question. Nationality was thus the historical vocation of the nation (Noro 1968, 31–3). The Fennoman notion that Snellman propagated was aimed at improving the position of the peasantry in Finnish society, establishing Finnish-speaking schools and giving the Finnish language an official status. Many activists in the Fennoman movement belonged to the upper class.

Snellman's aim was not to educate a new élite from the peasantry but to change the language of the old one. The proportion of representatives from the lower classes was in fact small compared with the situation in other Eastern European nations, where the key groups supporting nationalism originated from the peasantry. Alapuro and Stenius (1987) interpret this as being because the power of the upper classes was not based on land and property but on the strengthening role of the state, so that they had exceptionally good reasons to adopt or accept the language of the great majority and of the culture and thus integrate the peasantry into loyalty to the new state that was under construction. On the other hand, this linguistic and cultural integration and loyalty were necessary in order to build up a national consciousness in the face of the hegemony of Swedish culture.

Nevertheless, it was common in the discussions that took place in Scandinavia in the mid-nineteenth century to regard the Finns as inferior to the Swedes, as belonging to a lower race without the ability to become civilized. Scandinavian researchers were of the opinion that everything that had been achieved in Finland up to that period was based on the influence of Swedish nationality, even the creation of the *Kalevala*, and that the future of Finland was in the hands of the Swedes, not the untalented Finns (Puntila 1971, 41).

J. W. Snellman, together with J. L. Runeberg and Z. Topelius, represented the second generation of nation building in a Europe remodelled by the Congress of Vienna. Niemi (1980) labels them as the 'rebuilder generation', national idealists characterized by an affinity for a heroic, epic style. The writings of Runeberg and Topelius created the basic idea of a Finnish landscape dominated by lakes, and they also put forward a powerful, mythical interpretation of the 1809 war between Sweden–Finland and Russia. Much of the Finnish literature of the mid-1800s was characterized by Hegelian philosophy. According to Niemi (1980), Topelius and Runeberg can be regarded as having created the activist generation at the end of the nineteenth century and as having promoted the later aggressive nationalist mentality that characterized Finnish social consciousness during the 1920s and 1930s. Runeberg's *The Tales of Ensign Stål* (orig. 1848) depicted the braveness of the Finnish soldier, and Topelius outlined the features of the Finnish personality. Both provided a patriarchal view of Finnishness, but there was also a place for women in the form of the allegory of Finnishness, as will be discussed later.

An important force in nation building was the competition between the cultural movements emerging in civil society. Of special importance in the territorial shaping of the country was the organization of voluntary associations and the cultural and political organization that took place during the 1880s in particular. The institution of voluntary associations can have various roles in social integration. Such associations bind the individual to the totality by socializing and transmitting predominant philosophical values, for example, and can also mobilize the common interests of their members against other associations or the total system (Siisiäinen 1985). From the spatial perspective, associations integrate people concretely with the spheres of influence of various territorially based activities and ultimately with the national territory. The Fennoman movement in particular tried from the very beginning to integrate the organization of free associations with the state (Alapuro and Stenius 1987, 18), and the development of such associations has been crucial for the regional transformation of the Finnish provinces that has taken place since the mid-1800s.

As was pointed out above, organized mechanisms of physical force are one basic social element in the reproduction of the existence of a state in the system of states. A law requiring compulsory military service was passed in Finland in 1878, under which every 21-year-old male was eligible for national service. This independent army emphasized

Finland's position, since according to the law it was committed to defending 'the throne' and 'the fatherland', not the empire. As a result it became a thorn in the flesh of those Russians who were opposed to Finland's autonomy.

In summary, it has been argued that nationalism in Finland arose in the 1860–80s, when the possibility for developing political rights was denied and the rising Russian nationalism seriously challenged the existence of the Finnish state. Jussila (1979, 18) remarks that during the years of oppression, beginning in 1899, the spread of nationalism among the masses was promoted actively rather than passively. Hence the idea of a fatherland and a national consciousness spread slowly among the people during the early years of the present century, making the distinction between us and the Other more explicit. Finland was a deeply divided class society at the beginning of the twentieth century, both in the rural areas and in the towns. More than 70% of the population were working in agriculture and forestry, and almost half of them were landless, while the socio-economic position of the rising urban working class was weak. It was difficult for the peasants and workers to comprehend what was meant by the abstract idea of 'fatherland' (Jussila 1979, 19).

5.3 TOWARDS INDEPENDENCE

It has been argued that, even though there existed different views on the national question in Finland, as well as varying forms of activism, the Manifesto issued in St. Petersburg in February 1899 ultimately gave rise to a question among the large masses of the population as to what exactly was the 'fatherland'. Jussila (1979, 21) remarks that this event served to create national solidarity and an understanding of the danger of the oppression that faced the 'inherited society' and its institutions.

Ordinary people in many parts of Europe were invited at much the same time to identify themselves as members of a nation and as members of a class (Burke 1992, 294). The organization of the working class in Finland was slow. It was the upper classes who started the process, one of its aims being to promote solidarity between the classes. The need to organize the activities of the working class increased at the same time as labour markets were established as a consequence of the liberalization of the economy (Alapuro and Stenius 1987, 30), and thus the main task of the working-class movement up to 1895 was to persuade the state to set limits to the free market economy, and after that to obtain the right to collective bargaining on the labour market.

The nationalistic ideology and the internationalism of the working class are antithetical, but in Finland this basic line of thought was confused by the fact that Finnish nationalism at the beginning of the twentieth century was characterized by opposition to Russia (see Jussila 1979, 24). As a whole, the concept of 'fatherland' was now laden with new meanings by the workers. The turn of the twentieth century was a time of ideological revolution for the working class, and as Jussila (1979, 26) writes, the idea of the traditional 'nature-fatherland' was abandoned and the idea that was adopted was one of a workers' fatherland in which it was good to live.

The socialists were supporters of independence, for it was thought that the reforms could be achieved more easily in an independent country than under Russia. But there were different ways of thinking among the socialists. The nationalistic parliamentary group of the Social Democratic Party thought that independence could be achieved best by cooperation with bourgeois groups, while the supporters of the class struggle regarded a revolution as necessary.

In spite of the trends illustrated above, it must be remembered that until 1917 Finland was an autonomous part of Russia and the leading principle of Russian policy was to reduce the power of separatist movements and to strengthen loyal forces. Radical ideologies which were aimed at separating Finland from Russia were always objects of varying degrees of suppression (Konttinen 1991, 185). In particular during the 'years of oppression', 1899–1905 and 1908–17, Finland's separation from Russia and the changing situation in international politics caused the Russian government to plan a Russification of the Finnish administrative system and the introduction of the Russian language. The Russian authorities considered that language was the most important instrument to tie nations together (Polvinen 1984, 202–3). What was important from the viewpoint of territoriality and internal control was the fact that as part of this policy the Finnish army was to become part of the general Russian military forces, but after much resistance on the part of the Finns they were ultimately exempted from military service. This saved them from involvement in either the Russo-Japanese War or World War I. The Russification programme also created new military activity in Finland, and some 2000 Finns were sent to Germany secretly to receive military training from 1915 onwards. These members of the 'Jaeger movement' later played a crucial role in supporting full independence and in training the White Army in the Civil War. The ideal of Finland's independence was also supported by the Germans, who wished to isolate Russia.

Nevertheless, when Finland declared independence in early December 1917, the Russian goverment, led by Lenin, recognized the new state within a short space of time. This was a concrete manifestation of the thinking represented by Lenin, Engels and Stalin, for example, on the possibility of 'national self-determination' and the right of every nation to secede from feudal and bourgeois states, although they also looked forward to the day when 'the arrival of socialism would render the exercise of that right unnecessary' (A. Smith 1990, 2). Lenin was not supported unanimously in his own party and he based his arguments on his idea of the world revolution and the forthcoming return of independent states to Russia (Paasivirta 1984). In January 1918 Stalin was already putting forward the idea that the People's Commissariat had given independence to the bourgeois Finland against its will, since the proletariat had not taken power into its hands because of its 'shakiness and unbelievable tenderness' (Puntila 1971, 102–3). Sweden, France and Germany recognized Finland in January 1918, and Britain and the United States followed in spring 1919. The Western countries soon provided Finland with material help, which brought the country politically closer to them (Paasivirta 1984, 146).

5.4 TERRITORY AND THE IDEOLOGICAL CONTEXT BETWEEN THE WORLD WARS

The most serious conflict between the social classes took place soon after Finland had gained its independence, in the form of the Civil War in spring 1918 — a war which was supported by the Bolsheviks. This war has also been labelled as the 'liberation war' or 'class war', depending on the ideological connection of the commentator. In any case, it was a civil war in which the Finns fought each other. When the fighting broke out there were still 40 000 Russian soldiers in Finland, and some of them sided with the Red forces, while German troops arrived to fight on the side of the White army. More than 5000 Whites and almost 14 000 Reds were killed, and almost 12 000 Reds died later in prison camps when the war ended (see Upton 1981, 453–9).

The collective consciousness of the nation was divided. Upton (1981, 459, cf. Ylikangas 1993a, 24) is of the opinion that the working class, probably more than 40% of the total population, identified themselves with the losing side in the Civil War and hence became more or less embittered. The division of the nation was also manifested in the official histories, which aimed to represent the Civil War as a freedom war against 'misguided' workers (Upton 1965, 39). This prevailing view was balanced after World War II by a left-wing interpretation (Mäkelä 1947, Ylikangas 1993b, 93).

The Civil War was over by late April 1918, and the defeat of the Red forces led to a protracted division in the collective mentality of the nation and its various classes. For the traditional representation of the nation, i.e. the identity of the intelligentsia, the conflict was a deeply wounding affair. As Alapuro (1993, 8) writes, 'the nation stuck a knife into the back of its own intelligentsia' or 'one part of the nation was ridden with a red plague'. The division was strengthened by the fact that the activities of extreme leftist parties were proscribed soon after independence and the possibilities of the workers and rural poor for collective social action were restricted (cf. Hako et al. 1975). This dualism also existed to some extent in the field of culture, e.g. in Finnish literature (Paasivirta 1984, 283–4).

The reshaping of the geography of world politics proceeded simultaneously, and the victors of World War I developed plans to subdue the Soviet government. Finland also had a place in these plans, inasmuch as it was regarded as a barrier against the spread of Bolshevism. Furthermore the Head of State, General G. G. E. Mannerheim, was well aware of the state of affairs in Russia as a result of his long service in the Russian army (Puntila 1971, 126–7).

5.4.1   A new geopolitical code

The conflict in foreign policy also continued after the Civil War. Relations with the newly established Soviet Russia in particular became cool, for many reasons, and it can be argued that the 'geopolitical code', i.e. the set of political-geographical assumptions underlying Finland's foreign policy, changed (cf. Taylor 1988b, 24). A specific political and moral ideology, 'hate of Russians', emerged. Puntila (1971, 126) writes that the new Russia also inherited the sins of the old Tsarist Russia in anti-communist circles. The Lutheran Church associated itself with the White victors and treated the Red rebels as an anti-Christ, and the White forces were represented as an outpost of civilization against the 'spiritual and moral plague of the east' (Paasivirta 1984, 306).

After the Civil War the question of national integration emerged. It was found to be important to shape Finland's place on the world map in order to integrate the divided nation. The increasing success of Finns in international sports competitions in the 1920s provided one 'natural' basis for the rhetoric of integration, even though it was not — and never has been — considered that Finland can be equated with Finnish sporting life (Paasivirta 1984, 320–5). Success in sport nevertheless created one basis for the shaping of national feeling, surmounting the divisions of the classes, generations and geographical regions.

The Finnish case supports the arguments set out by the Swedish ethnologist Ehn (1989), who claims that sport is a profitable field in cultural studies. There is hardly any other area of activity in Sweden, Ehn points out, where expressions of love for one's country are made so strongly and so unanimously as in the field of sport. Furthermore, owing to its emotional expressions and nationalistic symbolism sport should have a key place in general research into nationalism and national culture. The Finnish nation-building process provides a fitting example of this. There is an illustrative spatial metaphor that some sportsmen 'ran Finland onto the map of the world'. This idea is basic among the myths constituting the collective mentality and Finnish identity. It has been argued in recent research that 'sport is an integral part of Finnish society, like the state church' (Reinikainen 1991, 91). In the nationalistic language and rhetoric of sports employed in journalism — which would probably become incomprehensible without these collective sentiments — it is common to exploit both negative and positive national stereotypes to depict the supposed collective character of other nations. Hence sport typically turns people living in various localities and nations at certain moments into one people, into one distinct, geographical and emotional 'we' with a collective memory (Ehn 1989).

In a broader perspective, two attitudes prevailed among the intelligentsia or ideological élite of Finnish society after the Civil War as far as the constituents of the social integration of the divided nation were concerned. Although all bourgeois circles were unanimous about the importance of achieving national integration, they were nevertheless not uninanimous about how to obtain it. The right-wing attitude aimed at creating a strong, unanimous national solidarity, i.e. to return to the time before the conflict arose (Alapuro 1973). The political centre, for its part, was of the opinion that to win the solidarity of the rebels, it was necessary to win the confidence of the people. This was the central background to the exceptionally strong, homogeneous nationalistic movement among Finnish students in Europe between the World Wars (1973, 172–9). The Finnish population did not provide any significant 'sounding board' for extreme right-wing movements but, as far as the construction of territorial ideologies in the area of social consciousness is concerned, the ideas provided by the students and academic intelligentsia became important. Among the circles of students it was commonly considered that Finnish patriotism had two sides: love of one's own country and hatred of Russia (cf. Upton 1965, 45–6).

As a whole, the Russian revolution contributed to the rise of Bolshevism as a powerful ideology in the international community. This created at the world level a set of boundaries which has acquired a particularly significant role in recent times, resulting in the line separating the communist and capitalist countries (Gottmann 1973, 141). From the early 1920s to 1945 this line coincided with the boundary of the Soviet Union, and it was strictly controlled on both sides and difficult to cross. The creation of a socialist state was a shock for the European political system. According to Immonen (1987, 72), the role of Bolshevism from the perspective of Western publicity rested on several premises. The first was its high level of abstraction and international character, the second was that the solutions based on this ideology were important in an international perspective, e.g. the question of private property, the third was its relations to religion, and a fourth was the concept of class, which provided both the bourgeoisie and the working class with an enemy.

Since the Soviet Union was the only representative of this ideology in the international community of states, this caused practically all comments regarding the Soviet Union to became comments on its socialist ideology, and as a consequence of these developments, Immonen (1987) writes, all discussions concerned with Soviet Russia/ the Soviet Union became ideological, even if this was not the original aim. Upton (1965, 45) is ready to conclude that in the case of Finnish/ Soviet relations, all publications dealing with Russia/the Soviet Union between 1920 and 1939 were propaganda put forward by one side or the other.

France and Britain encouraged the Finns to join to the fight against Bolshevism. There were influental persons in Finland who believed that Finland's independence was in danger as long as the Bolsheviks held power in Russia, and in spring 1919 relations with the East were becoming more active and aggressive. The Head of State Mannerheim was of this persuasion, and even though he lost the Presidential election, he was in contact with the interventionists among the Western powers and put forward a programme which committed Finland to take military action to remove all the dangers coming from the east (Nygård 1978, Vahtola 1988).

In 1919, to relieve the pressures building up from various countries, the Russian government agreed peace with those states that had become separated from it, and in September this also came to affect Finland. Nevertheless, the conflicts in border areas continued. In Eastern Karelia, beyond the Finnish–Russian boundary, there emerged counter-revolutionary uprisings in 1921 in which volunteer Finnish troops participated in order to help the Finnic people living there, although 'official Finland' did not support this activism. When the conflict ended in 1922, Finland and Soviet Russia signed a peace treaty regarding the border areas, which defined frontier zones stretching from Lake Ladoga to the North Sea, the number of troops allowed to be located there and the general mode of life in these areas. A demarcation of some points of territorial disagreement was carried out in 1925, and in this way Soviet Russia assured peaceful conditions along its Finnish boundary and succeeded in isolating the Eastern Karelians from the Finns. On the Finnish side, however, Soviet Russia emerged during the 1920s as the only possible threat to the independence of the new state (Selén 1987, 14–17).

The publication of the *Kalevala* had itself soon aroused a deep interest in Eastern Karelia, and both cultural and scientific Karelianism became important ideologies and motivators for scholars (Sihvo 1973). This interest continued after Finland became independent and the boundary was formally closed, and it can be said that although the scientific expeditions to Russian Karelia ended, interest in the area did not end. Old ideas from the nineteenth century regarding a Greater, natural Finland which would include the areas of Eastern Karelia beyond the boundary were revived (see von Hertzen 1921, Jääskeläinen 1961). Finally, in 1920, peace negotiations began in Tartu, Estonia. The question of boundaries was the central theme in the negotiations and there was a deep disagreement between the parties (Paasivirta 1984, 196). The Finnish expectations were for certain changes in the boundary and for an extension of their territory. The Russians were naturally opposed, and finally Finland received the Petsamo area but had to move its troops back from the communes of Porajärvi and Repola, while other stretches of the boundary followed the line established in 1812 (Poroila 1975, 81).

The nationalists and anti-Bolsheviks did not accept the territorial results achieved in peacetime, and it can be argued that this treaty became a primary element in the political unrest of the 1930s, being rooted in the ideology of Greater Finland. In particular the 'Academic Karelia Society', established in 1922, took as its point of departure the expansion of the area of Finland, especially towards the Finnic Karelian regions (Alapuro 1973). This Society was rapidly dominated by Finnish-speaking students, and many who later became leading politicians and scientists were involved in its activities.

In the course of the 1930s in particular the ideas of Greater Finland became so obscure that it is a highly complex matter to determine geographically what in fact was the shape of this imagined territory (Nygård 1978). The concept usually included Eastern (Soviet) Karelia (Dvina, Olonets), Ingria and the Kola peninsula, but the most utopia-minded activists also included Estonia, Western Bothnia and Finnmark in their idea of a great empire of the Finnic peoples. Thus the idea of building up 'Greater Finland' had a long tradition.

The ideology of Greater Finland reflected in itself the broader ideological connections prevailing in Europe paralleling the trends in several European countries, e.g. Germany, Rumania and Hungary (Joll 1990, 346–7). This ferment, together with the strains in the world political situation, strengthened the suspicions of Soviet leaders regarding Finnish aims (Puntila 1971, 128). Owing to active propaganda, the eastern boundary became a wall between Finland and an alien, hostile world. A typical example of the *Zeitgeist* is probably Räikkönen's (1924, 87) *Heimokirja* (Book of the Tribe), in which he wrote that a

Russian attack on Finland was probable and was only a matter of time. This forecast was to a great extent based on the 'long, unnatural border'.

As far as foreign policy was concerned, Finland was regarded abroad as an independent, viable and peaceful country during the 1920s. Its political and social development were not well known, but its anti-communist movements and actions taken against communism were noted. Communist activities were forbidden in summer 1930 and the Social Democrats became the only representatives of the left in parliament. Finland signed a non-aggression pact with the Soviet Union in 1933, but prevailing political attitudes were such that it was interpreted in various ways in political circles. The Social Democrats hoped that it would improve the development of relations with the East, while bourgeois circles regarded it as desirable but suspected whether Finnish political leaders could put trust in it. The extreme right resisted the pact. Economic circles of the country, for their part, hoped that the pact would open new possibilities for trade after the passage of a decade (Paasivirta 1984, 364).

According to Paasivirta (1984, 419, 425), it was characteristic of the end of the 1930s that the idea of national interest should become more prevalent and that the ideological premises were watered down as far as Finnish security policy was concerned. The strengthening of the military establishment was widely accepted, for instance, and one could argue that the content of the social consciousness changed in favour of national defence. There were several reasons for this: proximity to a great power, divided attitudes towards the German and Soviet systems, and the stabilization of class relations as a consequence of the rising standard of living.

5.5   THE WINTER WAR (1939–40) AS A
TERRITORIAL CONFLICT

It has been pointed out that the most striking influences on the border landscape and its inhabitants will frequently result from changes in boundary position, which transfer areas from one state to another (Prescott 1987, 63). This can be seen in the case of Finland. As Figure 5.3 displays all too clearly, Finland has for hundreds of years been a borderland between Sweden and Russia. Its eastern boundary has changed several times as a consequence of the territorial conflicts between the two states.

A boundary dispute was also among the background factors behind the Winter War. In October 1939 the Soviet government asked the government of Finland to send a delegation to Moscow to discuss 'concrete political matters', i.e. the Soviet leaders wished to improve the security of Leningrad in the prevailing international crisis, and according to their claims this could be done only by moving the Finnish–Soviet boundary farther away from the city. Leningrad was located 32 km. from the boundary and this was thought not to be enough by the Soviet leaders who were suspicious of the aims of the Finnish politicians. Furthermore, the Soviet navy demanded a base on the south coast of Finland.

In fact, this act concretized the credibility gap and the old suspicions of the Soviet leaders regarding the geopolitical threat coming from or through Finland and their old anxiety about the security of Leningrad, an anxiety that had persisted among Soviet leaders ever since Finland gained its independence. In the Peace of Tartu in 1920 the Finns dealt with the territorial problems from an economic point of view in the case of both the Petsamo region and Eastern Karelia and from a politico-ideological perspective in that of Eastern Karelia. The Soviet Russian leaders aimed in vain to make the Finns understand the aims of their security policy in the case of Leningrad, which was one of the most important industrial and commercial cities in the Soviet Union (cf. Korhonen 1966, 28–41, Apunen 1977). The broader international context for these arguments was of course the agreement signed in August 1939 by Molotov and Ribbentrop, which divided Europe into spheres of interest (Nevakivi 1987).

In return, the Soviet Union was ready to cede to Finland areas from the communes of Repola and Porajärvi, which had been objects of dispute earlier, after Finland gained independence (Nevakivi 1987, 53–4). The three Baltic States had already given in to demands regarding military bases, and the Finnish government was ready to move the boundary about 13 km. away from Leningrad and to make some other minor concessions, but these were not enough for the Soviet authorities and finally, in spite of the non-aggression pact that had been signed in 1933, the Red Army invaded Finland on 30 November and the Winter War began (Puntila 1971, 162–6). People living in the border areas were immediately drawn into the tumult of the battle.

During the next few days the attacks of the Russians continued. The inhabitants of those villages that had not been evacuated were escaping as soon as possible. Old men, women, children and animals. Frightened faces, crying children and lowing cows could be seen everywhere. The country people of the border area had left their homes and were going away. Nobody knew where . . . Fires were blazing behind them and beside them. Karelian villages were burning, throwing up a terrible blood-red glow against the dark sky. Their homes were destroyed there and then, and with them much more that is inseparable from home and native locality. (Ruotsi 1974, 132).

The military aims of the Soviet Union were based on a rapid solution to the war. Simultaneously with the Soviet attack, a rival, shadow government, 'the People's Government of Finland', was established on 1 December 1939 by a Finnish emigrant communist and Secretary of the Communist International, O. W. Kuusinen, to be located in the occupied area of Terijoki. The basic idea was that this goverment could be supplemented with Finnish representatives after the occupation of Helsinki. On 2 December the Soviet government signed an agreement with Kuusinen's goverment which ordered the 'People's Republic of Finland' to cede to the Soviet Union the areas it had claimed in the negotiations in the autumn and to receive as compensation an area including 70 000 sq. km. of Eastern Karelia (see Figure 5.5). According to Soviet propaganda, the time had arrived to 'fulfil the centuries-old hopes of Finnish people of connecting the Karelian people with relatives in Finland and thus creating a united nation' (Aminoff 1943, 459–60; for a broader interpretation see Paasikivi 1986, 123, Paasivirta 1992).

Alexander (1957), in his book on political geography, deliberates over the survival of Finland and puts forward a stereotypical argument referring to the 'perseverance' of the Finns, an essential part of their self-image. A more convincing explanation is probably their excellent knowledge of the physical surroundings in which the military activities took place, and a second important fact was the rise of a kind of class alliance in Finnish society, pushing the old class divisions of civil society into the background. There has been some debate over whether the story of the unanimity of the Finns during the Winter War is a myth (Upton 1965, 15), but it does seem that the division in the nation that prevailed after the Civil War had been largely neutralized, and there is even reason to speak of a national reconciliation which took place as a consequence of a conscious policy of unity that had its roots in the political activities of the 1930s. This was partly led 'from above', but arose partly from the international political situation (Soikkanen 1984). One active 63-year-old Social Democrat who worked in the Värtsilä factories before the war joined his personal history to this story in summer 1987 as follows:

During the war politics fused completely. There were no more Reds and Whites but Finns, or to be more accurate — people from Värtsilä. It was positive, the Finnish people were then unanimous. And when the negotiations were going on in Moscow, every worker who had been labelled as Red or an extreme Leftist before the war said that not an inch of the ground should be given to the Russians. (Interview).

This statement is interesting, inasmuch as it correlates so strongly with the idea of reconciliation and in fact represents an even more unyielding

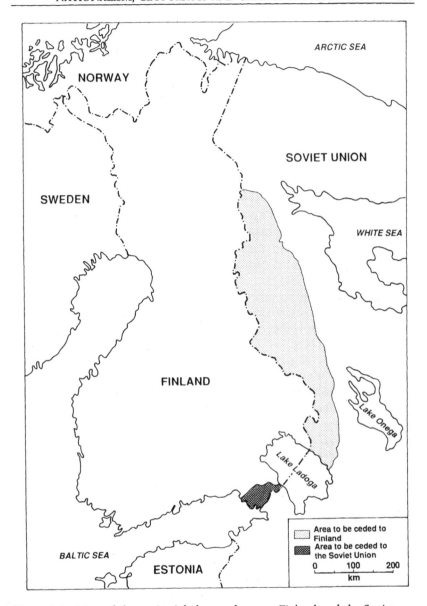

Figure 5.5 Map of the territorial changes between Finland and the Soviet Union as planned by the shadow government of the Finnish emigrant communist O. W. Kuusinen in 1939. Source: Redrawn from various sources, e.g. Aminoff 1943 and Kaukoranta 1944

attitude towards the negotiations than that displayed by the Finnish government or the military leadership.

The Soviet aims were thus much broader than merely to secure Leningrad: they included the occupation of Vyborg and Helsinki and an advance to the Swedish boundary. The order of the day given by Marshal Voroshilov ended with an encouragement to the soldiers to 'make a honorary salute to the Swedish border guard detachment' (Paasivirta 1992, 57). These efforts failed and after three months, in March 1940, the Peace of Moscow was signed. This process was accelerated by the fact that the allied forces let the Soviet leaders know of their intervention plans and that they would not accept the destruction of Finland (Nevakivi 1987, 58–64). According to the agreement, Finland had to cede to the Soviet Union large areas in Karelia and some smaller territories in Northern Finland. These areas were joined to the Eastern Karelia areas of the Soviet Union to constitute a Karelian–Finnish Soviet Socialist Republic, the 12th republic of the Soviet Union (Kaukoranta 1944, 30–1).

In addition, the peninsula of Hanko, located on the southern coast, was leased to the Soviet Union for 30 years. These boundary changes took place between the two states and thus constitute the formal history of what occurred. The reaction of ordinary Finns was another story. The terms of peace were a shock, especially because ordinary people had no idea of the military situation and what was going to happen. Censorship by the government had been very effective, and hence it is difficult to say how strongly the generalization of the psychologist von Fieandt (1946, 21) should be interpreted: 'When the terms of the Peace of Moscow became known in our country in 1940, the whole Finnish people formed a psychological group which adopted the same attitude.'

The loss of Karelia was a particularly difficult problem, since this area played a distinct symbolic role in the ideological socialization of Finnishness and the Finnish territorial identity. It was the borderland against the 'hereditary enemy', the border of Western civilization. The new border also radically changed the military position of Finland, as it was a land border, located in the immediate vicinity of important Finnish industrial centres and towns. The psychological effect was also significant. The official ideology of the period between the two wars had created anti-Soviet conceptions and many leading figures in political life, the Lutheran church, education in schools, newspapers and public propaganda maintained these persistently. It may be concluded that the war did not lead to a new evaluation of Finnish–Soviet relations. On the contrary, the feeling of an unjustified peace strengthened old attitudes (Upton 1965, 41–6).

5.6   THE CONTINUATION WAR AND THE IDEOLOGY OF TERRITORIAL EXPANSION

When Hitler initiated plans in 1940 for Germany to attack the Soviet Union the following year, Finland became a strategically important territory (see Manninen 1987, 84). The briefs that Hitler began to present to his military leaders from June 1940 to March 1941 contained a general programme according to which the Soviet Union was to be dispersed into separate areas: the state of Ukraine, the Baltic States, and a separate area of Belorussia should be established, and Finland should be extended to form Greater Finland (Manninen 1980, 15). It is not easy to delimit the boundaries of this Greater Finland from Hitler's thoughts, but he was evidently discussing a territory which would extend to the White Sea and also contain Northern Russia. Thus, in the German plans, Greater Finland was to be shaped so as to form a kind of shelter, as also were the other three large territories proposed which would then shift the actual boundary very much further to the east (1980, 24).

Finland's political and ideological connections with Germany became closer during spring 1941 and finally led to cooperation. One expression of these connections in Finland was the emerging idea of *Lebensraum* (living space), exploited effectively in the ideology of Nazi Germany, particularly from 1937 onwards. In Finland this ideology was partly based on the practical need to indicate to the Germans that Eastern Karelia belonged to the *Lebensraum* of Finland. In Hitler's original plans of 1941 Finland was to become a state within Greater Germany (Polvinen 1964, 9).

A more ideologically based argument revived the old idea of the Finnic tribes and the historical task of the Finns to join the separated areas occupied by the Finnic peoples into one territorial unit: to connect 'us there' with 'us here' (see Laine 1982, Paasi 1990). This task is put forward solemnly in Kaukoranta's (1944) book, for instance, where he expresses it in religious rhetoric: 'May the Almighty allow that the destiny of the Eastern Karelians should be shaped at the end of the present great time of tumult in such a way that it will be a full answer to the hope of freedom and national integration that has slumbered in the soul of these people and the whole of our nation.' When rumours began to spread concerning the expected war between Germany and the Soviet Union, this led to a marked revival of the spirit of Finnic kinship in Finland. People were sure that Germany would defeat the Soviet Union in the same manner in which it had succeeded in its previous offensives (Manninen 1980, 48).

The practical implications of this geopolitical ideology could be seen during the Continuation War, which began on 25 June 1941, Soviet bombings in Finland having begun on 22 June. The aim of the Finnish government was to return to the old boundary, but both President Ryti and the Commander-in-Chief, Mannerheim, proposed the construction of Greater Finland when deliberating over the aims of the war. Mannerheim was hence on the same track as he had been in 1918. The expertise of professional geographers as well as other scientists was effectively employed in drawing the boundaries of the new, 'forthcoming Finland', i.e. a marriage of practical geopolitical reasoning carried out by state élites (both civilian and military) and formal geopolitical reasoning by academics (Paasi 1990). The content of this expertise and the arguments that the geographers employed in this connection will be discussed more fully in the forthcoming pages, where the changing representations of the boundary that prevailed among Finnish political geographers will be analysed.

As a concrete manifestion of these ideas, the Finnish troops first re-occupied the ceded areas and subsequently swiftly occupied large areas of previously Soviet territory to create a 'Greater Finland' and to establish a 'safe boundary', as it was labelled (Paasi 1990, Paasivirta 1992, see Figure 5.6). The Petsamo area, the eastern 'arm' of Finland, was under German control, as were some Finnish troops in Northern Finland. Practical expression had thus been given to old ideologies and discussions of the fate of the Finnic tribes etc. (cf. Laine 1982). The crossing of the old state border was not accepted unanimously and masses of the soldiers refused to do so, pointing to the fact that the war was experienced as a 'war of the masters', i.e. that the idea of Greater Finland was strange to the majority of the Finnish population. This confusing situation has been illustrated in the most influental post-war novel *The Unknown Soldier*, written by Väinö Linna.

On the afternoon of the fifth day they began to notice that the road had begun to deteriorate. Before long it had become a mere forest track, and soon afterwards they arrived at a strip of land which had been cleared of all trees.
'Hey, it's the old frontier!'
The news swept down the line, cheering the men visibly. Hietanen, from his place in the column, called out: 'We're in Russia, lads!'
But Lahtinen, limping along sullenly, glowered at the others: 'So we are. And we've got no right to be here. We're no better than a bunch of bandits.'
'Bandits!' snapped Sihvonen. 'When we cross a frontier we're bandits. When the Russkies do, they're taking measures for their security. Bandits . . . Hah!'
Hietanen was busily examining their surroundings. 'There doesn't seem to be much worth stealing. Even the road's not as good as on our side. But the forest looks the same . . . You know, boys, this part of Karelia is supposed to be a land of song.'
'There'll be cause for plenty of sad songs over here.' Lahtinen took a swig of water from his canteen. 'From now on we're on Russian soil, so they'll really try to stop us. I think up to now they just figured, well, since you're always about it, take back that little bit of Karelia and welcome. But you know what happens when you go poking a bear in its lair.' (Linna 1957, 81–2).

Figure 5.6 'Liberating Dvina-Olonets.' A picture depicting the old border, with the caption 'Our soldiers were aware to which cultural realm Dvina-Olonets belongs. An indication of this is the sign "Uuteen Suomeen" ("to the new Finland") showing the way to the Uhtua area.' Source: SA-kuva (Finnish Defence Forces Education Development Centre, Photographic Section); the figure was printed with this caption in Salminen 1941, 49. Reproduced by permission

Nevertheless, the idea of Greater Finland also prevailed outside academic circles, where the ideology had been propagated since the 1920s. The Coalition and Agrarian Parties and a fascist-minded Patriotic National Front were in agreement, with some reservations, while the aim of the Social Democrats was independence and a democratic system (Manninen 1980, 129).

Crossing the border established in the Peace of Tartu had to be justified not merely to the troops but also to foreigners. The arguments were mostly based on the security of the state: the Eastern Karelians were again presented as belonging to the Finnish people and the area was regarded as forming a single entity (see Figure 5.7). 'A short boundary, a long peace', an idea that was put forward in 1941, soon became a slogan which was exploited to explain the territorial claims of the war (Polvinen 1964, 51). During the war years a number of Finnish scientists were involved, in the spirit of geopolitics, in the production of 'scientific evidence' for a Greater Finland which would include the Kola Peninsula and Russian Karelia — or even larger areas in some more idealistic plans. Here they effectively followed the metaphors of the state as a spatial organism that had been developing within geopolitical thinking. There were several important figures involved in Finnish geopolitics during the years of World War II, but the geographer and geologist Väinö Auer was the outstanding one, and he was the specialist consulted by President Risto Ryti as regards the definitions and arguments concerning Finland's *Lebensraum* (Laine 1982, Paasi 1990).

The Continuation War lasted more than three years, up to September 1944, and during most of this time it was mainly trench warfare. Even though the official foreign policy of Finland aimed at creating a low profile in the case of the occupied Russian Karelian areas, these areas were now commonly seen as belonging permanently to Finland. Originally the Social Democrats and the representatives of the Swedish Party in particular objected to the incorporation of these areas, but eventually even they changed their attitude (Laine 1993).

During the years 1941–44 the Finns established an occupation administration in Eastern Karelia and began to believe in their permanent possession of the region, so that they started to educate the local 'national civilians' (some 36 000 inhabitants with some kind of linguistic or other Finnish kinship) to become citizens of the forthcoming Greater Finland. Half of the 'non-national civilians' (some 24 000 of the total of 47 000 Russian people at one time) were mostly interned in concentration camps (Laine 1982, 116–56).

Thus an effective administrative system was created in Eastern Karelia, and one particularly important aspect was the 'national education' of the population, which was now started, with the aim of creating a feeling that the population of the occupied areas belonged to

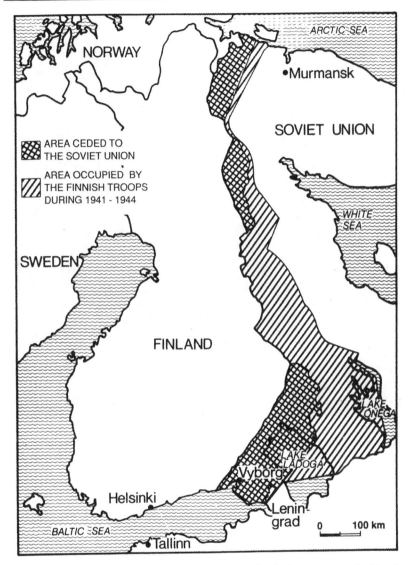

Figure 5.7 Map of the present boundaries of Finland, the area occupied by the Finnish troops in 1941–44 (and by German troops in Northern Finland) and the area ceded to the Soviet Union as a consequence of World War II

Finland and that they could understand the detrimentral effects of the Slavic culture on the whole Finnic tribe and on Karelia (Figure 5.8). It was time to change the content of the prevailing regime of 'truth' as far as the location of the state boundary was concerned.

National stereotypes were exploited systematically in the educational

Figure 5.8 'Shaping the disposition favourable for Greater Finland': a geography lesson in the occupied area of Eastern Karelia in 1943. Source: SA-kuva. Reproduced by permission

programme, the aim of which was to create a wall against the east that would hold. Some educational subjects were regarded as more important than others, and the most important of all — labelled as 'subjects shaping the disposition' — were history, geography, the Finnish language and religion. As for geographical education, it was claimed that the concept of 'Greater Finland' should be introduced as soon as possible, and in some instances it was emphasized that Eastern Karelia could provide something unique for Finland: a natural boundary, the only one which could permanently guarantee safety and freedom for the country. Finnish literature, including the works of Runeberg and Topelius and the *Kalevala*, was recommended reading, and a 'Book of our Land for Greater Finland' was compiled (see Laine 1982, 159-218, *passim*).

5.7   THE TERRITORIAL, ECONOMIC AND SOCIO-POLITICAL CONSEQUENCES OF WORLD WAR II IN FINLAND

Finally, the Soviet Union began a heavy military attack on the Karelian Isthmus on 10 June 1944 (Polvinen 1964, 263), and the Finnish troops were obliged to retreat rapidly from Eastern Karelia and the Karelian Isthmus. Again, these events were a surprise at the local level, as the following extract from an interview with a 62-year-old man displays:

The nobs of Karelia laughed, God damn. In the evening we heard that the time of departure had come and they were laughing. And at six o'clock in the morning, no, it was at four, the airplanes were above us and we had to leave. They said that the Karelian Isthmus will not break, but it seemed to break — and we still have memories of it. Well, now it has all been and gone.

The suspension of hostilities began in September 1944 and finally, as a repercussion of World War II, Finland had to yield to the Soviet Union — in line with the Peace of Moscow of 1940 — the possession of large areas of Karelia, and now the area of Petsamo in Northern Finland as well. The total area was 45 792 sq. km., 12.5% of Finland's land area (de Gadolin 1952). Finland also had to 'lease' the peninsula of Porkkala and surrounding sea area as a military base for the Soviet Union for 50 years, although in practice the Soviet troops, more than 20 000 men, left Porkkala in 1955 and the area was restored to Finland. Hence the changes that took place in the shaping of Finland as a consequence of this territorial transformation were considerable.

After World War II the state had to organize a resettlement programme for more than 420 000 inhabitants of the ceded areas — 11% of the country's population — to move them to other parts of the country, mainly to the southern and central areas. The evacuation and finally the resettlement of the Karelians was, according to Upton (1965, 77–9), an indicator of two important things. First it shows that the Soviet government wanted to clear the territory of its Finnish population and hence promote its military security, and secondly the resettlement of the population served as an indicator of the vitality of nationalism in Finland.

Owing to the fact that there existed a real possibility that the great mass of workers and refugees would create a 'revolutionary situation' (de Gadolin 1952, ix), resettlement was implemented by passing a land acquisition act in 1945, partly to promote the social integration of the resettled people in their new home areas and thereby in itself contributing to a peaceful movement to new social conditions (Paasivirta 1992, 306). Other refugees were located in towns to search for jobs according to their professions. It was important for the stability of internal politics and for the rise of a new foreign policy that no members of the Finnish population remained beyond the new border (see Tarkka 1987, 195).

The establishment of new social ties raised serious problems for the displaced people and for their cultural adaptation to new, at times even hostile, socio-spatial frameworks in various parts of Finland (Waris 1952). Von Fieandt (1946, 62) writes that in some circles the Karelian evacuees were regarded as symbols of the lost war.

The loss of the ceded areas also created a severe economic problem for an agrarian, slowly industrializing country, with 432 factories, over 25 000 industrial jobs and 25% of the country's hydro-electric power capacity having been left in the ceded areas, together with prominent towns such as Vyborg and Sortavala (Paavolainen 1958). Vyborg particularly was simultaneously an important traditional symbol of the Karelian heritage and a modern international city. In northern Finland, the area of Petsamo with its nickel mines had also been ceded. Furthermore, the transport routes to the east were broken off after the war, and Finland had to produce as war indemnities various goods, mainly metal and timber products, for the Soviet Union to a final value totalling about 450–500 million dollars, amounting to between 10 and 20% of the national income (cf. Virrankoski 1975, 224).

The war indemnities and the return of all the evacuated machines, locomotives, motor vehicles, historical monuments and so on nevertheless also marked the beginning of a new economic interaction between the two states after the war and the trade agreement for the period 1951–5 became a basis for a continuing bilateral trade. Many new metallurgical plants had to be built in Finland after the war so that the country could handle these debts, and as far as timber, pulp and paper exports were concerned, Finland was linked to the Western European economy. Soon after the war the Soviet Union and Britain became Finland's most important partners in trade, while until 1955 Britain was the most important partner for Soviet trade and Finland was the second (Heikkilä 1986).

Regarding further boundary changes, it should be noted that Finland sold the Jäniskoski power station area (176 sq. km.) in the Lapland border area to the Soviet Union in 1947 and that Finland leased the Saimaa canal in the southeastern border area from the Soviet Union in 1963 (Poroila 1975, 137–8).

Besides the cession of certain areas and the resettlement of their inhabitants, the war years and the final establishment of the new boundary led — as far as the everyday life of the inhabitants of the border areas was concerned — to situations which deeply affected the structures of the life-worlds of the displaced people. The transformation in local conditions was especially dramatic in the new border landscapes — in the regions where a new border was delimited, demarcated and finally symbolized after the years of war. The land boundary in southeastern Finland was now four times longer than the previous one located on the Karelian isthmus (Puntila 1971, 175). The boundary itself also became an essential part of the establishment of new relations between Finland and the Soviet Union, as several agreements were signed between the two states — in 1948, 1960 and 1969 — under which the maintenance of order in the border regions and the prohibition of various activities are defined in detail (Numminen et al. 1983).

A special case — and a problem for the social interaction of the people living in the border areas — was the establishment of a frontier zone along the Finnish–Soviet border. The frontier zone was established by the Finns, the Soviet Union never demanded this (Kosonen and Pohjonen 1994, 360). A protection zone had already been established on the eastern border after the Winter War which limited the scope of people's daily lives (Kirjavainen 1969, 250), and a new frontier zone was established under a specific law in 1947 to secure peace along the frontier and maintain and emphasize general order and safety. The zone is an area that runs along the boundary line with a maximum width of 3 km. on the ground and 4 km. at sea. Permits to move about and stay in the zone were originally granted by the police superintendent, and subsequently by the border guards of the district covering the area for which the permit was requested. Permits for foreign citizens to move about and stay in the frontier zone were issued by the Security Police.

The new — and present — boundary, established in 1940 and 1944 by the Peace of Moscow and confirmed by the Treaty of Paris in 1947, became in many respects a unique boundary in the field of world geopolitics. It was for several decades a border between the Soviet Union and a small nation-state which was, together with the Baltic countries and eastern Poland, apportioned to the sphere of interest of the Soviet Union in the secret additional memorandum to the *Ribbentrop Agreement* between Germany and the Soviet Union signed in 1939 (cf. Puntila 1971, 155, Paasivirta 1992, 25). Finland was the only state among those defined in the agreement which was able to remain independent after World War II. In fact, besides Great Britain and the USSR, Finland was the only European country involved in World War II which was not occupied, the only one to maintain the continuity of her constitution and political institutions and, finally, the only one to make the transition from war to peace without internal rupture (Jakobson 1984, xvii). It had to develop a new neutral foreign policy after the war, and it served for a long time as a 'model' for the Baltic states, particularly Estonia, in their efforts towards cultural cooperation, and later independence (cf. Parker 1988, 153–4).

A Control Commission of the allied forces supervised the execution of the truce during the period 1944–7. The allied forces had already developed the concept of political war crimes during the last years of World War II, and in this spirit several leading Finnish politicians of the war years were condemned for war crimes in 1946 and committed to prison (Polvinen 1981, 148). Vihavainen (1991, 32) remarks that the Soviet representatives on the Control Committee threatened repeatedly occupation and other acts of revenge if the claims they made were not granted, and this led to a cautious realism in politics.

A political turnaround also took place after the war, and the left-wing parties became strong political and ideological powers with the legalization of the Communist Party. In addition, anti-Soviet and fascist-minded organizations were disbanded (Paasivirta 1992). A coalition government between the three largest parties was established following the 1945 parliamentary elections, consisting of the Agrarian Party, the Social Democrats and representatives of the Finnish People's Demo-cratic League (SKDL), which was established after the war by the Communists and the Social Democratic opposition. This led to tensions in internal policy, particularly between the extreme left and the other parties, and the situation was changed by the 1948 elections where the SKDL lost a quarter of its votes to the Social Democrats and the Agrarian Party, thereby losing its place in the government.

The lost war also forced a change in official attitudes towards the Soviet Union, partly owing to the treaty of 1944, which forbade organizations with fascist sympathies. As Vihavainen (1991, 33) observes, this meant in practice that the Control Commission sought to eliminate anti-Soviet literature and publishing in Finland. The Ministry of Justice sent out a circular letter in 1944 which urged public libraries to dispose of hundreds of books which contained representations of the Soviet Union that might be regarded as hostile. Subsequently school books were 'cleaned up', with pupils having to cross out certain passages or glue pieces of paper over them. As far as the press was concerned, the censorship of the war years prevailed until 1947, with Finland's political leaders informing the representatives of the press how to act in the new situation (Immonen 1987, 19–20, Vihavainen 1991, 33–9). Once again, it was time to change the concepts of truth and inherent social practices in education and the media.

Urho Kekkonen, Member of Parliament and later to be President of Finland from 1956–80, who is today regarded by many as the key person behind this 'Finlandization', was an active propagandist during the years of the Continuation War, and he was also actively anti-Russian and pro-Greater Finland (Paasivirta 1992, 240). He changed his attitude in the course of the war, however, as is evident from a speech given in Sweden in 1943, in which he said that Finland had two alternatives open to her when deciding on the direction of her foreign policy at the beginning of the 1920s: one was to join with those who pursued an anti-Russian line, and the other was neutrality in relation to Russia. At first Finland chose the former, but he was now of the opinion that it would not be in Finland's national interest to join an anti-Russian front as long as Russia remained a great power (Kekkonen 1943). In this connection he recalled that a peaceful attitude towards Russia had prevailed in Finland in the nineteenth century. This distinction made by Kekkonen recalls the two powerful myths describing differing and conflicting qualities in the supposed national identity that have found favour in Finnish literature. According to Niemi (1980, 159–63), the first emphasizes Finnish valour on the battlefield and the second depicts a peaceful country made up of peasant communities living in the backwoods. Now it was time to move from war to peace.

In August and September 1944 J. K. Paasikivi, K. A. Fagerholm and Urho Kekkonen proposed that the aim of Finnish foreign policy should be good, harmonious relations with the Soviet Union. This argument soon became common in the press and was gradually established as the cornerstone of foreign policy (Paasivirta 1992, 284–7). At the same time, however, the Soviet press was painting a rather disparaging picture of Finland and the Finns — a fact which leading politicians were obliged to bring to the attention of the Soviet authorities (Vihavainen 1991, 41–6).

The Treaty of Paris, signed in 1947, also provided basic outlines for Finnish foreign policy. The point of departure for the negotiators was to establish and maintain permanent relations of friendship with the Soviet Union (Polvinen 1981, 210). Even though Prime Minister Paasikivi's opinion was that 'it is not the small nations which decide things now, but the large ones which draw the borders they want on the map' (cited in Polvinen 1986, 159), some hope of boundary changes still remained, e.g. the regaining of Vyborg. Particularly the question of the possible return of the resettled Karelians was a hot topic for the Finns, who were still prepared to propose some changes to the border line during the negotiations. Eventually the Soviet Deputy Minister of Foreign Affairs, Vyshinsky, threatened the Finns openly with a new war if they would not stop discussing boundary problems (Polvinen 1981, 220, Tarkka 1987, 199). Finland's fate was deeply dependent on its Soviet relations.

The basic tenet of Finnish foreign policy since the war became 'active neutrality', which meant that the country strove actively to keep itself outside of conflicts between the superpowers. The key idea of Finnish post-war foreign policy can be expressed as follows: 'Geography is something that the Finns can't help', an attitude usually attributed to J. K. Paasikivi, President from 1946–56. This expression aptly points to one of the fundamental ideas that the word 'geography' denotes. The

Treaty of Friendship, Cooperation and Mutual Assistance between Finland and the Soviet Union was signed in 1948, having aroused wide international interest, since Western European leaders were worried about alliances and the power balance between states (Nevakivi 1984). According to some commentators the treaty soon became comparable to Finland's constitution as far as the country's security was concerned (Saraviita 1989, 7).

It has been argued that the treaty had an effect on all spheres of Finnish life. The official attitude was that it was an ordinary international treaty, even though everyone knew that it was not: the parties to the treaty were not equal, and it differed only slightly from a military alliance. One significant feature of the treaty was the ideological aspect, which according to recent comments reflected the motives of the Soviet leaders (Tarkka 1992, 207).

Neutrality is a complicated, continually changing concept in international politics. Even though fundamentalist thought regarded neutrality and good relations with the Soviet Union as the basic principles of Finland's post-war identity, it must be noted that it was only during the period of Mikhail Gorbachev in 1989 that this was officially stated for the first time by a Soviet leader. According to Väyrynen (1990, 38) this was the first unreserved statement of the tenet that the Finns themselves had been propagating for a long time, and it was only then that the political gulf between the Finnish and Soviet definitions of neutrality — an eternal dilemma for the Finns — was removed. The key problem was that the Soviet interpretation of Finland's position and neutrality was continually being modified along with the changes in its own position in the world-scale geopolitical landscape. During the 1970s in particular, Soviet leaders became very disinclined to speak about the neutrality of Finland and instead wanted to emphasize the 'alliance of friendship' (Tarkka 1992, 56–8, 131–3).

The aim of official state policy after World War II was also to change the negative image of the Soviet Union and the Russians that prevailed in Finland and to change the openly hostile image of Finland that had prevailed in the Soviet Union before the war (cf. Hakovirta 1975, 120–3). The Soviet authorities wanted to alter the attitudes prevailing in Finnish society which prevented the development of good relations between the two neighbours, and their suspicious attitude towards Finnish internal policy, particularly against right-wing movements, can be clearly seen in the reports written by the Soviet ambassador Lebedev in the mid-1950s (see Tsernous and Rautkallio 1992). At the level of ordinary society these efforts did not imply any rapid changes, since the picture of an enemy that had been rooted in people's minds over the last 25 years continued (Raittila 1988, 21).

5.7.1   A crisis in national socialization?

When mapping the ideological forces integrating the Finnish political system, many authors have emphasized nationalism, the will for independence, certain traditions and national symbols — factors that have been instilled by the Finnish national ideology and all-round education. Communal and provincial allegiances have been equally strong (see Salonen 1970, 257–8). All these values were challenged during the 1950–60s, first by the structural transformation of Finnish society which forced people to move from their home areas, and secondly by international cultural integration.

In spite of the Treaty of Friendship, Cooperation and Mutual Assistance, the new political realities and strong Soviet connections in foreign trade, a powerful movement of integration with the economic and cultural bloc of Western Europe and the USA began to emerge in Finland in the 1950s (Paasivirta 1991). As far as the content of the social consciousness is concerned, it was obvious that certain important ideologies which had been significant in the national identity before World War II and during the war years had to be abandoned. Tuominen (1991), analysing the world views and ideologies of the generation born after the war, writes that the 'generation of the soldiers' children' was born into a world of crisis and change, in which national identities, political structures, culture and technology were all changing.

After the war the older Finnish generations still pursued the idea of national unity in their minds and activities, and this was also the basic idea in the model of national socialization. The war and the values it represented in Finnish society were still important constituents of the post-war world view, as can be partly seen from the fact that every sixth novel published in Finland during the 1970s dealt with war, and in most cases with World War II (Niemi 1980, 9). Also, several novels published after World War II, e.g. Väinö Linna's classic *The Unknown Soldier*, gave rise at first to much controversy because the criticism that they raised against the war was regarded as a criticism of the Finnish soldiers (Niemi 1980, 157).

It is interesting that whereas the masculine literary culture has produced innumerable novels and descriptions of World War II, the simultaneous descriptions of the experiences of the Karelian evacuees were in most cases produced by female writers. Sihvo (1981, 144) discusses the possible reasons for the role of women in this literature and points out that, whereas the men had to rebuild society, the women had the pressure of taking care of the elderly and the children, i.e. participating actively in reproduction, and also were probably more sensitive to the loss of the ceded areas.

All in all, literature has had a significant function within Finnish culture in transforming the history of the nation into a spectacle. Tarasti (1990, 191) argues that history becomes understandable in the form of spectacles, since people identify themselves with the roles played in these. Linna's novels are a typical example of making history into spectacles and this, Tarasti argues, is the reason for their success.

International cultural and political tensions, e.g. the Cold War, the Berlin wall, the Cuba crisis and the Vietnam war, were also reflected in the crisis of socialization in Finland, which was directed at several traditionally strictly controlled areas, such as attitudes towards militarism, patriotism, pro-internationalism, sexuality, drugs etc. Thus, during the 1960s in particular several new international identities arose in Finnish society which challenged the traditional territorial — and national — identities. Simultaneously politics became increasingly a matter of Finland's relations with the Soviet Union and their regulation (Tuominen 1991, 80). One important cause for this was the 'note crisis' of 1961, when the Soviet leaders, in accordance with the basic provisions of the Treaty of Friendship, Cooperation and Mutual Assistance, suggested military consultations with Finland. This act was part of the increasing international tensions of the early 1960s and it was a harsh reminder to the Finnish political and military leaders that the Soviet Union viewed Finland's position above all in military terms (Tarkka 1992, 97).

5.8 LOCATION ON THE IDEOLOGICAL BOUNDARY OF THE COLD WAR: FOUR DISCOURSES

Perhaps the most significant long-term effect of World War II on the world's geopolitical map and ideological landscape was the rise of the sharp distinction between the USA and the USSR and then between the Eastern and Western blocs after the war years. This became particularly evident with the founding of NATO in 1949 and the Warsaw Pact in 1955. Besides being a division in the global geopolitical landscape, it was also a division between Atlantic Europe and Eastern Europe, the Europe of the USA and the Europe of the Soviet Union (Taylor 1991a). The old concept of Europe and its sub-regions thus underwent profound alterations.

By virtue of its new economic and political relations and agreements, Finland was more than ever before 'something between' the great powers, and its relations with the Soviet Union were now the key element in its foreign policy. This also meant that Finland remained outside the Marshall Plan negotiations, for instance. In the international landscape, Finland was an independent state located in the Soviet sphere of interest, and its good relations with the Soviet Union were regarded as crucial in this landscape (Polvinen 1981, 274–80).

Finland's international position was an important background to the final creation of NATO. Even though it was a peripheral state, it was from the perspective of US foreign policy an important country for restricting the expansion of the Eastern bloc. Hence it was crucial to maintain Finland as neutral as possible (see Tarkka 1992, 34–5). Nevakivi (1984) writes that the Treaty of Friendship, Cooperation and Mutual Assistance that Finland and the Soviet Union signed in 1948 created a panic in Norway, and the question of that country's future became an important element in the establishment of NATO. As regards the background to the Warsaw Pact, Finland was asked together with other European states to participate in the meeting in Moscow in which a security organization for Europe was to be organized, but along with other Western states, it did not participate in the meeting that gave rise to the Warsaw Pact (Tarkka 1992).

Hence a new border, which Winston Churchill had already labelled the 'iron curtain' in 1946, extended across the continent of Europe, from the Arctic to the Adriatic. As a consequence, as Cohen (1964, 218) argues, the traditional idea of Central Europe lost its geopolitical substance. Europe, like the world ideological landscape as a whole, was divided into two parts. It was only after the mid-1980s that the re-institutionalization of this vague territorial entity began (Morley and Robins 1990, Väyrynen 1992).

Thus a new spatial scale for the international conflict emerged, and the conflict was transferred from the boundaries of territorially organized states 'to the minds of men' (Pounds 1972, 99). Even though academic 'geopolitics' lost its popularity after the war, the tendencies taking place in formal geopolitical reasoning led to the Cold War as the new geopolitical order and to the adoption of its specific rhetoric. Hence the world was divided into places of three types: 'ours', 'theirs' and various disputed spaces (an excellent analysis of the construction of the international landscape of the Cold War is given by Taylor 1990). The question of the Same and the Other became more acute than ever.

On the world scale, Finland was undoubtedly one of these disputed spaces, i.e. it aimed at neutrality even though it had specific agreements with the Soviet Union. It also had increasing economic and cultural connections, which led to an increase in individual interaction (tourism) and common construction projects in the Soviet border areas during the 1960–70s. As leading politicians declared after the war, 'geography' was still the cornerstone of Finland's foreign policy. The creation of the Warsaw Pact in 1955 made Finland's distinction from the eastern bloc more apparent.

The return of the leased Porkkala military area to Finland substantially increased the possibilities for pursuing a policy of neutrality
(Visuri 1989, 71–5). The question of the Porkkala area was part of the
general peace offensive of the Soviet Union, and reflected the fact that
Soviet military thinking was becoming increasingly global during the
1950s and hence its military lines were moving further away.
Meanwhile, Washington's foreign policy of the 1950s regarded Finland
as an independent, economically healthy state which was democratic
and oriented towards the West. The aim was to maintain this situation,
with Finland as a buffer against any possible Soviet attack on
Scandinavia (Tarkka 1992).

The development of arms technology and the changing locations of
military bases affected the position of Finland. During the 1960s the
strategic structure of Northern Europe changed and the Kola Peninsula
became a huge military base, the security of which was a crucial
question for the Soviet Union — and a key target of US military policy.
In Finland the question of Leningrad and the Karelian Isthmus had
previously been the main territorial problem for security policy, and in
fact the country's defence policy after the war had been based on the
threat of Soviet occupation (Tarkka 1992, 95). Now Lapland became
the key direction, mostly Murmansk and the border areas of Lapland.

The strengthening of the defence of the Lapland area directed the
front to the west, and Western attitudes towards Finland simultaneously
became more suspicious (Tarkka 1992, 96). The leading Finnish
politicians, for their part, put more emphasis on representing Finland's
international position as being that of a *mediating* state between East
and West. Finland was represented as being between them, and as
actively able to mediate or bridge the gap between them in international
relations. This also became one point in the national socialization
process, as can be seen in school geography, for instance (see Kalliola
1969, 143). The culmination of these ideas was probably the CSCE
meeting in Helsinki in the mid-1970s.

Particularly during the 1980s the technical development of strategic
weapons directed the focus of the military activities of the superpowers
to the north (Tarkka 1992, 168–9), and it was a shock for the Finns to
find that some Finnish localities were also targets of NATO missiles and
that the planned routes of NATO bombers passed over Sweden and
Finland.

Finland's cold war ended, Tarkka (1992, 193–5 *passim*) maintains, in
1990, when the Finnish leaders put forward a new, unilateral
interpretation of the Treaty of Friendship, Cooperation and Mutual
Assistance and the Treaty of Paris which abandoned the articles
restricting the nature and quantity of Finland's military power. The
arguments behind the declaration were first the fact that Germany had

been united and the situation in Europe was hence completely new, and secondly, that the treaties which restricted Finland's sovereignty were in conflict with the fact that Finland was a full member of the UN and CSCE.

5.8.1   West or East?

It was an undeniable fact after World War II that Finland's eastern boundary became the longest one between the leading socialist state and a Western capitalist state, over 1200 km., and it also became a fitting illustration of an *ideological boundary*, at times quoted in textbooks of political geography (see Pounds 1972, 99, Vuoristo 1979, 55–6). Whereas the Soviet Union gradually became the 'big bear' in Western Cold War humorous propaganda (Pickles 1992, 208–9), Finland was now literally 'tucked under its armpit'. We have already seen how Finland's new geopolitical situation directed its foreign policy activities. It became common to represent the Finnish case as unique in the international geopolitical system, particularly as compared with other countries in the 'Soviet orbit' (cf. Vloyantes 1975, Allison 1985), and as a result, it is appropriate to offer a brief discussion as to how professional political geographers have represented Finland's position on the geopolitical map.

Because of its location between the Eastern and Western blocs, and because of the centuries during which cultural influences have come from both sides, the border area has continuously been loaded with strong ideological connotations. It has been an essential constituent in the iconography of cultural and political 'Finnishness' and in the creation of the internal and external representations of Finland's place in the world. At times, especially before World War II and during the war years, these ideological efforts have manifested themselves in the endeavours of some Finnish authors to classify their country exclusively in the realm of Western Europe and the Western cultural heritage — even though the eastern connections have at times been identified as a background to these developments (Paasivirta 1992, 70 and 206–7, Sihvo 1992). This also manifested itself in Finnish geographical teaching and the textbooks of the 1930s:

The national culture, which is based on the spiritual heritage of the Kalevala, forms the immovable basis of Finnish cultural life. Long connections with Sweden and contacts with Central Europe have later linked Finland firmly with Western civilization. The strength of our own national culture and union with Western civilization once saved Finland from capitulating in the face of pressure from Russia. (Auer and Poijärvi 1937, 171–2)

A number of scientists also aimed to write Finland into Western civilization. Iivari Leiviskä (1938a), Professor of Geography, argued that Finland is not a cultural intermediary area between East and West but belongs to the Western cultural sphere, constituting its eastern border. Väinö Voionmaa (1933), a historian, wrote that the isthmus of Finland has been the contact area and 'battlefield' between the Eastern and Western cultures, and J. J. Sederholm (1923), a geologist, pointed to both the cultural and geographical connections between Finland and the Western countries. This emphasis continued after the war but the country's role as a bridge between West and East was also emphasized (Kalliola 1969, 143). Similar arguments were put forward by political geographers elsewhere (Bowman 1928, 446, Weigert et al. 1957, 178).

The dichotomy between East and West that was established as a consequence of the geopolitical division after World War II was based on political and economic arguments, and its basic idea was the division between capitalist and socialist states (cf. Arnkil 1984, 73). As far as the foreign representations of Finland's place in the world are concerned, it is interesting to note that in some Anglo-Saxon textbooks on political geography, Finland has now been placed in the Eastern realm according to certain unstated arguments, obviously partly owing to its physical location and partly on the grounds of its cultural or new political connections. In Alexander's (1957) *World Political Patterns*, for instance, Finland is located in Eastern Europe, as 'something of an anomaly among eastern European countries, since it has not been occupied and has maintained as neutral a position as possible with respect to the Cold War' since World War II. Paasivirta (1962, 187–8), in his study of the development of the image of Finland in the United States, argues that the 'American way of thinking' in the 1950s was characterized by scepticism of small countries that had close relations with the Soviet Union. This was also the case with Finland.

In the geopolitical frame of S. B. Cohen (1964, see also Cohen 1982), Finland's territory is mostly located in Eastern Europe and the heartland area (Figure 5.9), and Cohen labelled Finland together with Austria and Yugoslavia as 'buffer' states (cf. Weigert et al. 1957, 178). A Finnish political geographer, Vuoristo (1979, 18), pointed out, however, that since Finland is a politically uncommitted state, it would have been more natural to classify it in the area of maritime Europe. Cohen's schema is hence very much an American one (cf. Taylor 1988b, 21), obviously based both on Finland's physical location and perhaps also partly on his interpretation of its economic and political connections with the Soviet Union, particularly the Treaty of Friendship, Cooperation and Mutual Assistance which located it in the Soviet

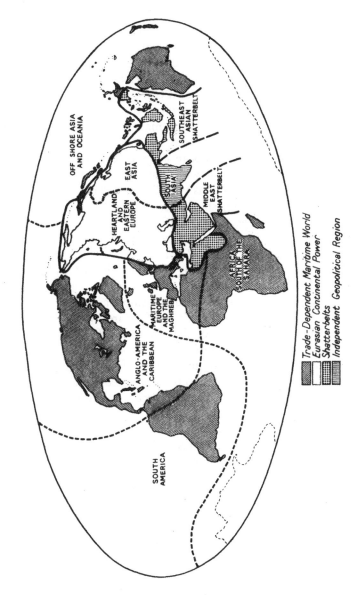

Figure 5.9 Finland represented as part of Eastern Europe: geostrategic regions and their geopolitical subdivisions. Source: Cohen 1964, 63. Reproduced by permission of Methuen, London

sphere of interest, and not on basic political or mainstream cultural ideologies. Vuoristo's comment for its part appears to be based on adoption of the official view of Finnish foreign policy which was typical of the post-war period up to 1990. Furthermore, Finnish representatives before the collapse of the Soviet Union would doubtless have liked to put more emphasis on Western institutional connections and forms of cultural and economic integration, such as membership of the Nordic Council from 1955 onwards, associate membership of EFTA from 1961 or the free trade agreement with the EEC from 1973.

Another example of Finland's location in the geopolitical mindscape is provided by Carlson (1958), who makes a distinction between Western, communist and neutral blocs, and locates Finland among the Atlantic basin states and the European neutrals. Carlson puts emphasis on Finland's cultural connections with Scandinavia, strong bonds with the West and historical connections with Finno-Ugrians. As regards the political context of the state, she writes that in the case of Finland 'neutrality is dictated by geography' (1958, 126). Furthermore, she is ready to conclude that '*Neutrality* is a necessity, however, for Finland is actually within the sphere of the Russians; they can do nothing but cooperate with the Soviets and along the lines dictated by the latter' (1958, 360).

This changing discourse on Finland's position among the Western/ Eastern realms and the representations put forward by political geographers form a fitting illustration of the argument of Foucault (1980b) that each society has its own regime of truth, types of discourse which it accepts and which it causes to function as if they were true. This accentuates the fact that texts have to be interpreted in their contexts — which appears to hold particularly well in representations of the world political order.

New representations of Finland's position on the political map are under construction at this very moment. Finland's Minister of Foreign Affairs labelled Finland 'as a member of the Euro-Atlantic Community' during his visit to the USA in October 1993 — a community which has been typically connected with NATO, the USA and its sphere of influence in Europe. Nevertheless, the recent interpretations of new features of the world geopolitical map do not inevitably correspond to these internal visions. The position of Finland on Cohen's (1991) new map of the world's geostrategic regions, for instance, is still connected with the Eastern realm in a way, since it is placed in the gateway area between East and West, together with a group of newly established states representing former Eastern European countries, e.g. the Baltic states.

Nevertheless Finland has now taken a long step into the Western realm with its membership of the European Union from the beginning of 1995. Accordingly, one of the key questions now appears to be the relations of the state with the emerging defence organization known as the Western European Union (WEU) and ultimately with NATO. At the moment Finland is an observer member of WEU and full membership would, according to many commentators, lead to membership of NATO as well.

5.8.2 Reconstruction of the Eastern social space and the rise of commercial Karelianism

In spite of the efforts to connect Finland ideologically to the Western world and to base its foreign policy on the geopolitical facts of physical location, Eastern cultural influences have become particularly significant in the cultural debate prevailing in Finnish society during the last few decades. The argument is based on Finland's traditional role as a border country between the Eastern and Western cultural realms, and the historical culmination of these relations in Karelia, its border province. As was indicated in earlier chapters, Finland's cultural connections with the East have been to a great extent mediated through Karelianism, which has a long tradition.

Pronounced interest in Karelianism arose in Finland at the end of the nineteenth century, and extended from the early, extremely lively romantic cultural enthusiasm of poets and artists to explicit political programmes in the 1920s which aimed at showing how the Finns and Russians are hereditary enemies. It also emphasized the connections between the Finns and the Finnic tribes in Russian Karelia, and finally led to geopolitical thoughts on Greater Finland (Sihvo 1973, Laine 1982).

On the whole, the discussion about Karelia, the Karelian people and the concept of Karelianism, and on the role of all these in Finnish cultural history, has been multidimensional. Karelia has been a battlefield, a bridge and a myth in the history of Finnish–Russian relations (Sihvo 1989). One reason for this ambiguity has been that the most significant symbols of Finnish national identity, the invention of a tradition and particularly Finland's national epic, the *Kalevala*, have in the course of time been firmly connected with Karelia (Wilson 1976). The loss of the Karelian areas was important in this respect, a fact which created new forms of Karelianism after World War II, forms which manifested themselves not merely in changing discourses about the Karelian social space but also physically in the reconstruction of the eastern Finnish landscapes and socio-cultural space. These tendencies have doubtless followed from the fact that as a consequence of World War II, the border of the 'East', as it was comprehended in Finland, was moved radically towards the 'West'. The 'proper' Karelian space was now beyond the border, as were many of its material and symbolic signs.

Nostalgic memory is, as Urry (1990, 109) deliberates, quite a different matter from total recall. It is a socially organized construct. When people live in a specific locality or region, they usually just live their daily routines, without reflecting on it. For the evacuated Karelians, the violent movement from their home regions became the cause for the emergence of a broader consciousness of a common Karelian heritage (cf. Karjalainen 1988, 100). During the last few decades, arising to a great extent from the ideological interests of the organizations for resettled Karelians, which in turn have been connected with the economic interests of certain local districts in eastern Finland and the emerging tourism industry, it has become popular to identify and stress the Eastern roots of Finnish culture. These activities have contributed to the creation of a mythical, exotic atmosphere for the border areas in the eastern parts of the country, tendencies that have taken place mainly beginning from the middle 1970s onwards and are manifested in a strong revival of Karelian and Orthodox religious culture. The 1970s and 1980s in particular witnessed certain new features which revitalized Karelianist thought.

One of the most prominent features has been that certain icons of Karelian culture, the Orthodox religion and the lost physical Karelian area, buildings and inherent forms of practice, have been reconstructed in the spatial context of the border areas to symbolize the Karelian past. As far as the Orthodox culture is concerned, the change since World War II has been a radical one. During the war years in particular it was not unusual to label the representatives of the Orthodox church as 'Russkies', and during the Continuation War, for instance, the ability of the Orthodox priests in the occupied Russian areas to create a Finnish national consciousness was seriously doubted (Laine 1982, 211, Paasivirta 1992, 216). The rise of the Orthodox religion to reach the core of Finnish cultural specificity is an expression of cultural activism inasmuch as the proportion of people belonging to the Orthodox church is a mere 1.1% in Finland as a whole and no more than 6% in Northern Karelia, where this 'Easternness' is most effectively exploited.

Accordingly, a kind of commercialization of the image of Karelia has occurred. This tendency has manifested itself partly in the form of numerous wooden buildings and houses in the old Karelian style, and partly in the organizing of various spectacles, e.g. plays and exhibitions depicting the features of a mythologized Karelian history and culture or the history of World War II. In general these interpretations of the past have been exploited to connect the present to the past and thereby to strengthen the present-day image. Such buildings and spectacles have become selected signs of 'Easternness', Karelianism and a lost Karelian past, and they have been actively marketed to tourists. It should be noted — as semiotists do — that the reality constructed through the spectacles does not emerge merely from the exhibitions themselves or the actors but also from the members of the audience, who reflect on their own expectations and beliefs in their minds and project them onto the stage (Tarasti 1990, 186). In this way they partly produce as well as reproduce the socio-spatial dimension of hegemonic cultural structures.

In particular the 'Road of the Bard and Boundary', which runs along the eastern border areas and villages, containing several new buildings (restaurants and hotels) designed in an old Karelian style, effectively exploits the heritage of the Karelian area (Figures 5.10 and 5.11), and several modern spas have been built along this route. The road also exploits the military heritage of the border area, and is said in its advertisements to provide a view of the 'nature and culture of Eastern Finland' and 'the good-humoured Karelian cultural heritage and hospitality' as well as modern facilities for travellers. The largest number of the almost 70 features, monuments and happenings provided for travellers along the road in 1992 were military (29), while 17 represented Karelian traditions and 23 explicitly Orthodox traditions (Runon ja Rajan tie 1992).

Thus it can be concluded that a new symbolic space with ideological and historical dimensions has been produced by various Karelian organizations and businesspeople since the war to revitalize and reproduce the Karelian culture and to promote the heritage business. One obvious aim behind these efforts has been to create both representations of space and spaces of representation, to employ the categories of Lefebvre (1991) discussed in Chapter 2. These point to symbolic substitutes for the lost Karelian territories and signs that denote both the past and the present — and finally enable economic exploitation of the idea. These trends are illustrative of the remark of Urry (1990, 120) on the role of architects and architectural practices in shaping the contemporary tourist gaze. Although these 'Karelian structures' are important in their signification of the lost territories and the eastern boundary of Finland, any profound evaluation of whether these buildings contain hints of critical regionalism or are merely rather simple-minded attempts to revive the hypothetical forms of a lost vernacular must lie beyond the scope of this book (Frampton 1985, Urry 1990, 120–6).

In any case, as Urry (1990, 109) maintains, heritage is often bogus history. The Road of the Bard and Boundary calls to mind the argument pointing to an increasing role of *simulations* in the constitution of our social world put forward mainly in the debates of French post-

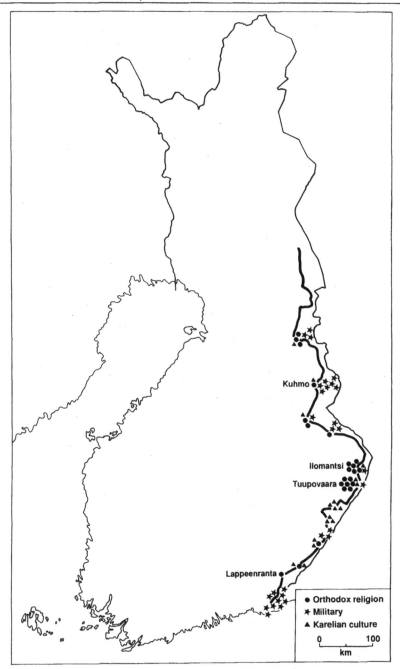

Figure 5.10   Reconstructing the Karelian socio-cultural space in Eastern Finland.
The Road of the Bard and Boundary and the numbers of military, Orthodox and
Karelian memorials and events featured in the various Finnish border communes
in 1992. Source: Based on information in *Runon ja rajan tie* 1992

Figure 5.11 Korpiselkä House — a typical example of the exploitation of Karelian tradition in the field of the modern heritage industry and tourism (photographed by the author)

structuralists. Some authors (e.g. Baudrillard) argue that simulations and models are to an increasing degree coming to constitute the world, so that the distinction between the real and the apparent becomes obscured (cf. Featherstone 1991). As far as the construction of history and heritage is concerned, simulations may also be copies which do not have any originals behind them (Jameson 1984). Hence, even though some of the buildings constructed along the Road of the Bard and Boundary may have real referents in the Karelian past, the spectacles that these 'signs' provide surely move much further towards simulations than the 'original' referents could ever do.

The attempts to revive the Karelian past have been criticized by cultural reseachers for exploiting the cultural tradition selectively and commercially, and even for transforming folkloristic mythology into concrete history. The reality that is represented, e.g. in the imaginary spaces of the Kalevala village at Kuhmo, Petrisalo (1989, 268, cf. Sihvo 1989) argues, is comparable to Disneyland, Donald Duck or the Sleeping Beauty. Knuuttila (1989) points out, however, that to emphasize the past as an essential feature of a regional periphery is not the primary purpose for selling the place, but is a result of the economic, social and cultural self-analysis made by the inhabitants themselves. The result, e.g. various written regional identities, is then transformed and exploited by various institutions and organizations in the field of tourism (cf. Paasi 1986a, 1991a).

All in all, 'East', and 'Easternness' are also signs that have various referents, both real and imagined, commercial and non-commercial. The discourse on the changing roles of Easternness is a fitting illustration of the disappearance of public spaces and experiences, which become incarnated in the pages of advertisements and simulated signs and rituals.

5.8.3 'Finlandization' and the changing self-understanding of the Finns

As the third illustration of the changing socio-spatial representations of Finland's location on the world political map, it is necessary to discuss briefly the interpretation that the international community put on Finnish–Soviet relations after World War II and how the Finns themselves interpreted these relations. Certain basic facts characterized Finland and its position in the system of international relations after the war. Finland was still a minor nation. It was a parliamentary democracy whose economy was based on a capitalist mode of production, and its foreign policy was based on neutrality (Väyrynen 1987).

When analysing the forms and manifestations of foreign policy and political culture in Finland after World War II, some commentators since the 1950s, and particularly since the 1960s and 1970s, have argued that these political developments have led to the rise of 'Finlandization', i.e. to a situation in which the Soviet Union was thought to have a major influence on Finnish internal political decisions even though there was no Soviet military takeover. Finland's international position and the agreements with the Soviet Union did not fit into the general models of international politics: Finland was simply an exception or even an anomaly in the international geopolitical landscape and her position and strivings towards peaceful co-existence did not comply with the rules and dictates of power politics in the post-war world. As a consequence of this, the country's foreign policy remained somewhat odd to many foreign observers (Väyrynen 1987).

The basis of Finlandization would then be conceived as rooted in a common boundary between a small state and its huge neighbour, geographical isolation, limited defence potential, weak national identity, economic dependence on the large neighbour and the strong position of the Communists in internal politics (Väyrynen 1987, Tarkka 1992). Where Finland was for some a model of how a state should organize its relations with its victor in a war, for others this state of affairs represented Finlandization in negative terms. The idea of Finlandization was consequently raised on both sides of the Atlantic as a spectre to frighten people into believing that Western Europe could also be 'Finlandized'.

In fact, Finlandization has been interpreted as a Western European replacement for the domino theory (Taylor 1988b, 17). Some commentators have also pointed explicitly to this idea when analysing the world political map and claiming that for the USSR to establish a sphere of influence on the basis of the model of Finland's role in Western Europe, it would have needed to replace the influence exercised at that time by the United States (cf. Vloyantes 1975, 177–80, Allison 1985).

Hence, according to the most extreme arguments, Finland had all the symbols and signs of national independence but the activities of its leading decision makers were restricted by the presence and pressures of its giant neighbour to the extent that it could be regarded as a province of the Soviet Union. On the other hand, this view was not accepted in some authoritative circles, and the location of Finland and its relations to its neighbour were interpreted as a unique solution on the scene of international politics. During the 1970s it was only the representatives of the extreme right in Finnish political life who regarded the debate on Finlandization as an illustration of the pursuit of an unsuccessful foreign policy since the War (Hakovirta 1975).

Even if Finland was probably never 'Finlandized' as much as the British Foreign Secretary believed after the war (Polvinen 1987, 140), the previous discussion briefly illustrates the character of the internal and external social and political boundaries of discourse that the Finns had to confront after the war. Some Finnish commentators have made a distinction between Finlandization and self-Finlandization. The former was not, according the commentators, a problem, but the latter, the political exploitation of foreign policy in internal policy to strengthen the ideological position of one's own group, has been discussed recently in much more critical terms (Jansson and Tuomioja 1981, 101–15).

Recent self-analysis, put forward since the dispersal of the Soviet Union, has been much more objective. Finlandization has been characterized as referring to a distinct change in the social consciousness of the Finnish people. According to a recent comment, self-censorship was strongest during the 1970s. Tarkka (1992, 203–4) aims at explaining these trends by referring to the changing Soviet 'tactics'. Hence, during World War II, the Soviet Union tried to solve the 'Finnish problem' by military means, while in the 1950s and 1960s it used political pressure. In the course of the 1970s the focus was moved to the 'mental field' and this was effective. A considerable proportion of Finland's political élite participated in this 'game', whereas civil society remained to a great extent outside this sphere.

Vihavainen (1991) is prepared to speak of the rise of a new political culture in which totalitarian elements become part of the pluralistic culture and the Finlandization of consciousness. It has been argued that this, for its part, created specific forms of political rhetoric and practice in Finland, which have only now begun to be revealed after the dispersal of the Soviet Union. In the course of 1992 and the early part of 1993 considerable 'sensations' came to light about the past activities of leading Finnish politicians in relation to Soviet leaders, and several authors have aimed to explore this new political situation to provide illustrations of 'Finlandization'. Even some leading politicians themselves have been active in this way.

At the same time new, somewhat controversial evidence is regularly being published regarding the actions of Soviet politicians and other participants and their intentions to influence Finland's internal politics. On the other hand, the aims of some leading politicians (e.g. Urho Kekkonen, who was President for 25 years) to exploit personal relations with Soviet leaders to prop up their own political aims, i.e. 'self-Finlandization', have also been discussed. The cautious relations established between Finland and the Soviet Union after World War II continued, with varying success, until the dispersal of the latter, and the most recent themes in these debates have concerned the motives of the Soviet military élite, who aimed, unsuccesfully, at establishing close military cooperation with Finland (Vihavainen 1991, Tsernous and Rautkallio 1992). Such efforts were part of the changing international military landscape.

Recent intensive discussion of Finland's entry into the EU also raised comments in which the idea of Finlandization was turned upside down. Vihavainen (1991, 286–7), for instance, wrote that in some respects the belief in the beneficial effects of integration resembles the traditional cargo cult: they will give us things cheap. He points out ironically that Finlandization is still going on; but the direction is now towards the West.

Thus, a new story — perhaps several rival stories — about Finnishness and Finland's place in the international community may be seen to be emerging. Whether wisely or not, this discourse promotes an understanding of the most radical change in social consciousness in the history of the independent Finnish state.

5.8.4 The discourse about returning the ceded areas of Karelia

The most recent feature in the symbolic and cultural struggle over the representations of the Eastern social space is the lively debate on the future and fate of the ceded areas of Karelia that has emerged in the sphere of civil society in Finland. Hence this discourse is the only one that is explicitly connected with the present boundary between the two states.

This public debate has risen in Finland only following the dispersal of the Soviet Union, and it is a typical example of the changes that have taken place in the socio-spatial consciousness of the Finns since the end of the Cold War. One source for these debates has been the opening of the doors to tourism in the areas that belonged to Finland before World War II. The current possibility to visit the 'real' or 'original' East, i.e. the areas of Russian Karelia that are in many cases a nostalgic and painful constituent of the personal history of the Finns, has raised the question of restoring these areas to Finland. Simultaneously the opening of the boundary has obviously reduced the impact of the reconstructed symbolic space of Karelia on the Finnish side of the boundary.

Since the war — in fact since the Peace of Moscow in 1940, and particularly since 1944 — there have been more or less continuous unofficial discussions in various circles of the possible return of the ceded areas, but not on an official political level (Laine 1992). Explicit, open discussions concerned with leasing or buying back these areas, or at least initiating negotiations with the Russian authorities about the possibility, have been formally proposed since the break-up of the Soviet Union, although it has typically been emphasized that these aims are not 'political' but rather historical, cultural and human. Anyhow, as far as the boundaries of sovereign states are concerned, these questions always turn out to be political even if they are not intended to be.

The idea of recovering the ceded areas has been put forward mainly by some leaders of opinion among the organizations of resettled Karelians, particularly the Karelian Union (Karjalan liitto) and certain private individuals, and certain former and even present leading politicians have recently taken up these themes in their discussions. Nevertheless, the official opinion represented by the Finnish state authorities is unchanged and hence does not support such claims. Accordingly, the official foreign policy view has been in agreement with the official attitudes of the Russian authorities. This view is still based on the fact that geography matters.

Nevertheless, some tensions still seem to exist between the two states, and surprisingly, one day before the first round of Finland's presidential election in January 1994 the Ministry of Foreign Affairs received an official Russian diplomatic note in which the (anonymous) authorities — it was not clear whether the note had come from the embassy or from Moscow — asked for information on the activities of two extreme nationalist associations formed since the disappearance of the Soviet Union which have been openly demanding a return to the boundaries of the Peace of Tartu. This note was sent only a short while after the extreme nationalist Zhirinovsky had been so successful in the elections in Russia. According to the Russian Ambassador, the Russian authorities were worried about the rise of extreme nationalism in both countries. Bearing in mind the practices of the Soviet period, the Finnish reaction was one of great annoyance and presentiment of whether this was one more expression of drawing lines between West and East.

6

THE CHANGING SOCIO-SPATIAL
CONSCIOUSNESS

6.1 REPRESENTATIONS OF FINNISHNESS AND ITS 'SYSTEMIC' BASIS

The discussion above made it clear how various forms of nationalism and their symbolic manifestations emerged gradually in the nineteenth century along with the process of the institutionalization of Finland and promoted the rise of the consciousness of Finnishness. This chapter will proceed to scrutinize the historical construction and the form and content of the spatial dimension in this consciousness, employing various sets of empirical material. It will discuss both the construction of the Finnish 'self-portrait' and the changing representations of the Soviet Union and Russia as the Other, and it will also scrutinize the institutional basis through which these representations have become part of the national socialization.

Even though nation building in Finland progressed gradually from the early nineteenth century onwards, the most important systemic, state-bound basis for it and for national socialization, that is the general education system, was still in its infancy. The move from being a part of Sweden to being a Russian autonomous state did not cause any marked changes in the education system. Education was organized by the church and it aimed at providing every person with a satisfactory knowledge of Christianity and an ability to read. In practice the latter was not usually very good.

Geographical education at the beginning of the nineteenth century, for example, provided a deficient picture of Finland. Stenij (1937, 2–3) wrote that little more was required than the knowledge of the name of the King or Emperor and maybe the number of administrative provinces, a few towns and some notable hills and lakes. It was only after the mid-1800s that the emerging liberalism substantially affected general education. The most important reforms were the founding of the primary school system, changes in the secondary schools and the establishment of adult education. The role of education as the main factor in the national awakening was thus assured (Klinge 1972, 73).

Nevertheless the spread of the school network, the 'material basis' of national socialization, was slow. The primary school statute of 1866 stated that a primary school had to be established in every town, but this was voluntary in the rural communes. In 1871–2 there were only 8000 children attending primary school and 5600 attending secondary school, although the figures had increased to about 29 000 and 6500 respectively by 1880 and 54 000 and 9500 by 1890. The russification policy at the end of the nineteenth century increased the national emphasis on education and there were already 112 000 pupils attending primary school by 1900. General compulsory school attendance was established in 1921, and thus the number of pupils in 1930 was more than 420 000 (Vehvilä and Castren 1972, 286–93). Hence it should be stressed that the stereotypical socio-spatial representations of Finnishness or otherness, as presented in textbooks, for instance, were certainly not familiar for most Finns during the nineteenth century. There were no common languages of integration and difference, and the internalization of these representations was only possible in association with the development of the education system.

Patriotic literature soon became an essential part of the national content of the curriculum, promoting an image of Finnish identity. Geography and history together with patriotic songs aimed at developing a national and religious spirit in education (cf. Wuorinen 1935, 140–52).

The construction of an identity on the national scale did not take place independently of the production of identities on other territorial scales. We saw above the role of spatial scales in the nation-building process, and it has been argued that scale provides a distilled expression of spatial ideologies (N. Smith 1990, 173). The axis 'home–home region–fatherland–other countries' has always constituted a unity in the ideological basis of Finnish nationalism and in the construction of a hierarchy of socio-spatial representations in this frame, and religion constitutes another key to this four-part scheme.

Material dealing with home regions has constituted a significant part of Finnish popular education (Oksanen 1982), as it has in Sweden, for Löfgren (1989) argues that the province or region has provided a micro-level model for patriotism: 'by learning to love your region — one part of the national whole — you prepared yourself for national feelings on a higher level'. According to Löfgren, this was the key idea of education in Sweden at the beginning of the twentieth century. Empirical evidence for this was put forward by Alsmark (1982) when studying the relations between regional and national identity in Swedish school books. Hence, even if regionalism may pose a threat to the nation-building process, Löfgren (1989) argues that in the Scandinavian countries it has often functioned as a stable factor in the national landscape and more of an integrating one than a threatening one.

We will now examine the basis for this integration in the Finnish case, looking at the material and functional basis of nation building in this chapter and the cultural and ideological constituents of the process in the chapters which follow. It can be argued that before the rise of the social division of labour, Finland hardly manifested itself territorially to its inhabitants at all. A parish was the territorial frame where the activities of most people were organized in eighteenth-century Finland, and even up to the mid-nineteenth century (cf. Jussila 1979, 17). This is of course not uncommon in the history of nation building, as it has been pointed out that far from having any concept of what it meant to be French, the peasantry and rural artisans of France, for instance, inhabited not one but a number of lesser Frances, such that the inhabitants of one were hardly aware of the existence of those in the other (Mac Laughlin 1986, 303). The content and plot of nation building are typically constructed from the viewpoint of the present, to create a past along with its representation.

In the process of state building, systems for managing local government and the territorial organization of the state become necessary along with the construction of socio-spatial consciousness (cf. Agnew 1987, 38). In Finland, the administrative communes and the church parishes were formally distinguished from each other in 1865 and the communes now took care of many important social functions necessary for social reproduction. This did not change the boundaries, however. In the countryside the area of a parish became that of the commune as well, and every town became an independent commune (Tiihonen and Tiihonen 1983, 182). The philosopher J. W. Snellman, for instance, greatly respected the communal self-determination and regarded it as a distinct moral sphere in the action of civil society (Pulkkinen 1989, 17). This agrees partly with the argument regarding the key roles of the local state, according to which the function of the local state is concerned more with the securing of a social consensus and integration than with ensuring the conditions for production. These are of course not independent of each other (Johnston 1989, 301). Between 1831 and 1921 eight administrative provinces provided the regional context between the communes and the scale of the state as such (Paasi 1986b).

The spatial and social mobility which spread through Finland in the second half of the nineteenth century and the construction of roads and railways could, from a logical point of view, have implied a more frequent severing of local connections and a weakening of the regional consciousness. 'Logical' refers in this connection to the discussions of classical sociologists, according to which the role of local traditions will diminish through the increasing division of labour, while various general, 'common', non-local features will gain ground in the social consciousness. Nevertheless, where the division of labour differentiates between people, it does so in a way that impels them to cooperate with one another (cf. Abrams 1982). One key to understanding the flexibility of territorial identities lies in their hierarchical nature and their connection with various institutional practices which are not unambiguously local or non-local. This is the case with religion and the media, for example.

The emergence of the division of labour is generally the time when the *anonymous authority* — to employ the ideas of Fromm (1962) — comes into play. Hence the responsibility for communication and control over the socialization process is to an increasing degree transferred to anonymous social institutions, and the power relations in society become increasingly concealed (Paasi 1986a). National symbols become crucial in nation building and flags, national anthems, national poetry and literature begin to shape the identification.

Brennan (1990, 49) puts emphasis on the role of the novel and the newspaper as the major national vehicles of the printed word, which helped to standardize the language, encourage literature and remove mutual incomprehensibility. In many cases the manner of presentation set forth in novels allowed people to imagine the special community that was the nation.

Aleksis Kivi, a key figure in Finland's national literature and reformer of the descriptions of nature (Tiitta 1982, 20), aptly describes the Finnish 'nation building' process from the perspective of Eero, the youngest brother in his classical *Seven Brothers*, the best-known, and according to many commentators, the most revered work of Finnish literature, originally published in 1870. The gradual extension of the world view of the nineteenth-century Finnish peasants and the establishment of its spatial boundaries is clearly illustrated:

On Sundays and holidays, he either read the newspaper or himself wrote about parish news and affairs for some newspaper. The editors always welcomed the pieces he sent in. They were all highly pithy in content, and sharp and clear, sometimes even brilliant in style. This kind of activity broadened his view of life and the world. To him, his native land was no longer an indefinite part of a vague world, its kind and location completely unknown. He knew where it lay, that dear corner of the world where the people of Finland lived and built and struggled, and in whose bosom lie our forefathers' bones. He knew its boundaries, its seas, its quietly smiling lakes, and the woody picket fences of its piney ridges. The complete picture of our homeland, with its kind, motherly face, was forever imprinted deep in his heart. (Kivi 1991, 340–1)

The realistic perspective provided in *Seven Brothers* was not accepted by the upper classes at first, and Klinge (1975, 17–18) argues that it came into its own only after the 1905 general strike, when the old Runebergian heroic image was discarded.

The contemporary process of nation building did much to reinforce the role of the region, i.e. province. The number of historic Finnish provinces in the mid-nineteenth century was nine, and these areas did not correspond to the administrative provinces. At the end of that century several new provinces were gradually institutionalized through regional transformation eventually to become established parts of Finland's regional structure and socio-spatial consciousness. This was a product of the emerging modernity and capitalism, through industrialization, increasing mobility, the rise of centres and spheres of influence and the replacement of local identities by mobile, anonymous, literary identities (Paasi 1986b, cf. Bleicher 1989).

The vast increase in production, consumption and trade promoted by the emergence of industrial capitalism, brought about a trend towards concentration, as a result of which population mobility increased and was directed towards the towns and industrial communities. The formation of the provinces was further promoted by improvements in transport connections and the development of a regional press (Paasi 1986b).

The organization of voluntary associations also supported these tendencies. Even though the members of these associations organized their activities according to their ideological premises and aims, the network of associations spread simultaneously across various spatial scales, i.e. national, provincial and local (Salonen 1970, 238–9). Of particular importance was the role of the student fraternities at the University of Helsinki, as their activities were organized along provincial lines, appealing to the mythical traditions of the original Finnic tribes and exploiting these notions to construct a national identity.

The notion of 'tribe' usually presumes notions of a descendency group — mythical or historical. Giddens (1987a, 117) points out that such concepts, together with religious symbols, have always been the main sources of group identity and exclusion. Genealogical myths seem to have been the most common means whereby actual descent and kin connections become solidified with group identity. They have been as much part of the history of the ruling classes within class-divided societies as of overall cultures. In Finland the establishment of the distinction into 'tribes' contributed to a progressively more normative and pervasive nationalism (cf. Gellner 1983, 86).

6.1.1   The rise of the Topelianic self-image of Finnishness

One of the most outstanding creators of national and provincial identity in Finland was Zacharias Topelius, Professor of History at the University of Helsinki, whose books — above all *Maamme kirja* (Book of Our Land), published in 1875 specifically for use as a school textbook — fashioned the concept of Finland and its landscape in the public consciousness and provided stereotyped descriptions of the characteristics of the Finnic tribes and the Finns as a whole. Topelius was, together with Runeberg, an 'unofficial' creator of national symbols. The government realized the political role of their writings and they were accorded important positions in recognition of this (Tiihonen and Tiihonen 1983, 190). The books of Topelius and Runeberg were regarded as dangerous by the Russian authorities, however, since they made a clear distinction between Finland and Russia instead of representing Finland as an autonomous part of Russia (Polvinen 1984, 194).

Alongside history, Topelius also lectured in geography, reflecting the close connections between the two which existed in many European universities up to the end of the nineteenth century. At that time geography was commonly regarded as an auxiliary to history. Hegel's introduction to the philosophy of history, for example, contains a whole chapter on the geographical basis of world history, in which he speculates at length about the relations between various geographical environments and their role in shaping the character of the inhabitants (Hegel 1978). A typical feature of Hegel's historical thought is its Eurocentric approach. Nevertheless, Topelius regarded geography and history as being of equal value, inasmuch as they are interdependent. Furthermore, he regarded geography as an independent science.

Topelius was inevitably an important figure as regards the national movement in Finland. He had originally been inclined to interpret history in a Hegelian manner, regarding the history of Finland as having begun in 1809, but in the face of criticism he eventually pronounced that the history of Finland was the 1000-year struggle of the spirit of the Finnish people on behalf of their own conscience and moral freedom (Noro 1968, 82) and that one could not conceive of a history of Finland without it being grounded in the Finnic 'tribes' (Tommila 1989, 55).

Topelius became Professor of History in 1854, and lectured on the geography of Finland for eight terms before starting his lectures on history. He discussed such topics as the Finnish state, its boundaries and the inhabitants of its provinces (Stenij 1937, 6). His chief influences were Hegel and the Swedish historian Geijer, who was himself influenced by Hegel. Topelius tried to connect geography both with his religious and romantic philosophy of life and science and also with his patriotism, and he literally constructed a picture of Finland in the national consciousness with his books (cf. Tiitta 1994). The following citation from his *Book of Our Land* gives a fitting illustration of the territorial distinctions and religious elements that Topelius constructed in his book and the metaphors he employed for this purpose.

This nation has grown up as one, like many trees form a large forest. Pines, spruces and birches are different timber species, but together they are a forest. The people in one province differ from those in another province as regards their appearance, clothing, character, habits and standard of living. We can easily distinguish a person from Häme from a Karelian, a person from Uusimaa from an Ostro-Bothnian; we can even distinguish the inhabitants of one parish from those from neighbouring parishes. But when we travel abroad and meet our fellow countrymen from other parts of our country, we will find out that in many respects they are similar. And when a foreigner travels to Finland, he finds that the inhabitants of this country have much in common as regards their character. For those who have lived in the same country since their childhood, subordinated to the same laws and living conditions, must have, with all their differences, much in common . . . Such a sense of belonging or quality (*national character*) is much easier to feel than to explain. It is the stamp that God has impressed on all nations by letting their inhabitants live together for a long time. (Topelius 1981, 124)

Nature has occupied a crucial position in determining how Finland and the Finnish identity have been represented, being present both as a factor explaining the Finnish character and as a crucial element in the rise of the national identity, mediated by folk poetry (cf. Tiitta 1982, Suutala 1986, Lehtinen 1991). Both nature and the idea of landscape, which had become popular early in the autonomy period, also formed significant structural elements in the image of Finland as idealized by the landscape painters.

Meinig (1979a) points out that 'every mature nation has its symbolic landscapes' and, further, these landscapes are usually an essential part of the iconography of nationhood. Landscapes become important constituents of social integration, since they also become part of the shared ideas, memories and feelings which bind people together — a part of the prevailing ideological structures. Landscapes can be particularly important in this respect, since they often provide a visible, materialized basis for this abstract symbolism. Topelius' book *Finland framstäldt i teckningar* (Finland presented in pictures) played a major part in establishing the view of Järvi-Suomi (the lake region of Finland) as the primary symbol for the whole of the country, and many artists contributed to this idea (cf. Tiitta 1982, Saarela 1986, 210–11, see Figure 6.1).

Topelius' exploitation of the metaphor of trees and forests in his text

Figure 6.1 'Metsälampi' (Forest Lake) by Lennart Forsten: A prototype of a stereotypic Finnish landscape in Topelius' *Finland framstäldt i teckningar.* Source: Republished in Topelius 1981, 178. Reproduced by permission of Werner Söderström Osakeyhtiö, Helsinki

is hence not unique. Buttimer (1978) writes that in many countries patriotic songs referring to the native soil and forest have built up the spirit of nationhood. Throughout America and Australia the national histories have consisted of creating a country from the forest and the grassland (Short 1990, 19–21). Daniels (1988, 43), for his part, suggests that 'trees and woodland have proved as rich a symbolic resource as a material one, frequently being exploited to represent ideas of social order'.

As a conclusion to his essay, Daniels (1988) is ready to talk about a 'political iconography of woodland' in England, and his conclusions also hold good in the case of Finland, as can be seen from the conclusion of Lehtinen (1991) that forests are an essential part of the national landscape, providing a source of inspiration for artists aiming at building a national identity (cf. also Burke 1992). Hence it can be argued that in nation building some local landscapes are chosen and transformed into abstract symbols of nationhood, regardless of whether these symbols actually 'correlate' with the local landscapes and physical environments which ultimately constitute the basis of local identity and of the life-world of the people.

It was argued above that nationalism aims at homogenizing the national space and making it a source of identification. Nevertheless, as the case of Topelius indicates, a nationalistic ideology can also effectively exploit regional variation in the construction of national representations — to concretize and put life into the empty space of the nation-state, as it were. A typical example of the exploitation of the hierarchy of socio-spatial identities is the employment of various features of the Finnish provinces — the 'character' of their inhabitants and various features of nature — in the construction of the national identity since the nineteenth century (Paasi 1992). But the ideological structures of nationalism change in content along with political relations between states, and thus they are not permanent (Anderson 1988). A fitting illustration of these changes may be obtained from the contents of geographical teaching in Finland, where the relations between various states, commonly represented in the form of national stereotypes, have varied greatly in relation to the changes in foreign policy.

In the *Book of Our Land* Topelius puts forward a long list of desirable characteristics that he regards as typical of Finns, and his interpretations of the influence of nature on the Finnish character remind one of Geijer's corresponding descriptions in the Swedish case (Noro 1968, 83). The *Book of Our Land* sold 310 000 copies during the nineteenth century, a substantial figure when one remembers that the whole population of the country only exceeded two million in the 1880s (Lehtonen 1983, 57)! Also, the stereotypes created by Topelius were not found significant merely in early nation building, for it seems that they were still being exploited in Finnish geography textbooks in the 1960s and contributed to the language of national integration.

6.1.2 Identity and otherness in school geography textbooks

The importance of the role of geography was realized in the emerging Finnish state. Geography as a precisely defined subject has been a part of the Finnish school curriculum only for about 100 years, although the map of the world, extended by the voyages of discovery, had featured in the school ordinances as early as 1724 (Rikkinen 1989, 23). H. G. Porthan, an influential academic figure in the late eighteenth century, linked geography and history together in the school system, geography occupying largely an auxiliary function with respect to history, and they remained combined in the official curriculum up to 1914, even though teaching positions in geography began to be combined with those in natural history by the beginning of the current century (Leikola 1980).

The construction of social integration and difference is part of the production of space and spatial representations, and this is particularly the case with cartography. The cartographic representation of Finland began to take shape even when the country was still part of Russia. The first map of Finland with names written in Finnish (1846) already clearly depicted the boundary with Russia, i.e. the role of maps in constructing and not merely reproducing the world was obvious (see Figure 5.2). The representation of the boundary became even more explicit in the first edition of the *Atlas of Finland* (1899), which the Geographical Society of Finland produced partly in order to heighten the self-consciousness of the Finns.

By the end of the nineteenth century 'geography' had also become important in various institutionalized forms, not merely as a physical or cultural basis for nation building. School geography was of crucial importance when the subject became an institutionalized academic discipline in many European countries (Capel 1981, Taylor 1985), and it was also essential for the rise of academic geography on account of the fact that it established the first labour markets for professional geographers and thus rendered the maintenance and reproduction of academic geography possible (cf. Vartiainen 1984). Its national function in Finland was to provide a background for academic geography — as can be seen in the establishment of Finnish geographical societies before the institutionalization of geography (cf. Mead 1991). The first major national task for the discipline of geography was the education of teachers.

The teaching of geography mostly took place in the junior schools, and with the founding of the elementary school system geography also established its position in the curriculum at this level. Whereas elementary school teachers received their training at seminaries, secondary school teachers attended the University of Helsinki, where geography remained a part of history until its establishment as an independent discipline. Thus geography was taught in all types of school in the nineteenth century. By the turn of the century it was being connected with the natural sciences in schools rather than with history (Rikkinen 1989, 25–6), and this combination together with the important status of the teaching profession as a source of employment for geographers subsequently influenced the content of university geography, leading to a powerful naturalistic approach.

The publication of books in Finnish was a somewhat rare phenomenon in nineteenth-century Finland, less than 200 titles being published between 1544 and 1809 and only 900 between 1809 and 1865. There was only one bookstore and two printing houses in operation in Finland at the beginning of the nineteenth century (Peltonen 1992, 69).

The first geography textbook to appear in Finnish, in 1844, was a translation of a book written by the Norwegian pedagogue Platou, and the first original textbook in Finnish was produced in 1862, an extremely early date bearing in mind that Finnish did not become the second official language of the country alongside Swedish until 1882, and that it was only in the 1890s that the number of Finnish-speaking pupils attending secondary school, the level leading to university entrance, exceeded that of Swedish speakers (Konttinen 1991, 198).

The role of language in the construction of the Finnish nation is apparent from the above analysis. Language also had more delicate meanings, in which the role of power became apparent, the crucial motive for various groups and classes being control over language in order to control the content of social consciousness. Language was also a weapon for the Swedish-speaking élite and civil servants in their effort to maintain their positions against the Finnish-speaking majority which was struggling for economic and social improvement (Rommi and Pohls 1989, 90). As was seen during the years of active Russification, the language connected with geography and history teaching was resisted by the Russian authorities (Polvinen 1984), and hence language — both the language of integration and that of difference — became critical in the representation of relations between nations and states.

The earliest geographical textbooks published in Finnish were largely lists of facts presented in a neutral manner, and it was only towards the end of the nineteenth century that attempts at illustration and causal explanations were made, frequently in the spirit of environmental determinism. The role of school geography from the viewpoint of nationalism was thus at first one of defining the spaces of nations and the boundaries between states — and between us and the Other — in a technical sense and setting out and delimiting facts regarding the natural environment, culture and administration of each in a neutral manner. In addition, the maps that were exploited to support geographical education represented Finland formally as a territorially restricted entity. In this way they served to construct the formal distinctions necessary for the production and reproduction of the territoriality of the state. Stereotyped descriptions of nations and the people living in them began to abound by the 1880s, and normative, usually favourable, descriptions of the character and moral codes of the Finnish people appeared in books at approximately the same time. It was common to explain the national character in a determinist vein, in terms of the physical conditions (Paasi 1992).

A more subjective body of teaching material spread into the textbooks in the late nineteenth century, and at the same time the collective descriptions provided of the Finns seem to have become more normative in character, i.e. their role in the construction of ideologies and hegemonic structures became more crucial. The depictions of Finland were mainly positive and thus contributed to the construction of the self-image of the Finns. The books usually iterated clearly, in the manner of Topelius, how Finland is defined, who the Finns are, what territorially based sub-groups, emerging from the historical 'tribes', exist within them and how the Finns are to be distinguished from the Other.

Simultaneously, colourfully embellished travellers' tales derived from the voyages of discovery were used as a background when presenting corresponding stereotypes for other countries.

As was customary at that time, the textbooks did not mention their sources (cf. Hakulinen 1984), but a number of teaching guides suggest that they were based on travellers' reports not only of Finnish but also of Swedish and German origin (Paasi 1984a, 62). Lehtonen (1983, 122) in turn observes that close attention was paid in Finland over the period 1870–84 to developments taking place in foreign textbooks, from which points illustrating the Other were adopted. Thus these depictions were not exploited systematically in relation to the picture of Finnishness. It should also be remembered that from the point of view of the dissemination of geographical knowledge there were still many blank areas on the map of the world in the nineteenth century, and the colonial branch of geography was actively engaged in gaining possession over this unknown territory at the same time as academic geography and social science inadvertently contributed to nation building (cf. Mac Laughlin 1986).

The merits of geography as a science were ascribed in quite a matter-of-fact manner in many works precisely to the voyages of discovery and their achievements (cf. Lehtonen 1983, 98). Concerning the relation between text and context, it should also be recalled that these narratives, which now strike us as wholly unscientific, represented the highest level of spatial knowledge available at that time, even though they were a mixture of information, imagination and fantasy.

The principal emphasis in elementary school geography at the turn of the century, as expressed in the stated aims, was to be on the pupils' own country, and attention was to be paid to other countries only when they were of importance for Finland. One sign of a broadening of the image of the world, however, was that it was stated as early as 1908 that attention should be paid to those areas that represented major non-European cultural trends, India, China, Japan and the United States being recognized as such. In practice, however, minority nationalities and representatives of more distant cultures tended to be presented in a gradually more critical light, and any conditions that departed from the Western (European) cultural sphere were explained from a Western viewpoint as attributable to the peculiar nature of the peoples discussed, with no attempt made to outline the cultural or historical contexts for these conditions and customs (Paasi 1992).

These accounts were typical Euro-American representations of 'third world' Others, who were regarded as lazy, ignorant, traditional and less rational than 'us'. Duncan (1993, 45) writes that representations like these helped to provide the ideological superstructure for the spreading European exploitation of the world.

The early Finnish textbooks thus contain typical evidence of the doctrine of social evolutionism, in which the advance or destruction of 'weaker' communities or societies was seen as a manifestation of the general laws of development. The subjugation, exploitation or destruction of underdeveloped, 'non-viable' societies was condonable as a part of this same natural development. The main period when evolutionism dominated Finnish research in the social sciences was between the two world wars (Haavio-Mannila 1973). It had already been pointed out in 1914 that the inspection of textbooks was to be tightened up in order to make them more objective (Makkonen 1984, 33).

School geography occupied a powerful position after Finland gained independence in 1917 (Rikkinen 1978a). Patriotism was heavily emphasized in the curriculum, but geography was also seen as a subject of some importance for increasing international cooperation (cf. Makkonen 1984, 34), in that it taught about countries with which it would be essential for Finland to establish contacts in the period following independence. The way of thinking was based on a national self-consciousness, establishing a routine in things that were considered to be 'Finnish'. Paasivirta (1984, 227) writes that at the same time as 'Finnishness' was connected with traditions and conservative attitudes, foreign countries could be experienced as strange and foreigners, in fact, as 'Others'. A conservative-national attitude regarded what was 'international' as being in sharp opposition to itself and not as natural phenomena providing influences and points of comparison. This dualistic attitude between national and international — the languages of integration and difference — also continued in Finnish society during the 1930s (1984, 351).

A committee appointed by the Finnish Geographical Society issued a public statement in the late 1920s emphasizing the importance of an understanding between nations (Rikkinen 1978b), and again in the early 1930s the contribution of geography from the point of view of the problem of internationalization was discussed officially in a government committee when the role of geography teaching in fostering international understanding was once more brought to the fore (Komiteanmietintö 1932).

Particular attention was paid to those countries and peoples which were of educational or economic significance to Finland or otherwise merited special treatment on account of their geographical proximity or historical connections. The aims set out by these committees did not find their way into the textbooks, however, where the same stereotyped, subjective elements continued to be prominent. Suspicion continued to be greatest in the case of peoples whose culture differed from the European model, these being felt to be alien and regarded in the spirit of evolution as representing a lower level of development than the manifestations of European culture.

Finnish political circles had their own fixed points abroad during the 1930s, fixed points that they admired and evaluated on the basis of their own ideologies. Right-wing circles were oriented towards Germany, the extreme left towards the Soviet Union and the Social Democrats and Swedish-speaking circles towards the Scandinavian countries, etc. (Paasivirta 1984, 360). Simultaneously, the impending world crisis was reflected in the geography textbooks of the 1930s, for example. The books adopted a typically critical attitude towards the Soviet Union and a sympathetic attitude towards Germany. National defence also emerged as a topic to be dealt with in schools (Rikkinen 1978a, 1978b).

The representation of more distant cultures continued to be couched in highly critical terms during the 1930s, in spite of the fact that the continuation of colonialism had altered the relations between the cultural realms, so that the black population, for example, had begun to adopt lifestyles and cultural features belonging to their subjugators. The attitude towards the black peoples was essentially different, however, for where the old textbooks had claimed in the evolutionistic spirit that the problem for the members of alien cultures was that they had not 'mastered' Western and European culture, the line was now drawn so as to imply that Western culture was utterly inappropriate to them (Paasi 1992).

Questions of the relations between peoples emerged in quite a new light in international politics after World War II, especially following the approval of the Declaration of Human Rights by the UN General Assembly in 1948. Stereotypes now became an object of scientific study, once scientists became aware of their role in international under-standing, and a major project for their investigation was implemented by UNESCO in 1947–51 (see Klineberg 1950, 1951).

Again a committee was formed in Finland to consider the aims of geography teaching in schools, in the light of the new awareness of the importance of the subject for increasing international understanding. The resulting report indicated implicitly that warped views of some countries 'might exist' but did not specifically state that the traditional teaching of geography had failed in this respect (Komiteanmietintö 1952). It was mentioned, however, that the focus of the descriptions should be on mankind rather than on differences in human lifestyles as manifested in cultural form.

The Finnish geographical textbooks published after World War II and even in the 1960s continued the tradition of providing stereotyped descriptions of the peoples of the world, but the subjective element with its deprecatory overtones became less marked. The Other now became divided into three spatial realms: the positive (Western) European one, a relatively neutral Eastern European one, and representatives of the most distant cultures, to which the most prejudiced comments continued to apply. There were still references to the 'black' and 'yellow' threats to the 'white' labour market, and racist attitudes were still very much in evidence, even though the harmful stereotypes had disappeared. Thus the accounts of the relations between white and black populations shifted from the field of explicit stereotyped descriptions to *indoctrination* in the form of affectively loaded photographs, which clearly identified the relative social positions of white and black people (Paasi 1984a). As has been argued above, the 1960s produced a major change in the international attitudes of the Finns. The Vietnam war in particular has been regarded as an important background for these changes (Tuominen 1991).

The passage into the 1970s was marked by the introduction of the 'comprehensive school' system in Finland, the declared aims of which were to promote an attitude of understanding with regard to all nations, races and religions, to foster a problem-centred attitude towards all forms of social system without labelling them as good or bad, and to encourage a democratic view of government and support for United Nations initiatives in the maintenance of peace (Makkonen 1984). The aim of school geography nevertheless continued to be to provide pupils with a clear picture of Finland as an independent state and to arouse a sense of identity between its natural environment and its people. Although the language of integration in principle remained the same, the language of difference had been removed from the textbooks.

At the same time, specific mention was made of developing an understanding and appreciation of other peoples and encouraging pupils to promote cooperation between the Finns and other nations. It was at this stage that the textbooks made a clear break with disparaging descriptions of different cultures and set out in a more objective manner to lay emphasis on the characteristic features of cultures. This meant a general reduction in subjectivity. The aims laid down for the teaching of geography in schools in the 1980s stated explicitly that education in international understanding requires the creation of favourable attitudes towards foreign countries and peoples as early as possible, though inherent in this was a lurking danger of stereotyped descriptions (Rikkinen 1989, 36).

6.1.3 The rise of the allegory of Finnishness

The stereotypic image of Finnishness had to be structured in the form of personalities, human beings, and this points to the question of the role of metaphors and allegories in nation building. Lakoff and Johnson (1980, 33) write that perhaps the most obvious ontological metaphors are those in which the physical object is further specified as being a 'person'. The personification of various physical and social entities is an essential part of human discourse. Similarly, it is typical to raise social groups as subjects of action and cognition (cf. Gilbert 1989, 206). Examples such as 'the USA beat Germany,' 'Finland's attitude to the EU', 'the relations between Sweden and Denmark' are typical expressions of the language of anthropomorphism, typical both in everyday speech and in international politics. Metaphorical expressions are very useful in linguistic practice on many occasions, and it is often extremely difficult to specify the subject of the action under discussion. This has become obvious above, for example, where 'Finland's' attitudes or opinions, 'Soviet' ways of thought, 'actions' of the state, and so on were discussed. But who is the subject when one speaks of war and peace? Who is the subject of collective representations and stereotypes?

These metaphorical expressions referring to collective subjects are fitting illustrations of the potentially ideological roles of language, in which the power relations involved in the construction of these groupings and in their control are eclipsed. The implication of these personifications is that total peoples have a joint attitude towards certain themes. Even though their academic integrity and intellectual credibility may be weak, these practices in the use of language are doutbless as common among academic as non-academic people and are effective in constituting national identity, being narrative accounts for establishing an idea of the distinction between us and them (cf. Bloom 1990). Thus they are also expressions of ideologies and inherent power relations.

Much of the following discussion will be based on the development of the allegories of Finnishness as explained in Reitala's broad account of this theme, cursorily picking out topics that are important for the present purposes. Reitala (1983, 157) comes to the conclusion that the personification of Finland in various forms was not a temporary poetic allegory but rather a symbol which was constructed consciously and which changed gradually with the ideological motives behind it. Further, these allegories were not merely literary and artistic creations but were politically significant.

Geographical personifications and mythological expressions referring to places and countries became common in the Roman Empire, particularly in coinage, in which occupied provinces were also represented in the form of female figures. Regional personifications then underwent a new renaissance during the seventeenth century in the form of winged allegories. Geographical entities were also represented as divine women in both pictures and poetry. Even though coats of arms and flags became the official symbols of states, personifications were employed in festive and poetic genres, pictures and eventually caricatures (Reitala 1983, 11).

The personifications of states played their most important role during the second half of the nineteenth century and early twentieth century, the background for this lying in politics and nationalism. Reitala (1983, 13–14) writes that personifications attained their most important role where the purpose was to strengthen solidarity among the people and to create a national consciousness. As a consequence of this tendency, most European countries have had their own anthropomorphic symbols, muses or at least some patriotic symbolic meanings attached to feminine figures (1983, 14).

All Finland's neighbouring states have had their own 'mother figures'. *Matuska Rossija* has a long tradition, which was originally connected with the Emperor and later with the people, while *Svea mamma* (Mother Svea) is the typical personification of the Swedish nation. The Soviet Union, for its part, was prodigious in its use of huge patriotic personified statues (Reitala 1983, 17), and both these and paintings depicting the background figures of socialism constituted an important ideological landscape and were a concrete and constant manifestation of state power at the level of everyday life (cf. Pickles 1992, 201). A fitting illustration is Figure 6.2, which depicts two views of Vyborg, the main Finnish town ceded to the Soviet Union as a consequence of World War II. In both pictures the symbols of socialist ideology are an essential part of the iconography of landscape and social space: above in the shape of Lenin, below in the form of the traditional hammer and sickle.

The stereotypical Finn described by Runeberg and Topelius was above all a brave soldier and hardworking man coming from the forests of Central Finland. Aleksis Kivi, for his part, spoke about the 'motherly face' of Finland in his *Seven Brothers*. The first female personifications originated in the seventeenth century, and the subsequent idea of 'Mother Finland' took several forms. Löfgren (1989) also draws attention to this specific gender bias in national stereotypes, noting that whereas the typical Swede, Dane or German is usually a man, the *father*land is usually symbolized with a national mother.

During the period of Russian autonomy, and particularly during the 1870s, the emerging nationalism and nation building began to manifest themselves in various personifications. Simultaneously the move from mother Finland to an attractive young maiden began to take place. The concept of *Suomi-neito* (The maid of Finland) became common, and it was again Zacharias Topelius' *Book of Our Land* that was a key instrument in spreading this concept (Reitala 1983, 59).

During the period of autonomy, and especially during the years of

Figure 6.2 Two views depicting the socialist ideological landscape in Vyborg, the city ceded by Finland to the Soviet Union as a consequence of World War II (photographed by the author in 1983)

oppression, much political art was created to illustrate Finland's relation to Russian hegemonic power. The relations were often illustrated in the form of a contrast between a young Finnish woman and ugly, bearded Russian men. In this way the derogatory stereotypes of Russians began to appear. The most prominent political expression of this relation was probably Eetu Isto's painting *Hyökkäys* (The Attack, 1899), which depicts a symbolic struggle over the law between the two-headed eagle and the maid of Finland (Figure 6.3). It has been argued that the effect of this painting and the pictures of it that were distributed among the Finns was considerable as far as the formation of public opinion was concerned (Immonen 1987, 64). In some cases it also became typical to present the Topelianic–Runebergian national landscape together with the Maid of Finland (Figure 6.4). When the Finnish state became independent, the role of these allegorical personifications declined in response to the rise of the official symbols of state. Nevertheless the female allegory of Finnishness continued its existence in numerous tourism advertisements as well as in humorous 'national' cartoons (Reitala 1983).

National symbols form a hierarchically organized textual entity in the language of integration and difference. The oldest 'layer', representing the symbolic language of the oldest societies, consists of coats of arms, flags and decorations and, subsequently, 'great men', commemorative days, national anthems, poems and epics. As Klinge (1981, 280) points out, national symbols simultaneously express and direct the forms of national identification.

The red and yellow coat of arms depicting a lion and the white flag with a blue cross became key symbols of Finland, even though those who chose them were not unanimous at the time. The symbolic and ideological role of these can be seen in the fact that the coat of arms, for instance, is protected by law and cannot be used as a symbol for any private enterprises, political party or association etc. (Klinge 1981, 242–4). Finland's national symbols are typical examples of abstract symbolism which does not directly represent the history of the state. Rather, they are metaphorical expressions of power. As Duchacek (1975, 40–1) writes, few countries lack national symbols — such as lions, eagles, tigers etc. — whether or not such strong wild animals are to be found in the country. This is also the case with the Finnish lion.

6.1.4   The representation of Russia and Russians in Finland

In spite of numerous national symbols that express Finland's internal ideologies, it is an undeniable fact that the relations between Finland and Russia have been an essential constituent in the story of Finnishness, and in many cases the boundary between the two states has been regarded as an icon of these relations. An attempt will now be made to trace the content of the images of Russians and Russia as the Other, as they have been represented in various texts. These are expressions of socio-spatial consciousness, a part of the collective mentality of the nation, and a complex of values and attitudes. It is

Figure 6.3 An example of the use of allegories in the construction of symbolic space. The maid of Finland fighting with the two-headed eagle symbolizing Russia, in Eetu Isto's painting *Hyökkäys* (The Attack), 1899. Source: National Museum of Finland. Reproduced by permission

Figure 6.4   Two views of the allegory of the Maid of Finland. Left, the Maid and a stereotypic Finnish landscape in 1905; and right, the Maid of Finland in a popular poster circulated in 1948 for promoting tourism. Source: left: National Museum of Finland, right: Helsinki City Museum. Reproduced by permission

usual to regard collective mentalities as being fairly stable and to believe that they change relatively slowly. In fact, as far as the stereotypic socio-spatial representation of these attitudes is concerned, the discourse can change quite rapidly. This is owing to the fact that the structures and their representations are never 'floating' above people and the power relations constitutive of them, so that when political passions require it the representations can also change, as indicated by the discussion on the nature of stereotypes.

Thus, when the old Soviet system still prevailed, Raittila (1988, 11) argued that the image of the Soviet Union that the Finns carry in their social consciousness was firmly connected with the ideological and political side of their world view. Hence the words 'Soviet Union' would stir up many more political connotations in the mind than Sweden, for example. All this was because of the fact that the Soviet Union was the undisputed figurehead of the Eastern bloc and a concrete manifestation of all its ideological and symbolic implications.

The ideological and political aims in the construction of the world view and stereotypic attitudes towards others have usually arisen from the internal political motives of states. This was the case with the Soviet image in Finland, and it also holds good in the case of the images that have prevailed in the United States in the course of time, for example. The threat associated with the Soviet Union was emphasized above all at times when the political situation in the United States demanded it, i.e. internal causes may have been behind this image building. The ideological implications become obvious in the illustrations given by Dalby (1988, 1990) of how a series of 'security discourses' (Sovietology, realist literature in international politics, nuclear strategy and geopolitics) were exploited ideologically in the USA to construct an image of the Soviet Union as a dangerous 'Other'. This is a fitting illustration of the claim put forward by Neil Smith (1990, 177) that the struggle for space in an area ranging from popular to the most philosophical discourse is acutely political.

Hence, it would be a serious simplification to believe that there is a single 'image' or representation in a society which the power-holding élites can change when necessary. In fact, there are always competing representations and ideologies put forward by rival social groups. As indicated in a number of investigations, the Soviet Union was a model and source of inspiration for the extreme left in Finland between the World Wars, while for the extreme right it was a manifestation of all possible evil and an enemy. As regards the potentials of these ways of thinking for shaping the content of publicity, Immonen (1987, 429–32 *passim*) writes that it was the anti-Soviet groups that largely defined the boundaries of the publicity and hegemony of the images of the Soviet Union. The largest publishing houses supported these ideas and had a powerful influence over the circulation of publications. The most important barrier for alternative publicity was that pro-Soviet comments would be the most effective way of inducing unfavourable publicity. The facts needed in foreign policy and commercial policy maintained a low profile, and the aims of military policy were in keeping with the aggressive view of the Russians.

The representation of Russia, and later the Soviet Union, has been mainly derogatory in Europe (and also in the United States) for hundreds of years, although more favourable appreciations, amounting to Russophilia, have been put forward in some countries at times (Luostarinen 1986, 54). Many of the derogatory stereotypes that have characterized the representations of the Russians have emerged from experiences of Russia's expansionist policies. Such representations are thus expressions of Russophobia, fear regarding the aims of Russia (or the Soviet Union) as a state and of its means of achieving these aims (Immonen 1987, 40). The basic idea of Russophobia can be see in a book by Essén (1941) from the years of World War II, in which Russia is described as vague and different. This attitude persisted throughout the time of the Cold War, and lasted in the United States, for instance, until the end of the demise of the Soviet Union, as it was this that was understood in the security discourse as the expansionist Other (cf. Dalby 1988, 422).

The Finnish image of the Russians during the Grand Duchy period was generally favourable. Immonen (1987, 48–59) claims that it is possible to detect different interpretations at different periods, but no hatred of Russia or the Russians as such can be found before Finland gained independence. This is owing to the fact that the early Finnish identity was partly constructed on the basis of opposition to the dominant Swedish culture (Paasi 1992). Finnish national poetry aroused anti-Russian feelings at times, but Topelius tended to depict Russians in fairly favourable terms in his *Book of Our Land*, even though his aim was to make a clear territorial distinction between the two states.

Nevertheless it was only during the 'years of oppression' in the early part of the twentieth century, when the Russians introduced new laws aimed at Russification of the autonomous Finnish culture, that anti-Russian feelings began to be directed at their administrative practices, e.g. in underground publications (see Immonen 1987, 66). All in all, there existed a number of attitudes towards Russia and the Russians in Finland during the Grand Duchy period and they varied in time and space as well as from one group to another (1987, 69).

A radical change took place in this representation after Finland gained its independence in 1917, based on the longstanding European Russophobic tradition and partly reflecting Swedish–German influences in which Russophobia was strong (cf. Korhonen 1966, 34–41). The early social integration and intensification of publicity in the independent Finnish state was to a large extent based on an 'enemy image'. Finnish–Soviet relations before World War II were characterized by mutual hostility, distrust and a minimal amount of political, economic and cultural interaction. Klinge (1972, 57) writes that the hatred of Russia which emerged in the 1920s continued throughout the 1930s up to the war years. This hatred was partly systematic action organized by the Academic Karelia Society and its secret group known as the Brothers of Hate. Klinge (1972, 70) argues that the most important incentive for this attitude was the fact that Finland had become independent and there was consequently a *boundary* between the two states, something that simply had not existed during the nineteenth century: 'Finland did not have boundaries as it now has. People did not think of a winding red line across Lake Ladoga and extending northeast from it.'

Strong anti-Russian and anti-Bolshevik feelings were instilled into the consciousness of citizens between the wars in the form of a myth of an eternal struggle between Good and Evil (Hakovirta 1975, Luostarinen 1989), and if the Soviet Union was experienced as a threat to Finland, the feeling was the same in the Soviet Union. These attitudes were maintained by the international political situation, the press and other opinion builders (Vihavainen 1988), and it is revealing that during this period the Soviet ambassador claimed that the Finnish press was the most hostile in the world as far as relations with the Soviet Union were concerned (Puntila 1971, 158).

In brief, the Soviet Union as a whole was depicted as the eternal hereditary enemy of Finland, which posed as a bastion against this threat to Western civilization and Christianity. These attitudes became especially apparent during the 1930s, and were instilled into the consciousness of the Finnish people through various forms of publicity, not least novels and poems. This was also the case with the teaching of geography in schools (cf. Paasi 1992). According to Immonen (1987), the Soviet Union was the number one question in Finland during the 1920s and 1930s. In a peculiar way it was entangled with everything possible — foreign policy, religion, the church, the question of the Finnic tribes, national defence, etc. The essential feature of the representation of the Russians that was created in 1918–44 was its mythical side, the myth of a Russian threat and the myth of the Russian character (Raittila 1988, 18–19).

Territorial attitudes are commonly shaped through what can be called territorial indoctrination: the use of a specific territorial iconography to reinforce integration and create distinctions (Bergman 1975, 44). Since this typically takes place through the education system, it is instructive to follow the changes in these attitudes as they were presented in school geography textbooks, since these were among the most effective systematic vehicles for national socialization — much more so than today, with the existence of multimedia. There were doubtless many other significant media involved in the construction of these images (e.g. books of adventures written in a strong anti-Russian style, see Immonen 1987), but the textbooks reveal the development of such representations (and the social consciousness) more systematically than these more intermittent publications.

It was found in the above analysis of Finnish school geography textbooks that the self-image represented and reproduced in them emphasized the cultural homogeneity of 'us' and of 'our' distinction from 'them'. The following discussion will concentrate explicitly on the presentation of the Soviet Union/Russia in these books. It will also attempt to interpret the context in which such references are found.

On the whole, it is easy to conclude from an analysis of the textbooks that until Finland gained its independence, i.e when it was still under

Russian rule, stereotypic attitudes towards the Russians were in general fairly favourable:

Russians are persevering, hard-working people. They are happy, they love social life and are hospitable and fond of singing, music and dancing. (Dannholm 1889)

This form of statement was partly a consequence of the fact that the education system and geography as a subject were objects of more or less effective 'Russification' and control during the period of autonomy under Russia (Rikkinen 1978a, 10, 84). Specific textbooks on the geography of Russia were employed in schools and they were generally favourable towards the Russians as well as actively aiming at changing some of the older deprecatory stereotypes, although such stereotypes were still not uncommon in these books (see Aro 1913, for example). This was a constant problem for the Russian authorities during the years of oppression.

Bobrikov, the Governor General of the Grand Duchy of Finland, was especially aware of the power of geography and history in the construction of social consciousness and identity — and of Otherness — and aimed to control the content of history and geography teaching, since he argued that education would produce generations that had become alienated from Russia and were even hostile towards that country. Since logically the common 'fatherland' was Russia, the works of Runeberg and Topelius, like geographical textbooks, were particularly harmful from the viewpoint of active Russification. These books usually represented Finland as a separate state, with its political relation with Russia dismissed in a few words. A committee was set up in 1903 to alter the content of textbooks to show Finland's political position correctly, but its activities did not change the educational situation (Polvinen 1984, 193–212 passim).

After gaining independence, and especially during the 1930s, when the 'enemy image' became more common in the representation of the Soviet Union and Russians in Finland, the attitudes put forward in books also became more aggressive, and a more explicitly anti-Soviet and anti-communist atmosphere became evident. The attitudes may be said to have contained some evidence of the 'hatred of the Russians' that had arisen among the extreme right when Finland gained independence and that had prevailed in the country throughout the 1920s and 1930s, a hatred that was in effect an ideological phenomenon directed against communism, but it was at the same time an attempt to translate the internal class struggle manifested in the Civil War of 1918 into a struggle between Finland and the Soviet Union. The principal promoters of this way of thinking, the Brothers of Hate, even planned to revise Topelius' *Book of Our Land* and the accepted histories of Finland in accordance with their views (Klinge 1972, 66). Thus highly disparaging descriptions were written of both the Soviet social system and its inhabitants:

> The Soviet Republic is not a real republic. All the power is held by a small group which governs absolutely. Only one party, the Bolsheviks (communists), enjoys the rights of citizenship. . . Russia is enormously rich in its natural resources, but because of the ignorance and laziness of the people many of the sources of income are left unused and the people are often starving. (Hakalehto and Salmela 1936, 126–7)

The geopolitical argument, which was influenced by ideas prevailing simultaneously in German geopolitics and which reflected the old Ratzelian heritage in its employment of organismic analogies, is also evident in the pages of the textbooks. These ideas had already been present in Finnish political geography in the middle 1920s, but it was only at the end of the 1930s and during the war years that they became more explicit features of geographical education (Rikkinen 1978a, cf. Paasi 1990). The most explicit geopolitician in the history of Finnish geography, Väinö Auer, was also famous for his school textbooks. Auer and Poijärvi, for instance, wrote as follows:

> The Republic of Finland is thus, as far as its structure and function are concerned, a gigantic uniform organism, just as a human being is a uniform organism. A specific human being, animal or plant, which has a living organism, is called an *individual.* In the same manner a state can be regarded as an individual, since it is in many respects similar to a living organism in its structure and function. (Auer and Poijärvi 1937, 12)

The majority of the Finnish-speaking cultural intelligentsia were favourably disposed towards the new Germany after 1933. Critical attitudes were scarce and were underlined by a fear of the communists. Scientific contacts with Germany were also close (Paasivirta 1984, 433–4). Even though the official political connections with Germany were limited, Finland's cultural connections with Central Europe, and with Germany in particular, were clearly evident in some circles (1984, 417). These connections also manifested themselves on the pages of the school geography textbooks. Geography teachers had connections with Germany during the 1930s (Rikkinen 1978a), and thus Professor Iivari Leiviskä depicted the English in a somewhat unfavourable light but was more enthusiastic about Germans:

> Germans are talented, hard-working and energetic, and thoroughness and method are among their strong sides, too. These qualities characterize German economic life, administration, education and science. Germany is among the

first civilized countries and all kinds of civilizing influences have spread from it
to other countries, including our country. Due to its geographical location,
Finland has always been in closer contact with Germany, being located on the
shore of the Baltic Sea, than with other leading countries such as England and
France, and as regards the principal languages, the Finns have most often
studied and have been able to use the German language, which has had deep
roots in Finland since the days of the Hanseatic time. (Leiviskä 1937, 43)

The following description of Poland is probably among the most
extreme indications of the *Zeitgeist* of the late 1930s:

There are a lot of Jews in the towns. They push themselves to the foreground
everywhere and occupy business and industry. The education of the people is
poor with the exception of the old German areas. (Hakalehto and Salmela
1936, 125)

After the defeat by the Soviet Union the content of Finnish school
geography textbooks became neutral, the markedly derogatory stereotypic material referring to different nations and peoples vanishing from
them for the most part (Paasi 1992). This was among the radical social
and political changes that took place in post-war Finland and a consequence of the censoring of anti-Soviet content in school geography
textbooks. All in all, the fundamental changes in the relations between
the two states led to new ways of representation in geographical education. Whereas in the case of the Western countries the descriptions
usually contained favourable stereotypic elements, the attitutes towards
the Soviet Union and the Eastern bloc presented in textbooks in general
became principally neutral after the war (Paasi 1992). Discussion of the
Finnic tribes living beyond the border soon disappeared entirely from
the pages of the books. It is only in recent years, since the collapse of
the Soviet Union, that information on the Finnish-speaking populations
of the former Soviet areas appears to have returned to the pages of
school geography textbooks. Compared with the expansionist, tribalist
tones of the 1920s and 1930s, however, the 'Finnic tribes' are now
represented in these books merely as a matter of historical fact.

As far as the aims of Soviet geographical education and the textbooks
used in education are concerned, an obvious fundamental ideology after
the war was to turn to the basic principles of dialectic materialism. The
purpose of education was to convince students of the superiority of the
socialist economic system over the capitalist one, to create Soviet
patriotism and proletarian internationalism, etc. (Heikkinen 1987, 15–
17). One basic idea in education was to create a picture of the uneven
conditions prevailing in capitalist states. Simultaneously the books
criticized the fundamental premises of capitalist societies and pointed
out the advantages of Soviet socialism and socialist states in general

(Mihailov 1950, Heikkinen 1987). Whereas the capitalist countries in general, and particularly the United States, are criticized strongly in the textbooks, Heikkinen (1987, 33–4) maintains that attitudes towards Finland were fairly neutral and the good relations between the two states were emphasized.

6.1.5 Changing representations in civil society

It has been argued above that social consciousness cannot be reduced to the sum of the ideas that individual actors bring with them. To complete the picture provided by various documents, the representations of the Finnish people regarding the former Soviet Union will be analysed. This will be done on the basis of survey material published by geographers, political scientists and sociologists, and thus the discussion will proceed from the socio-spatial consciousness manifesting itself in various documents to an analysis of the aggregates of individual consciousness (the latter being both constitutive of and constituting the former).

It is easy to conclude on the basis of the previous analysis that politico-ideological relations between Finland and the Soviet Union and now Russia have changed markedly since World War II. The period of the 1920s–30s, which was characterized by mutual hostility and 'enemy images', has given way to a situation of peaceful co-existence (Raittila 1988, Luostarinen 1989). This change first took place at the level of the two states, where the official 'language of friendship' was established after the war and has manifested itself in the unanimous character of the rhetoric of political parties, in which official foreign policy has been the point of departure (Borg 1970). What was at times questioned, e.g. by some parties which remained in opposition, was not the basic line of foreign policy but its use as a vehicle in internal politics, i.e. 'self-Finlandization'. A number of well-known politicians were already commenting on this in the 1970s (Vennamo 1970, 53–63, Salonen 1977, 16–17, Westerholm 1978, 199).

Relations were much more complicated in civil society, however, where the consciousness of individuals and social communities was strongly coloured in a time–space-specific manner, and where, after a certain delay, the content of the social consciousness followed the decisions and interpretations of official state policy. This was owing to the fact that social representations, reflecting a more general social consciousness, were typically galvanized by old, deeply embedded visions and judgments (cf. Moscovici 1981, 190). This delay has also been noted by many important politicians (Virolainen 1971, 26, Salonen 1977, 15). There were also differences in the attitudes prevailing among the various political parties, e.g. the anti-socialist ideologies of the right-wing parties (cf. Raittila 1988, 26).

It should also be remembered that the socialization of various generations before World War II into a world of derogatory stereotypes and anti-Russian feelings probably affected many Finns who will still be alive in 2010–20 and whose children will have received an influential education at home, in spite of the fact that the official aims of national socialization have changed radically as a consequence of the war!

Present attitudes towards the ceded Karelian areas are indicative in this respect. At the level of official state policy they were taboo until recently, but the idea of re-acquiring (leasing or buying back) these areas has emerged from time to time in the sphere of civil society. The recent radical changes in Eastern Europe, the dispersal of the Soviet Union, the possibility of visiting former home areas in Karelia and the 'rediscovery' of the Finnish-speaking peoples of the Karelian areas have revealed the hidden attitudes of many Finns towards Karelia. Nevertheless, a recent survey suggests that most Finnish people have an attitude towards the Karelian question that is in agreement with the official state policy: 68% of the Finns do not want the state to engage in any negotiations for the return of the ceded Karelian areas (Suomen Gallup 1992).

Surveys carried out just before the break-up of the Soviet Union indicate that the hostile attitudes of the Finns towards it had changed markedly since the war (Luostarinen 1989, Raittila 1988), and the authors put forward several reasons for this: the disbanding of many anti-Soviet organizations after the war, the increase in economic and political contacts between the two countries, cultural interaction, etc. Probably the most important fact has nevertheless been the contested 'Finlandization of consciousness'. Although there was no official criticism of the Soviet system, some individuals — sometimes very significant figures — made public statements at times. As has been pointed out, stereotypic, antagonistic attitudes have probably been much more evident at the level of civil society and everyday discourse (Klinge 1972, 58, Raittila 1988, 26).

In any case, the attitudes of the Finns became neutral or favourable after the war, whether through forced friendship or otherwise (Raittila 1988). It was also significant that the majority of Finnish people were not afraid of the Soviet Union as were those in many Western countries (Luostarinen 1989, 124). Although attitudes towards the Soviet Union in general had become more favourable, comparative space-preference surveys carried out in 1983, in which Finnish students were asked to rank countries according to their desirability as places of residence, tended to place it clearly as the least preferred. This was considered to be based on the simple fact that the Soviet Union was still the leading socialist state, the social system and culture of which were felt to be somewhat strange among Finnish students — even though tourism had brought the connections closer since the 1970s (Paasi 1984b).

Attitudes have changed again since the disintegration of the Soviet Union. A recent — rather technical — Finnish Gallup Poll indicates that according to the opinions of the Finns the most obvious course of future development in Russia will be a continuation of the present chaotic situation. Environmental problems are becoming more important in the socio-spatial consciousness of the Finns, inasmuch as Russian nuclear power stations and environmental conflicts appear to be the features of which they are now most afraid in Finnish–Russian relations. Secondly, the Finns are disturbed by the threat of the increasing invasion of Finland by Russian professional criminals (Suomen Gallup 1992). This organized crime seems at present to be the most concrete 'threat' from the East perceived by the Ministry of Internal Affairs.

It could even be argued that to some extent the Finns look forward to the future with excitement. Administrators and enterpreneurs in many communes located along the eastern border hope and believe that they can open communications with Russian areas, and in the case of some communes several actual roads, and initiate cooperative activities. The military leadership has recently been worried, however, about the strategic changes in military geography that will probably take place after new road connections have been built over the Finnish–Russian border, particularly in Northern Finland. It seems that no explicit threat from Russia is experienced in military circles, but old enemy images still exist, even though not explicitly directed at Russia (Joenniemi 1993). In any case, a future theme in this respect will be Finnish attitudes towards integration into Western Europe and possible forms of cooperation in the spheres of economics, politics and military activities. It is also possible that membership of the EU may cause stricter boundary control.

7

SIGNIFYING TERRITORIALITY: THE CHANGING ROLES OF THE FINNISH–RUSSIAN BOUNDARY

7.1 THE FINNISH–RUSSIAN BOUNDARY AND CIVIL SOCIETY

The rise of the territorial principle as a part of Finland's nation-building process and the factors contributing to it have been sedimented into a number of social practices (politics, economy, administration, education etc.). The institutionalization process will now be discussed explicitly by concentrating on the territorial shaping of Finland through an analysis of the specific representations of the Finnish–Soviet boundary and how these have been exploited to signify territoriality. The representations also illustrate the changing ideas of geography, whether put forward by professional geographers, geography teachers or other actors in society.

It has already been seen how the boundary between Finland and Russia has changed several times as a consequence of strategic aims and wars, and it is clear that its role in civil society and the socio-spatial consciousness of Finnish society has varied greatly over the years. Up to the eighteenth century the question of the existence of a boundary was a very vague one even for those who lived in the area: in fact there were people living in the frontier zone. It can be argued that a formal border was created when the Finnish Customs Department was established in 1812, and particularly when some customs stations were established to control goods traffic between Finland and Russia (Hämynen 1993, 168). Nevertheless, these stations were local manifestations of the increasing control over space rather than expressions of boundaries dividing the space. In a way, they were an expression of the role of state interaction in the gradual development of a frontier into a boundary.

In civil society there were many contacts across the boundary. During

Figure 7.1 The Finnish–Russian border was open during the autonomy period. The well-known Finnish photographer and ethnographer Samuli Paulaharju (second from left) and his colleagues resting at the border post on their way to the Dvina in 1910. Source: National Museum of Finland. Reproduced by permission

the nineteenth century many students went to Russia to continue their education and stayed there to work after graduation. Migration to Russia expanded markedly after Finland became a part of Russia in 1809, and as many as 1000 inhabitants a year migrated there. During the period 1852–8 as many as 100 000 Finns visited Russia, which was a considerable number since the whole population of the country had only just passed 1.5 million during the 1840s. Cultural cooperation between Finland and Russia was also strong during the Grand Duchy period, and the Russian authorities made an active effort to integrate the Finnish intelligentsia. Many scientists, for example, worked in Russia during that period (Immonen 1987, 56).

In general, people moved freely and had commercial contacts across the boundary (Mead 1991, Laine 1992, 1993, Figure 7.1). An orientation towards Russia was natural for the Border Karelians since there were people on the other side of the boundary who spoke the same language and professed the same religion. People moved over the boundary when searching for a job and, particularly in the Orthodox areas of Karelia, it was common to marry according to religious rather than formal state boundaries (Hämynen 1993).

The boundary in Karelia was in a sense located functionally in the 'wrong place', for as far as cultural and social activities were concerned, the spheres of influence of Russian culture extended into the Finnish territory (Hämynen 1993). The boundary was not a significant barrier to the activities of local daily life, and this relatively open situation continued during the early years of independence, as a book written by Sakari Pälsi, an anthropologist and explorer, demonstrates, as he describes how, as a war correspondent, he 'slipped' over the boundary almost unnoticed (see Pälsi 1922). When the border was closed, many Border Karelians moved to other areas in Finland and above all to Russia, since it was no longer possible to search for a job or buy grain in Russia (Hämynen 1993).

The nature of the Finnish–Russian boundary was already being discussed by some authors during the Grand Duchy years, however. Ignatius (1890, 60), for instance, in his *Geography of Finland*, considered the country's boundaries in detail and pointed out how Finland's eastern boundary was very sinuous and did not follow the 'dictates of nature'.

When Finland became an independent state, this inevitably produced a change in its territorial strategy, so that it had to secure its boundaries and use them to signify the territoriality of the state. The boundary confirmed in the Peace of Tartu (1920) followed mostly the same line as the boundary of the Grand Duchy, but its practical role was completely different. State authority was established along the border areas of Finland and Soviet Russia through a system of border guards. The army and civil guards undertook this task during the Civil War, and a formal system was established in summer 1918. A specific Border Guard Detachment was established in 1919, and the patrolling of the boundary became more efficient, but particularly on the part of Soviet Russia this was somewhat theoretical for a long time (Peltoniemi 1969, Vahtola 1988, 319). The Border Guard Detachment was permanently established only in 1931 (Kosonen and Pohjonen 1994, 157).

Whereas the previous boundary may be called a 'formality', the new boundary was above all a political construct (Hämynen 1993). While it had previously been possible to interact and communicate across the boundary, this was no longer allowed. The new boundary was in principle closed — but only in principle, inasmuch as illegal interaction across it continued and at times caused diplomatic conflicts. A lot of Finns and some of the workers from the border areas moved into Russia during the Civil War, and the Russians claimed that the Finns should close the boundary to avoid these conflicts. In 1922 an agreement was signed whereby peace was secured in the boundary areas, and this finally closed the boundary. Nevertheless it did not entirely prevent violations, which continued in 1923 and particularly in

1930, when the representatives of the nationalistic movements forcibly ejected some Finnish communists over the border into the Soviet Union (Korhonen 1966, 73, 203–4). A second wave of illegal immigration took place during the economic depression of 1929–32, and as many as 12 000–15 000 people defected illegally from Finland to the Soviet Union, more than 7000 doing so in 1932 alone (Hämynen 1993, 313).

On the whole the closing of the boundary meant a radical change for the less developed areas of eastern and southeastern Finland. Commercial connections with the east were severed, especially with the economic sphere of influence around St. Petersburg.

The special character of the Finnish–Russian border had already been realized at the end of the nineteenth century, particularly during the years of intensive Russification. These actions stirred up a wave of Finnish patriotism, which also sought to strengthen national feeling in the border areas, and finally turned its attentions towards gaining independence (Palosuo 1983). As we have seen above, the generations of the 1920s became well known for their activities in the Academic Karelia Society, and their relations with Finnic tribes and the concept of a Greater Finland. 'Suomi suureksi — Viena vapaaksi' (Finland greater — the Uhtua region free), the title of a poem by Ilmari Kianto (1918), became a well-known slogan in these affairs.

During the 1920s the specific role of the boundary itself and the areas located near it became important in activities in civil society. Immonen (1987, 317) is prepared to conclude that the boundary became a symbol through which both the distinction between Finland and Russia/the Soviet Union and their dependence on each other were expressed. Waterways, and particularly the River Rajajoki beside which a railway connection ran to Russia, became symbols of the boundary in the 1920s and 1930s, and of the role of nature in the formation of boundaries (see Figures 7.2 and 7.3). Fig. 7.3 shows how the signs of the socialist ideology were gradually taken into use to symbolize the Soviet side of the border landscape at Rajajoki station. The Rajajoki bridge became an internationally famous 'border between two worlds', the place that Western tourists often visited. Russian refugees labelled the place as the 'gate to hell' (Kosonen and Pohjonen 1994, 189). Paavolainen (1930, 40) characterized the border area of Rajajoki as follows:

A kind border guard gives advice to the traveller as to where he is allowed to go. From here he can see a piece of another world, he can see on the other side of the bridge a genuine Red soldier walking with a rifle on his shoulder but not with his finger on the trigger, as good relations prevail nowadays between the two neighbouring states. The boundary in nature is not broad but the distinction between the worlds located on either side of the river is greater. On one side is the west, and on the other the east. On the one side chaos prevails, and on the other an organized, legal society.

Figure 7.2 The idea of a 'natural boundary' became important in the nineteenth century, and this was later exploited by expansionist state ideologies. The picture depicts the Finnish–Russian boundary in the realm of nature in the form of the River Suojoki. Finland is on the right bank, Russia on the left. Source: Enckell 1939, 64. Reproduced by permission of Patrick Enckell

The relations between the two states soon engendered a boom in literature about the Karelian question, including the Finnish tribes beyond the border. The ideas of Greater Finland became an essential theme in such books (Räikkönen 1924), and in many cases the literature revived the myth of the eternal conflict between Western civilization and Oriental barbarism (Immonen 1987, 338). Together with other material, such as school textbooks and newspapers, this doubtless shaped the social consciousness of the Finns. Patriotic poetry also referred continually to the eastern danger. Probably the best-known poem of this kind is 'On the Boundary', written by Uuno Kailas in 1931. The following extracts from the poem together with its illustration (Figure 7.4) in Kailas (1941) are also indicative of the spirit of the socio-spatial consciousness of the 1930s and the role of Finland's eastern border in this consciousness:

The border opens like a crack.
In front of us is Asia, the East.
Behind us is the West and Europe;
I will guard it as a sentry. . .

. . . A nightly, howling wind brings
snow from beyond the border.

Figure 7.3 Transformation of the ideological landscape. The River Rajajoki between Finland and the Soviet Union became an important symbolic and ideological landscape during the 1920s and 1930s. Sources: above: SA-kuva; below: National Museum of Finland. Reproduced by permission

Figure 7.4 The ideological confrontation between East and West was characteristic of Finnish society between the world wars. The illustration to the poem 'On the Boundary' from Uuno Kailas' collection of patriotic poems is a typical example of this. Source: Kailas 1941, 64. Reproduced by permission of Werner Söderström Osakeyhtiö, Helsinki

> — My father, mother, Lord, let them
> sleep with solemn dreams. . .
>
> . . . Gloomy, cold is the winter night,
> the East breathes ice.
> There is slavery and forced labour there;
> which the stars behold. . .
>
> . . . But fathers from their graves
> are riding with their phantom mounts:
> with bear spears in their hands,
> they charge towards the border. . .
>
> . . . The iron sole of the enemy
> won't step insulting
> the place of your heroic rest —
> I will protect the border of my country!

The closing of the border also generated systematic work on behalf of the border regions on the part of some activists: to lift them to the level of the other Finnish regions and by this means to provide a single ideological basis for social integration in the country (Figure 7.5). This development activity was regarded as important by the activists because the boundary was partly a symbol of fear and disappointment, partly a reflection of national romanticism and partly an important inspiration

Figure 7.5 'Doing the washing in front of the boundary marks' at Rautu. The establishment of the border between Finland and Soviet Russia made everyday life complicated in the border areas. Source: Photo Archives of Otava (published originally in *Suomen Kuvalehti* 1933). Reproduced by permission

for organized social action (Korhonen 1973). In brief, the role of the boundary areas in the nation building process was a crucial one.

After the Finnish state gained its independence, the most important motive for this activity was to guarantee its security. The efforts to develop the border areas, and particularly agriculture there, were also supported by the Council of State, beginning in 1923 (Palosuo 1983). These aims were obvious expressions of an intention to 'nationalize' the peripheries of the social space that Finland was now defined as constituting (cf. Augelli 1980, Kliot and Watermann 1983). In principle this could be implemented in three ways: through efforts to strengthen internal identity, to lessen external influences and by the pursuit of both of these strategies. The territorial motives to improve the living conditions in the border areas of Finland were originally based firmly on the idea of promoting the national level of social integration. This called for reducing people's consciousness of the distinctions between rich and poor, owners and workers, while simultaneously making a clear external spatial and social demarcation between 'us' and the 'Other'.

According to the well-known slogan of Benedict Anderson (1991), a nation is an 'imagined community', which has to be constructed, symbolized and legitimized. During the historical development of nations the idea of community has been indissolubly linked with the land on which such 'communities' have developed, and as Taylor (1994) points out, this has completely changed the nature of territory and particularly the integrity of its borders: it became the state's duty to defend the national homeland, as was also the case in the integration of Finland's border areas after the gaining of independence.

The voluntary work could not at first provide much for the people in the border areas. At Christmas 1934 a candle and a miniature Finnish flag were delivered to each of the inhabitants of the border areas to remind them through these abstract symbols in which country they lived. The abstract character of the symbols of national identity in relation to concrete everyday life is characterized by the following citation: 'The Christmas candle twinkling on the table, and the flag beside it with its blue cross, the symbol of the nation, directed their thoughts, at least for a moment, away from the everyday cares with which they were burdened' (Vainio 1958, 14). During the second half of the 1930s the development work became more significant, anticipating, in a sense, the creation of a systematic regional policy, which was to take place in Finland only in the late 1960s (Koljonen 1985).

As an indicator of the changing representations of the boundary, an analysis will now be undertaken of the discourse put forward in the journal *Rajaseutu* (Border area), which was established in 1924 by 'Suomalaisuuden liiton rajaseutuosasto' (the Border Area Department of the Association of Finnishness). The journal was to provide an organ for promoting the work on behalf of the border regions. *Rajaseutu* probably offers the best systematic material for analysing the representation of the boundary as manifested in action explicitly connected with Finland's boundaries.

The journal *Rajaseutu* has been published regularly 6–12 times a year since 1924, and its idea of border areas has embraced not merely the eastern boundary but also Finland's boundaries with Sweden and Norway, as most of the articles deal with general development problems in border areas. Attention is particularly directed to the representation of the Finnish–Russian/Soviet boundary, however, and to the changes in representation associated with broader ideological and social considerations. The representation of this boundary is in any case completely different from that of Finland's other boundaries.

During the first two decades of the journal's life almost every number contained a leading article or other articles in which a persuasive rhetoric was employed to emphasize the role of boundary areas in *national integration*, i.e. these areas were seen not only as a local-scale problem in the national territorial structure, but essentially as a national question which was deemed to be of crucial importance for the identity of the newly established state. This attitude seemed to be based on the argument that the nation building process calls for complete occupation of the national space, which ends expressly at the border line. Thus the boundary was considered to constitute an essential element in social integration, one which was necessary for the nation building process. In practice this implied above all the Finnish–Russian boundary, though this was not often discussed explicitly. The challenges of social integration and the identification of the national space were clearly expressed in the leader of the first number of the journal:

The question of border areas is an urgent one for us. It is an important question for the nation and the country. It is not a matter merely for areas adjacent to boundaries, but is a common matter for the whole nation, the independent Finnish state. The emergence and development of the boundary areas is particularly important as far as the maintenance of independence is concerned (1924, 2).

It is even possible to talk about 'boundary based state-regionalism', which aims to develop boundary areas to the level of other parts of the country, not for the sake of these regions themselves but for the sake of the whole state. The population of the border areas was in many cases regarded as 'morally rotten in their political opinions' (Kosonen and Pohjonen 1994, 49). The authors firmly emphasized in several articles how the boundary area should be inhabited by reliable, patriotic Finnish people — as this was seen to be the best way of preventing the 'danger' arising from 'strangers' (see Paavolainen 1925). It is recounted, for example, how the Finnish Service Club arranged instruction for the border guards to stimulate their patriotism, social integration and will to defend their country, and at times the national question was 'illustrated' by pointing out that it is artificial to reduce the question of border areas to the interests of parties, classes or language. Thus the aim was to emphasize national interests and to push some of the key conflicts of the 1920s into the background.

All in all, the ideological perspectives provided in the 1920s were highly *internally* coloured. They concentrated on creating social integration and to some extent on establishing a clear distinction — even hatred — between Finland and Russia. There was no room for international perspectives on the pages of the journal during the 1920s and 1930s (cf. Paasivirta 1984). The task of the border area was interpreted as a crucial one: to stand as a 'defensive wall against armed attack and spiritual plague' (pseudonym H. L. 1927, 211). The spirit of the 1920s is probably summed up well in a poem the 'Goalkeeper', published in 1925, of which the following verse is an extract:

Like a goalkeeper you stand, the people of the Isthmus,
against you is the threat of danger from the former oppressor.
He pushed, trod down, demoted you in his time,

he is not tired of being on the prowl for his prey.
Goalkeeper, goalkeeper, people of the Isthmus,
when the time comes, Finland will cast her eye on you —
you can bring her victory or defeat!
Go with honour, Isthmus, to your watching place,
strain your eyes and stretch your tendons,
and step with a gallant forehead to meet the danger —
if you can hold out, so will the happiness of your native country.
Goalkeeper, goalkeeper, the people of the Isthmus
you stand as a guard over the freedom of your land —
you can bring us victory or defeat! (pseudonym Tyyne P. 1925, 66).

A new social policy for work in the border areas was introduced in
the 1930s, but the principal aim was still the same as in the 1920s — to
increase the welfare of those who formed the outpost for the country
and by this means to increase the credibility of the inhabitants living in
these areas. This was necessary in order to strengthen the social
integration of all the Finnish people. Brander (1932), for instance, was
worried about the 'less developed population' along the border areas
who were still not able to understand why the boundary was closed and
why they had lost their privilege inherited from time immemorial to
trade across this curious boundary which divided members of the same
tribe. There were many reasons why these areas should be developed.
Infant mortality was five to six times higher in the border regions than
in the more developed areas, for example — mainly because of the lack
of nourishment and care. Tuberculosis was still common in these areas,
while it was disappearing in more developed areas (Vainio 1958, 8–9).
    Some authors were deeply satisfied that the boundary, the gateway
between East and West, was now closed. Its symbolic role became even
more explicit in some areas during the 1930s, since a 10 metre clearing
was cut into the forests and a barbed wire fence built along the border
line itself (Hämynen 1993).
    Although the central theme of social integration within the Finnish
nation was to emphasize its internal strength, a new feature appeared in
the rhetoric. What some writers were now worried about was that it
was not a boundary between different states but that it divided the
Finnic tribes into two groups. Religious aspects also became more
common, so that some authors expressed their anxiety about the
religious situation of their tribal brethren beyond the border, and about
whether it could prevent 'infection' crossing the border, behind which 'a
happy world was under construction without God' (Kärnä 1932,
Kukkonen 1933, Sormunen 1936, 1940).
    The unsettled world situation also manifested itself in the articles
published in Rajaseutu during the 1930s. The questions of the internal
consolidation of the nation-state and of the relation of work on behalf

of the border areas to national defence were discussed during the second half of the decade, and attention was concentrated more on defence problems particularly in 1939, the year that the Soviet Union attacked Finland. According to some authors there was a continual need to strengthen the 'living wall' located along the boundary (pseudonym A. E. K. 1939, 86). As far as the political connotations of the discourse have been analysed, the journal emphasized the neutrality that Finnish politicians had been putting forward between the two world wars, and the hope was that relations between the Finnish and Soviet leaders would improve.

As far as the outcome of the Winter War was concerned, the journal continued to follow its basic approach, i.e. stressing the development of the border areas but now more explicitly in cooperation with the military authorities (pseudonym A. E. K. 1940). It was stated that 'the boundary is a boundary at all times, under all conditions, but the point of time in which we are now living is particularly significant for the border areas' (Leading article no.2/1941, 21).

The question of the Finnic tribes in relation to the boundary became significant in autumn 1941, when Finnish troops moved into Eastern Karelia and occupied it. The journal was favourably disposed towards the occupation. Its leading article emphasized that the aim was to correct the historical injustice which had confronted the Finnish nation in the east and southeast and divided the Finnic peoples in two halves with a boundary line. Furthermore, it was emphasized that the old boundary was difficult to defend and the aim was to move it to an area where it was easier to defend. The journal was also disappointed with the reaction of British politicians who had objected to the occupation.

When the Continuation War ended and Finland had to cede both the occupied territory and her existing Karelian areas, along with meeting other economic and political conditions, the content of the rhetoric towards the Soviet Union changed drastically. Hopes of regaining the ceded areas were gradually lost but geopolitical facts and the aim of national integration were still the same, e.g. in the work that the voluntary associations undertook after the war (e.g. Vainio 1958). Likewise national self-analysis still continued in civil society. Niiniluoto (1957, 206), for example, discussing the altered position of Finland, wrote that the Finns no longer argue about being the outpost of the West but neither are they a long-distance patrol behind the lines of the enemy. He wrote that culturally Finland was still on the borderland of the West but as regards its military position, it was now an outpost of the East towards the West — because of its treaties with the Soviet Union. Niiniluoto has probably found here the key reason why some Anglo-Saxon political geographers, for instance, located Finland fairly firmly in the Eastern realm after World War II. As we saw earlier, he also anticipated the Soviet ideological strategy that was to emerge during the 1970s and 1980s, which used all possible means to restrict Finland's military, political and economic relations with the West (Tarkka 1992, 200).

A remarkable change also took place in the discourse of the journal *Rajaseutu*. As in the case of Finnish political life in general, the journal changed its tune radically. The extreme nationalistic attitudes, the defiant discussion about the Finnic tribes, the discourse on the hereditary enemy and the extreme militaristic tones ceased completely. It was emphasized that now, more than ever, it was necessary to work for the boundary areas and their development. The same original principles still prevailed, but with more restrained forms of rhetoric and practice.

It is possible to argue that, whereas the kernel of the representation of Finnishness in the rhetoric put forward before World War II was strictly the principle of territoriality and the Finnish–Soviet boundary as the key manifestation of this ideology, the representation put forward in the journal during the late 1940s and particularly the 1950s changed completely. It is no exaggeration to say that in fact the boundary and what it demarcated was disregarded by the journal. The representation of Finnishness was no longer based on the gloomy enemy of pre-war territoriality.

Nevertheless the peculiar nature of the boundary areas was recalled in some rare cases by employing religious metaphors : 'People of the borders! Be glad that you are on the right side! Be glad, that you can internally pass over the boundary of sin, as the Gospel calls on you to do' (Heikinheimo 1953, 84). The journal now contained many contributions which aimed to teach people the correct habits, and to avoid 'materialism' in their lives. Similarly, it started to deal with the problems of regional development. This theme became particularly strong during the 1960s, the decade when official regional policy was established in Finland and when the effects of the structural change going on in Finnish society began to be seen.

The border landscape was not merely a neutral context after World War II. According to a recent historical study, the border area in the 1950s witnessed intense espionage activity, particularly on the Soviet side, and Western states used Finnish territory for their spying activities (Kosonen and Pohjonen 1994, 415–18). The authors conclude that Finland was a through passage area for both power blocs in the 1950s. Local people were in many cases recruited by the Soviets as informants in a search for knowledge regarding the organization of the border patrol system, the numbers of guards, local political opinions, evaluations of the Finnish political system and so on.

Where the above analysis shows that the boundary almost completely disappeared from the discourse of the journal, now, after the dispersal of the Soviet Union, it has returned. The boundary is becoming more open, and many Finnish communes located in the border areas are looking forward to a new and more prosperous future.

At the moment there exist altogether nine border stations or passport checkpoints. Five of them are official stations where it is possible to cross the border for various purposes, two serve the purposes of cultural exchange, group tourism, economic activity or friendship, and two temporary checkpoints allow crossing of the border related to the diversifying economic activities of the adjacent areas (Ervasti 1993). The communes of the border area are looking forward to a future in which this will really become an international boundary in Europe, an open boundary which does not prevent the movement of people, capital, goods or ideas.

7.2   FINNISH GEOGRAPHERS AND THE REPRESENTATION OF THE BOUNDARY

So far we have analysed the changing geography of the Finnish–Russian boundary as part of the transformation of the international system of states and in the context of emerging Finnish nationalism. A detailed examination will now be made as to how professional geographers and geography textbook writers have presented the 'geography' of the Finnish–Russian boundary and in what rhetorical, ideological and conceptual contexts they have grounded their ideas. Genealogically — i.e. when looking through the sequence of ideas that geographers have put forward — it seems that the central problem has been the relation between human societies and nature, one of the classical themes of geographical thought.

It is clear that regions or territorial communities — whether they are nation-states, provinces or other spatial units — always have their physical or natural basis. This has been one of the key points of traditional regional geography, but it also is a significant point of departure for the 'new regional geographies'. Hence Markusen (1987, 18) argues that the region, perhaps more than any other conceptual spatial unit, is the meeting ground of humanity and nature. Thrift (1983), for his part, discusses the physical basis of regions extensively under the general heading of 'topography'. It was found in the earlier chapters that nature and the environment in general can also be exploited for the ideological construction of national landscapes and identities.

The discussion on the physical basis of regions leads to a portentous theme as regards the history of geopolitical thought, i.e. to the question of the relation between physical geography and a territory and, hence, to the question of 'natural boundaries' and 'artificial boundaries'. These themes were particularly popular between the wars. By tradition, there has been much debate within political geography over the unsatisfactory nature of this division, since all political boundaries and frontiers require some selection and for this reason are artificial or arbitrary solutions to problems (Prescott 1965, 41, Glassner and de Blij 1980, 87). As Giddens (1987a, 51) proposes, inasmuch as borders are in principle nothing other than lines (or in fact vertical levels) drawn to demarcate state sovereignty, it is irrelevant what types of terrain or sea areas they pass over. In spite of this basic fact, boundaries may of course be important for the fortunes of a state in territorial conflicts.

The idea of natural boundaries originates from eighteenth-century France and its use of the new rationalist philosophy to claim a larger 'natural' territory (Pounds 1951, 1954, cf. Taylor 1993b). Hence the idea of natural boundaries has by tradition contained a strong ideological aspect. Glassner and de Blij (1980, 87) point out that the arguments on behalf of 'natural boundaries' are usually advanced by representatives or academics of states that wish to expand their territories. As was seen above, the Finnish–Swedish boundary was negotiated by the Russians in 1812 to be a 'natural boundary' as far as its location is concerned, but Glassner and de Blij ask whether anyone can imagine a state offering to move its boundary *back* to a 'natural' line deep inside its own territory!

President Paasikivi (1986, 6–7) stated after World War II that the idea of Greater Finland can be partly reduced to the national ideologies of J. W. Snellman, who argued that people talking the same language will sooner or later find each other. Snellman's argument is hence strictly a cultural one (cf. Vahtola 1988, 22). By contrast, the Finnish geographers grounded their arguments in facts from the realm of nature. Zacharias Topelius discussed Finland's boundaries in his geographical lectures during the mid-nineteenth century, long before political geography had an institutionalized form. He began his analysis by making an explicit distinction between the state of Finland and 'geographical Finland', pointing out that the 'natural boundary' in the east does not correspond to the state boundary. A 'complete Finland' was an entity that was bounded by water, by the Arctic Sea in the north and east and by the White Sea, Lake Onega and Lake Ladoga (Stenij 1937, 6, Tiitta 1994). Topelius also referred to the connections between the Finnic tribes on both sides of the boundary, and he remarked that 'even if the geography of state would never accept this idea, physical geography has already done it, since Finnish nature with its fauna and flora does not recognize any other boundary for its natural area' (Stenij 1937, 6).

Topelius regarded boundaries as important on the basis of his organistic philosophy and believed that a state needs 'natural boundaries' to establish itself and similarly 'human beings need bounds around them to maintain their soul as an entity'. Through these ideas Topelius popularized the notion of 'Greater Finland' which was later exploited in the field of geopolitics (Tiitta 1994). These thoughts were not unique in the Finland of the autonomy period but rather were an expression of a more general tendency to define natural, i.e. the botanical and zoological, boundaries for a 'natural Finland'. Therefore the question of the eastern border of 'natural Finland' was a significant topic in certain academic circles throughout most of the nineteenth century.

Topelius points to the fundamentally important fact that the territorial space of states is always absolute as far as their sovereignty is concerned. Geographical time and space, for its part, is a dynamic concept that is connected more with the stage and development of human practice and consciousness. Hence functional space is continually changing along with human efforts to develop communication networks, for instance, or along with the tendencies for economic, cultural and political integration and disintegration. The territorial space of the nation-states, for its part, is continually being maintained and symbolized by similar abstract means to embody the group solidarity of its inhabitants. When the idea of geographical space is employed in the case of nation-states, this is very much prone to ideological exploitation and persuasion (cf. Paasi 1986a).

A fitting illustration of ideological activity in defining geographical space is the book *East Carelia and Kola Lapmark: Described by Finnish Scientists and Philologists*, published in Finnish in 1918, an English version of which was published in 1921 in *Fennia*, the periodical of the Finnish Geographical Society. Of the Finnish geographers, J. G. Granö and J. E. Rosberg contributed several articles, together with the representatives of other disciplines. Particularly in the description of the Eastern Karelian areas, the authors discuss the boundaries of the area and the relations between the natural boundaries (floral, faunal) and those based on settlement history and culture. The conclusions were in general based on the idea of the legitimacy of geographical space over political space: 'Geographically, physically and ethnographically, this country forms with Finland one natural and continuous whole, while it is clearly and sharply divided from Russia . . . The Finnish race ought at last to be united into one whole, a single people . . . The portion of that people which dwells in Finland has, under the influence of a common historic fate, coalesced with the Swedish inhabitants of that land into *one* people, the people of Finland. Now at last the hour seems to have come when it could be united with the Finnish people of East Carelia into a single whole' (Homen 1921, 254, 255, 264).

Early Finnish political geographers were well aware of the international trends prevailing in political geography and of existing concepts and ways of thought. From a present-day perspective, their rhetoric was in some cases explicitly 'strategic' and violent. Such expressions as 'dangerous', 'possible war', 'defence boundary' etc. were engendered specifically by the existence of the neighbouring eastern state (Numelin 1929, Leiviskä 1938b). Hence their discourse partly reflected the ideological debate that some activists had aroused regarding the connections between the Finnic tribes and the 'non-equivalence' of the geographical space and the space of the nation-state (e.g. von Hertzen 1921, Räikkönen 1924).

After independence, Finnish political geographers formed a clear opinion of the character of the Finnish–Soviet-Russian border. It was regarded as an *artificial* boundary. Here the ideas of Finnish geographers partly reflected a more general discussion that had existed in the emerging field of political geography since the publication of Friedrich Ratzel's works. This concerned the dichotomy between artificial or constructed boundaries and natural boundaries based on seas, rivers, lakes, watersheds, nationalities and religions. In fact these themes had been dealt with earlier in Finland by Väinö Voionmaa (1922, cf. Voionmaa 1919), Professor of History, in his *Economic Geography of Finland*, when discussing the relations between Finland and 'natural Finland' (the Finland of the natural sciences) and suggesting that Greater Finland should include large areas beyond the existing eastern border. Furthermore, the question of Karelia was presented as permanent and indelible, a problem which 'would not cease to exist until it was solved in one way or another' (1922, 15–16).

The first Finnish political geographer, Ragnar Numelin (1929, 75), devoted much space in his *Political Geography* to the nature of boundaries. To some extent confusingly, he regarded the Finnish–Russian boundary as an absolute cultural boundary — not a geographical, linguistic or religious one. Here he was partly borrowing the ideas of Rudolf Kjellen, who considered cultural boundaries to be natural ones (Numelin 1929, 75). Numelin pointed out that the usefulness of various definitions of boundaries depends on the point of interest that lies behind the choice. Natural boundaries can be bad strategic boundaries and good strategic boundaries can effectively prevent cooperation — thus 'good' and 'bad' boundaries are relative concepts.

The boundary between Finland and Russia was, according to Numelin (1929, 88) an artificially constructed one, since it did not take into account geographical or national claims:

Even though it follows small rivers and natural waterways in some places, the boundary between Finland and Russia on the northern side of Lake Ladoga, i.e. in Central Finland and to a great extent also in Northern Finland, is a complicated artificial line which does not take into account the claims of nature but leaves some of the Finnish people, the inhabitants of Repola and Porajärvi, on the Russian side (Numelin 1929, 88).

Here Numelin was referring to the debatable areas of Repola and Porajärvi which the Finnish government claimed in the Peace of Tartu in 1920, because these Finnish-speaking communes had been left under Finnish control in 1918 and 1919 following the counter-revolutionary activities that had taken place in them. Nevertheless the Finnish troops had to be withdrawn from them and they became part of the Soviet Union, an integral part of Eastern Karelia (Jääskeläinen 1961, Vahtola 1988).

Iivari Leiviskä (1938b), Professor of Geography at the University of Helsinki and an eager activist in promoting Finnish as opposed to Swedish culture, also discussed the nature of boundaries extensively in his book on political geography. He emphasized the Western roots of Finnish civilization and created a long, 'Swedish-free' history for the country:

Finland received its original population from the southern side of the Gulf of Finland during the centuries before Christ was born. The Christian religion came here by sea and with it western civilization, and the later cultural influences also arrived by sea, Finland having received them directly, without the mediation of Sweden (Leiviskä 1938b, 288).

Leiviskä made a distinction between natural or structural boundaries and artificial or non-structural boundaries. The former are based either on the physical features of geography or on national and economic features, while the latter are boundaries that are not based on the structural features of geography but are drawn regardless of them. Leiviskä also indicated that this distinction is vague and even fallacious. He discussed the nature of the Finnish–Russian border in military terms and pointed out that the threat to the Finns had always come from the east and southeast. He also speculated about a possible attack from the Soviet Union and in a way anticipated what actually occurred a year later (Leiviskä 1938b, 290–93). This discussion also partly reflected the Zeitgeist and Finnish–Soviet relations at that time. Hence Leiviskä (1938b, 111), citing Coudenhove-Kalergi, wrote in Political Geography that the Soviet Communist Party 'was no longer a common party but rather a religious group whose Bible was Marx's Das Kapital, whose founder and prophet was Lenin, whose pope was Stalin and whose outlook on the world was materialism'.

Leiviskä (1938b, 18) also stated that it was a pity that the Finns had not achieved a more secure boundary in the Peace of Tartu by moving the southern part of the eastern boundary to the isthmus between Lake Ladoga and Lake Onega, so that the eastern boundary would have remained as a hinterland boundary, located away from major military actions. He also indicated that the state must see to it that the border areas are settled with 'reliable people' who must be supported both economically and psychologically.

The discussions of Numelin and Leiviskä appear to be highly aggressive and the intention to make social distinctions sticks out strikingly. Their language was undoubtedly a reflection of a more common social consciousness prevailing in Finnish society, although this view evidently matched the Soviet vision of the nature of the Finnish–Soviet border. The point of departure in this thinking was that the western boundary of the country was nowhere a natural one: every neighbour beyond the zone was a potential enemy, inasmuch as the governments of neighbour states were clearly anti-Moscow. For the Soviet authorities, Finland was the northernmost, geopolitically important link in this chain of enemies. It was regarded as aggressive and anti-Soviet (Korhonen 1966, 30–31).

7.2.1 World War II and the activities of geographers in the representation of the boundary

Finnish academic geographers were exploited by political and military leaders during World War II as never before or since, and geography as the spatial structuring of the world, as academic practice and as a conceptualization of the world in spatial terms became inextricably fused in the process.

As was noted above, Finnish political geographers employed the naturalistic terminology of traditional political geography and regarded Finland's eastern boundary as partly artificial, since it did not correspond to the border which was written in the realm of nature. It is difficult to say to what extent the use of this argument had ideological implications, i.e. attempts to create an image of the artificial nature (or *wrong location*) of the boundary, since the authors usually failed to present any grounds as to why in the case of nation-states a natural boundary is more legitimate and acceptable than an artificial boundary.

These ideological purposes are in any case explicit at least in several articles that were published during 1941 in a special number of *Terra*, the Journal of the Finnish Geographical Society, dealing with the area of Eastern Karelia. This was edited by Professor Auer and was published in an ideological situation in which Finland's political and military leadership were influenced by the 'Greater Finland' ideology and professional geographers and other scientists were partly involved in promoting this. The corresponding material was published in German in 1943 (Laine 1993). The explicit aim of its several articles was to prove, particularly to the Germans, that the Finnish territory did not correspond scientifically to the 'real' territory which was written in the realm of nature. The real boundaries should be located deep in Soviet territory, in so far as geological and other natural features extended there (Paasi 1990). Hence the basic aim of these articles was rhetorical, especially since the ideological basis of this argument can easily be extended to suggest that on geological grounds Finland's territory should also include most of Sweden and large areas of Norway, although none of the authors considered this point in their papers (cf. Poroila 1975, 42).

Terra No.4/1941 consists of seven articles, six dealing explicitly with the area of Eastern Karelia and each of them trying to prove — implicitly or explicitly — that these areas form a part of Finland. Four articles deal with the bedrock, flora and fauna and economic connections of the 'forthcoming Finland', and two contain descriptions of parts of Eastern Karelia. The Finnish geographers Väinö Auer and Leo Aario were among the authors, and both explicitly employ the terminology of geopolitics and have an organismic concept of the state (Paasi 1990).

The Finnish Geographical Society, and in particular Väinö Auer, were active both in making concrete proposals and in their publication policy in their concern with the occupied regions of Eastern Karelia. On the initiative of the Geographical Society, the Ministry of Education established a Scientific Committee for Eastern Karelia, which consisted mostly of members of the Society. Auer, who was joint author with the historian Eino Jutikkala of *Finnlands Lebensraum* (1941) and special adviser to President Risto Ryti on such problems, acted as chairman of the committee from 1942 onwards (Laine 1982, 1993).

The authors of *Finnlands Lebensraum* argued that since the Finns had the highest birth rate of all the Nordic people and the country had the highest population density for its latitude, it also most needed to broaden its Lebensraum (Auer and Jutikkala 1941, 102). These ideas were also popular among other scientists as well as among the political leadership, as Poroila (1975, 126) has indicated. Among the clearest indications of this connection between Finnish academic geography and the politico-military élite is the fact that the Society introduced a Golden Fennia Medal, which would be given to Finnish scholars who had broadened geographical perspectives and in this way brought fame to Finland (*Terra* 1941, 4, p.292–3).

The first medal was given to Field Marshal Mannerheim in late 1941 for 'his contribution to Asian studies', although he had actually achieved these merits in the course of his service in the Russian army, when mapping on a reconnaisance mission in Asia (Halén 1989). As far as the scientific basis is concerned, Mannerheim (1951, 56) wrote in his memoirs that the scientific background of his exploration was very slight: 'Ultimately, thanks to a couple of English handbooks, I managed to contribute something to the domain of science'. In any case, he collected valuable ethnographic material side by side with his military duties (Granö 1941).

The basic idea of establishing the Scientific Committee for Eastern Karelia was to create a total picture of the occupied region and prove 'scientifically' that it belonged to Finland (Laine 1982, 56, Paasi 1990). The Geographical Society was also active in standardizing the area's placenames. The retiring chairman of the Society, Professor Heiskanen, said in a speech at the beginning of 1942 that:

We are living in great, heroic times, when a new Europe is being created and our sons are drawing natural boundaries for our country and guaranteeing industrial peace for our nation for a long time to come (cit. in Laine 1993, 129–30).

The Society also strengthened contacts with Germany, which had already been the most important source for the exchange of publications and general cooperation before the war. Now German geographers also attended meetings of the Society, among them Albrecht Haushofer, son of the famous geopolitician Karl Haushofer. He was also invited to be a corresponding member of the Geographical Society (Paasi 1990). It is interesting, too, that during the war years (1939–45) the proportion of publications written by researchers at the University of Helsinki in English diminished substantially (Laine 1994).

Väinö Auer had already been invited to be an honorary member of the Academic Karelia Society in 1927. His research field was not primarily in Karelia, and his interest in the area was based essentially on ideological arguments and 'practical geopolitical reasoning' (Paasi 1990). Auer was a supporter of a small group which had sympathies with Nazi Germany. According to Laine (1993, 108), some Members of Parliament representing the Patriotic National League organized a group called the Finnish National League, which started at the turn of 1941–2 to gather names of people who would begin explicit activities as soon as Germany had won the war — for a German victory was already envisaged in the minds of leading politicians (Upton 1965, 403). Support for the national socialist ideology in Finland was principally connected with the activities of this organization, although its operations were made difficult by the attention of the security police. Auer was invited to be an honorary doctor of the University of Bonn in 1942 (Laine 1993, 108).

As was indicated when defining various referents of the idea of geography in Chapter 2, maps of geographical entities are representations of power, and they also have a metaphorical and rhetorical character, which can be exploited in the social construction of reality (cf. Harley 1989, Wood 1992). The most explicit geographical examples of Finnish propaganda maps with strong ideological connotations were those made by Väinö Auer (1941, cf. Auer and Jutikkala 1941), who also discussed explicitly in *Terra* the location of the state boundaries of the 'forthcoming Greater Finland' (see Figure 7.6):

Before we can examine the forthcoming Finland as an economic entity, we have to know where, approximately, the boundaries will be located. If we assume that they will be drawn at the line where Finland can best be defended and where the Finnish natural sphere ends, somewhere between the White Sea and the Gulf of Finland, we will probably be on the right track (Auer 1941, 206).

Arguments for this legitimation were consequently sought in the realms of nature and military interests, and not from the history of boundaries, which reflects changing social relations on various territorial scales. Typical of the ideological background of the whole 'scientific Eastern Karelian movement', in which scientists from various disciplines participated, was the desire to show the connection between Finland and Eastern Karelia. This was not, however, explicit in all the published work. As Laine (1993) demonstrates, many contained concrete plans regarding the future of agriculture, forestry and fishing in the occupied areas.

Nevertheless, not all Finnish academic geographers were supporters of the prostitution of science to promote political aims. Hustich (1983, 107), for instance, wrote two newspaper articles during the war years on the tasks of science, directed against the attempts by geographers and other scholars to show 'scientifically' that Russian Karelia belonged to Finland. According to a personal communication, he was occasionally accused of an unpatriotic attitude in the war years. He nevertheless undertook botanical research in Eastern Karelia while on his military service (Hustich 1983, cf. Laine 1993, 180).

Certain other geographers were not involved in these tasks but were much more sympathetic than Hustich. After the occupation of parts of Eastern Karelia, Professor Leiviskä (1941), for instance, expressed the hope that the whole of Karelia might be freed. Leiviskä was an honorary member of the Academic Karelia Society and he applied for a large grant to complete research into 'The Land and People of Eastern Karelia', but the application was unsuccessful. Professor A. K. Tammekann, whose purpose was to analyse the distribution of Finnic people and find arguments for the political leadership to claim their areas from the Soviet Union, did receive a grant, but there is no evidence that he completed any of this research (Laine 1993, 156, 160).

The point of departure in these geopolitical speculations was certainly

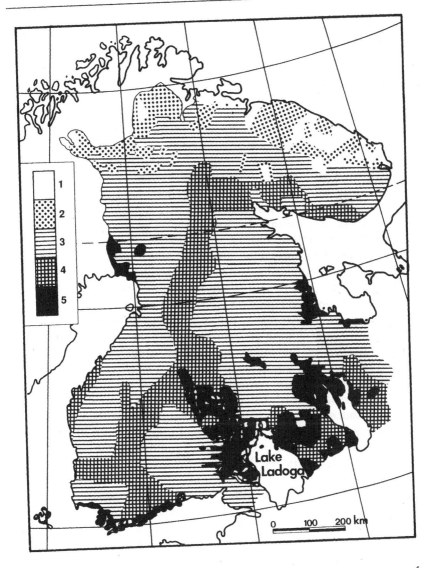

Figure 7.6 An example of Finnish war propaganda maps: the forest types of the 'forthcoming Finland' as outlined by Professor Väinö Auer (1941). Key: 1=bare mountain top; 2=mountain birch; 3=pine forests; 4=spruce forests; 5=forests with broadleaf trees. Source: Auer 1941. Reproduced by permission of the Geographical Society of Finland

not the fact that nation-states are man-made and always reflect past victories and defeats in social conflicts (cf. Taylor 1982, 27). These speculations were verbal expressions of geopolitical 'scripting' or of 'geo-graphing' as discussed by Dalby (1990, 39). Such practices refer to an ideological exercise, which in the case of war or the preparation for war pits geographically delimited political organizations against one another. The map in Figure 7.6, besides being a fitting illustration of geo-graphing, a propaganda map and 'persuasive cartography', is also an example of the power external to maps and mapping that serves to link the map to the centres of political authority. Most cartographers have a patron behind them (Harley 1989, Pickles 1992). In this specific case, the map was supporting the military aims set forth by the leading Finnish politicians (cf. Paasi 1990). In a way it is an ideological mirror image of that put forward by Soviet leaders in the early days of the Winter War (see Figure 5.5).

The geographers mentioned above — Auer, Leiviskä, Hustich and Tammekann — continued as academic geographers after World War II, although Auer was among a group of scientists who left Finland 'for reasons of caution', as Paasivirta (1992, 267) puts it. He worked for six years in Patagonia, Argentina, before returning to Finland.

The activities of Finnish geographers during the war years are thus a clear indication of the connection between geography and power. It should be noted that geographers were only a small minority — but an important one — in the research work directed towards the occupied Eastern Karelian areas. Laine (1993) observes that altogether 125 scholars undertook research in these areas, many of them among the leading representatives of their disciplines. About a third of them were members of the Academic Karelia Society, although the proportion of members was larger among those who planned and led the organization of the research.

As was found above, another referent of the word 'geography' became particularly important after the war — Finland's location in relation to the Soviet Union. This also caused a change in the social consciousness of the Finns and is probably one background factor explaining why Finnish geographers have not involved themselves in political geography very much since the war. Questions of political geography have been discussed explicitly in a few publications, these again mainly referring to the problems of the Finnish–Soviet boundary (Vuoristo 1979, Paasi 1990, cf. Seppälä 1993). Hustich (1974) and Vuoristo (1979) discussed the nature of boundaries and classified them as natural or artificial, e.g. on the grounds of physical and other factors, but they did not take any stand on the 'naturalness' of the Finnish–Soviet boundary.

A certain amount of research which can be labelled as political geography in the sense of the 'internal' development of the country has been undertaken since World War II, and similarly work regarding the relations of Finland in the international system of states (Hustich 1972, 1974). Part of this, e.g. that dealing with developed vs. underdeveloped areas, has been an explicit continuation of that done on behalf of the border areas from the 1920s onwards. The spirit, however, has been completely different.

7.3 THE FINNISH–RUSSIAN BOUNDARY IN GEOGRAPHICAL TEXTBOOKS

Education in geography was important in the construction of images of the nature of the boundary in successive generations. The representation of the Finnish–Russian boundary in school textbooks during the Grand Duchy period was generally neutral — as was the attitude towards Russia. It was typical simply to note this boundary as a fact among Finland's other boundaries without emphasizing its special features. What was important, however, was the fact that the authors made a clear distinction between Finland and Russia, i.e. they represented Finland as a clearly distinct territorial unit, even though its role as part of the Russian empire was noted. Some books provided a very detailed description of the course of the boundary and employed a naturalistic terminology, pointing out how Finland was partly shaped by 'natural boundaries'. It was also indicated how Russia was bounded in the northwest by Finland — in a way to strengthen the image of the independent character of these political entities (e.g. Dannholm 1889, 1894).

It became more common to emphasize the artificial nature of the Finnish–Soviet boundary in school geography textbooks during the 1920s and 1930s. Hence the content of education followed the content of academic political geography, supporting the ideologies of Greater Finland.

The maritime boundaries are Finland's natural boundaries. The other boundaries — with the possible exception of some river boundaries — are not natural. The Eastern border with Russia is especially rather an arbitrary political boundary line, similarly the greater part of the border between Finland and Norway (Auer and Poijärvi 1937, 11).

In some cases naturalistic language continued to be employed after World War II and even into the 1960s and 1970s (Hustich 1974), but in general it became more common to define the nature of the border in more realistic socio-political terms. Nevertheless, suggestions of the main naturalistic interpretations still prevailed in school geographies, for instance. 'Finland has natural boundaries in the south, west and partly in the north. In the east the boundary is artificial and could not be observed if there were not a border cut through the forest' (Hakalehto and Salmela 1951, 6). Again, 'A border is called natural if the boundary line follows a river, lake or sea' and a question at the end of the chapter asks: 'What can you say about the naturalness of Finland's eastern and southeastern borders?' (Myrsky and Ulvinen 1957, 35), or 'the largest part of Finland's long land boundary is merely a line that has been cut through the forest and which divides the terrain by winding arbitrarily and capriciously' (Kalliola 1969, 8).

It appears that the dichotomy between artificial and natural boundaries has been in part a reflection of the naturalistic language and environmental deterministic thought that has prevailed in Finnish geography. It is perhaps also in part a reflection of an ideological attitude as to how boundaries 'should be organized'. Nevertheless it is in both cases a question of the *naturalization* of the boundaries. Naturalization implies a deterministic emphasis on nature as the real basis of international relations and boundaries. Such an emphasis eclipses their historical and social character.

When we compare this naturalization with naturalization as an ideological strategy by which socially and historically created states of affairs are treated as natural events or as the inevitable outcome of natural characteristics (Thompson 1990, 65–6), naturalization in the case of the Finnish–Russian boundary is a time–space-specific phenomenon. In the first place, it is obvious that these ideas represent a specific language that geographers had adopted before World War II. Secondly, these representations of the boundary probably also reflect the disappointment of generations who have experienced defeat in the war and the ceding of the Karelian areas, and for this reason have normative concepts of where the boundaries should be located. It was not possible in this analysis to find any arguments to suggest that Finland's boundary with Sweden should be located elsewhere. In any case, after the 1970s the dichotomy between artificial and natural boundaries was put aside from Finnish geographical textbooks, and it has now been replaced by making a distinction between water and land boundaries.

Finally, it is interesting as an example of the 'pedagogy of space' and the inherent creation of spatial representations prevailing on the Soviet side of the border to scrutinize the content of some Soviet textbooks. Whereas the border area has for a long time been represented in the Finnish ideology as 'the outpost of the West', the interpretation given by the Soviet historians to the history of the border area has been at least as ideological: in that Karelia was for a long time regarded as the outpost of Russia against attacks from Sweden (Sihvo 1992).

The new boundary with Finland and the areas that Finland ceded as a consequence of World War II were also objects of redefinition in the Soviet Union. One Soviet geographical textbook, published in Finnish in 1950, contains several ideological comments which aim at creating representations of a certain kind for its readers. According to the interpretation presented in this book, the areas located to the west and north of Lake Ladoga were now *returned* to the Soviet Union in 1940 and the boundary of the Soviet Union was moved some 20 km. to the west (Mihailov 1950, 39). Actually the border line was moved about 100 km. to the west and the ceded Karelian areas had never been part of the Soviet Union — though they had been part of Russia several times in the past. Furthermore, the book does not tell the reader anything about the Soviet attack on Finland in autumn 1939 or the Winter War and Continuation War in general.

A more recent example of the political and ideological exploitation of cartographic representation for the construction of a specific reality is an atlas of the history of the Soviet Union published in 1990 (*Atlas Istorii SSSR* 1990). This contains a map of the most notable industrial building during 1928–40, but with the Finnish–Soviet boundary in Karelia located on the map as it has been since World War II. The areas that belonged to the Finnish territory in Karelia during the period concerned are thus represented as if they had belonged to the Soviet Union. Similarly, the territories of the Baltic States are represented as if they had been a part of the Soviet Union during the period of their independence. Boundaries make a difference in space, and as far as their cartographic representations are concerned they also make a difference in time.

7.4 BOUNDARIES, RELIGION AND METAPHORS: AN EXCURSION

It has been argued that language and religion are probably the key elements in the ideology of nation building (cf. Glassner and de Blij 1980) and that religious diversity still plays a part in the political organization of space (Gottmann 1973, 137). The role of language in the construction of the territories and inherent struggles about power has been made clear above. Now, at the end of this part, we will consider the relationships between nationalism, territoriality and religion. These represent a combination typical of many present-day conflicts all over the world (cf. Mikesell 1983). Before entering the local scale, one transcendental dimension of state formation, nation building and boundary construction — religious rhetoric — will be discussed. The so-called 'communitarian' ideologies of nationalism in particular have by tradition emphasized the role of religion and assigned significance to the community, based on shared beliefs and sentiments (cf. Smith 1983).

The affiliation between politics and religion has generally been called 'civil religion'. This can be a 'political religion' without explicit belief in any god, or also be explicitly based on conventional religious premises. The former case is more a question of a collective belief in specific affairs and ideals that are labelled as holy. Hence the idea of 'holiness' can also be laden with more secularized connotations, as was the case in Soviet geographical education, for example (Heikkinen 1987, 45). In the latter case religious beliefs and rhetoric are the fundamental point of departure for collective ideals and action. The Bible, for instance, contains many references to boundaries and their pre-ordained and historical nature. On the whole, it has been argued that modern nationalism took a number of its concepts from Old Testament mythology, beginning with the idea of a 'chosen people' (Brennan 1990, 59).

The ideas set forth by Olsson (1994, 120) probably hold good in the context of nationalism. He discusses the realms of thought and action which are so natural that we do not even notice that we think and do them: the territories of taboo. Olsson argues that for a 'heretic cartographer', a researcher who aims to map the invisible landscapes of power, a taboo-ridden term like 'Holy' is a synonym for 'taken-for-granted', and the name 'God' a pseudonym for 'power'.

It has been said that in almost all instances nationalism has been interwoven with religious predicates, while the nation is consecrated and is regarded ultimately as a holy entity (Alter 1989, 9–10). The significant role of religion in the history of nationalism is apparent in a number of investigations which show the dominant role of religion in the determination of state boundaries and even in the creation of states. The case of Northern Ireland continually indicates the power of religion in territorial conflicts.

Religion is often significant in the construction of socio-spatial distinctions, even though these distinctions are not the basic point of departure for religious discourse. Religious language nevertheless commonly spatializes the distinction between good and evil, the powers of light and darkness (Harle 1990, 2–3). National ethnocentrism has throughout history also effectively exploited the dualistic and even paradoxical role of religion, which can both create and remove prejudices. The universal rhetoric of the connection between religion and the state makes this explicit: 'cross and flag', 'chosen people', 'Gott mit uns', 'home, religion and fatherland' are all typical expressions of this rhetoric (cf. Eskola 1963, 155).

Nevertheless, where the relations between nationalism and religious language are concerned, the former does not inevitably legitimate power through some specific religion. Rather it defines the state, nation or boundary itself as a holy entity and the aggressive activities of the state as holy in themselves, i.e. it exploits religious discourse politically (e.g. the idea of a Holy War).

Whereas in its earlier forms the culture of a community was usually more or less synonymous with its religion, i.e. church and religion were the most important sources of norms in society, religion has subsequently become more of a hidden ideological constituent of social action, which has 'spread' into various institutional practices even though the specific religious feelings have diminished.

As regards the institutional role of religion in modern societies and its role in the construction of territoriality, the connection between church and state still prevails. The 'home–religion–fatherland' axis has by tradition been important in Finland, and has been particularly significant during periods of crisis such as wars. In fact a specific territorial rhetoric exists for these purposes. These forms of rhetoric in the case of Finland will now be briefly discussed, for the representations of territories and boundaries in general and of the Finnish case in particular contain features that are laden with religious metaphor. Hence the aim is not to analyse in detail the role of religious elements in the construction of Finnish nationalism or their changing relations.

It has been argued that secularization is a common feature of Finnish society, which means that the role of the church in the construction of social norms has diminished (see Allardt and Littunen 1975, 239–43). Religious rhetoric has nevertheless had a crucial role in the representation of Finnishness and the Finnish–Russian border.

As the previous analysis indicated, one specific feature in the rhetoric of symbolizing boundaries in Finnish geographical education, and also within the church, newspapers etc., has been the idea of their *holy* nature. Topelius in his *Book of our Land* already based his discussion of the fatherland fundamentally on religious rhetoric and began from the Book of Proverbs (22, 28): 'Don't remove the primeval boundary, that your forefathers have set'.

Similarly, religious expressions full of rhetorical eloquence have been exploited in various contexts. After independence, Finnish activists soon discovered that God was also in favour of Greater Finland (Wilson 1976, 148). It was seen, too, how between 1920 and 1939 in particular the representations of the Finnish–Russian/Soviet boundary in Finland were laden with religious metaphors aimed at convincing people that it was the historical duty of the Finns to stand against evil and the East. Hence religious definitions of the Other were exploited geopolitically (cf. Dalby 1990, 23). A curious synthesis emerged from the fact that the border areas between Finland and the Soviet Union were also essentially a meeting place for the Lutheran and Orthodox religions.

The ideology that depicts Russia and the Soviet Union as the Other has a long tradition in Finland, originating from the period when Finland was a part of Sweden and in practice a battlefield between Sweden and Russia. When Finland was a part of Russia and the period of active Russification began at the beginning of the twentieth century, it was again believed that the old Russian culture was a threat to Western civilization (see Upton 1965, 42). This is indicative of the fact that no specific religion is needed to declare the basic principles of civil religion! In fact, it can be generally argued that in nationalism the 'religious' is secularized and the national 'sanctified' (Alter 1989, 10).

The traditional view of Finland as a holy country was common between the World Wars. Paasivirta (1992, 47) writes that independence and territorial integrity have been understood in Finland as absolutist entities since independence was gained. The idea of the holy nature of boundaries was emphasized in school geography books particularly during the years of World War II and this continued in some cases up to the 1960s.

The boundary is holy, whether it is the boundary of home, farm, village, commune or state. Nobody may go into his neighbour's field to cut hay or into somebody's forest to take wood for himself. A foreign state similarly may not come over the boundary of another state to exploit something. Border guards are needed to protect Finland's boundaries during peace time. . . If a stranger aims at going into the protected area of someone's home with bad intentions, all the members of the family are ready to defend it. Similarly all states defend their rights. (Auer and Merikoski 1951, 36)

Thus, one aim of geographical education was to connect religious metaphors with national ethnocentrism (Paasi 1992, Eskola 1963). This argument is naturally connected with a more general discourse that aims at the personification of the state to dispel the distinction between individuals and collectives, which is one of the key arguments in stretching personal identity into a part of national identity (cf. Bloom 1990). Thus this argument represents a typical application of the connection between a strong nationalist ideology and the hierarchy of territorial consciousness, since the national ideal is not something original or natural to human beings, like their physique or their family, nor is the state the direct outgrowth of family, tribe or local community (cf. Smith 1979, 1). It has been argued that the rise of capitalism and the modern state changed this hierarchy inasmuch as it changed the context of socialization: the control of education and authority over children was transferred from 'fathers to governments', from 'private families to public agencies' (Brown 1987, 40–44).

Religion has a dual role in that it works at times to support the existing structures of power and sometimes as a channel for social protest (Allardt and Littunen 1975, 258). The link between the state and the church is constitutive for the former. There are several illustrations of this connection. For example, since Christianity became a state religion, military metaphors have become a part of the dominant narratives. The discourse of the 'righteous war', for instance, was established. The main ideologies of the church accept violence as a means of defence (Niemi 1980, 23).

Nisbet (1990, 91) remarks that if there is any single origin of the institutional state it has arisen from the relationship and circumstances of war. This, for its part, points to the political existence of human beings, as discussed by Ricoeur (1965, 234–46), for instance. He argues that 'the political existence of man is watched over and guided by violence, the violence of the state which has the characteristics of legitimate violence'. One of the cornerstones of this connection has been the link between religion and (state) authority, which can be argued to be laid down already in the Bible. Religion, territory and power are thus often close to each other, even though not the same thing (Klinge 1981, 280). In the tradition of Finnish political geography, Numelin referred in 1929 to the almost religious meaning that boundary signs have had among most nations.

Hence the ideological role of religion in nation building seems to be particularly important during the years of war. A good illustration of its social role in Finland concerns events at the beginning of the Continuation War. When England allied with the Soviet Union in the war against Germany, which for its part had connections with Finland, the Archbishop of Canterbury in August 1941 wished 'the Soviet people and brave Russian armies' good luck in their fight. In Finland this statement was interpreted as a contention for Bolshevism, and Archbishop Kaila interpreted it as an abandonment of Christianity and the gospel. This kindled anti-British attitudes in Finland and in the extreme case the expression 'Anglo-Russkies' was taken into use to depict the English during the war years (Paasivirta 1992, 190, 212–14).

As far as the role of the Finnish–Soviet boundary is concerned, those of a religious disposition considered that to guard the border was a special challenge set by God, and that the Finnish people were a chosen people. These ideas were not merely typical of clerics, but newspapers also reproduced them and, symbolically, elevated the Finnish people to the place of Christ on the cross. This doubtless shaped social consciousness, and even though the mass of the population did not engage actively in such speculation, the long political and national socialization made this rhetoric seem reasonable (Upton 1965, 44).

In the jingoistic circumstances of the Continuation War the Bible also provided strong arguments for the ideology of Greater Finland. This is a fitting illustration of the argument put forward by Klinge (1981) about the relation between political power, religion and territory. Forsman (1941), for instance, argued that 'wrong boundaries were not eternal' and found support for this in the Acts of the Apostles (17, 26–7): 'He created every race of men of one stock, to inhabit the whole earth's surface. He fixed the epoch of their history and the limits of their territory. They were to seek God.' Wrong boundaries were temporary and, Forsman (1941) continued, now was the time when God was sweeping them away for the Finnish people and the Karelian tribe. The fight of the Finns was 'God's fight against the powers of darkness and in this fight the powers of light will win', Lilja (1942, 136) wrote about the struggle of the Finns. 'The common enemy of humanity' and the 'beast of the east' was to be beaten far back beyond the old boundaries. The following excerpt from a book published in 1942 by Bishop Eino Sormunen is almost poetic:

The boundary between East and West has during the whole of our history divided Karelia into two parts, whose governmental stages have been completely different, but whose civilizing connection the foeman was not able to prevent until the period of Bolshevist power. . . By fighting the whole time against the East, our people have created the peculiar Nordic peasant culture. When guarding the heritage of this, the faithful Finnish soldier has surprised the world with his heroic deeds, independence and readiness to make sacrifices. In this culture the Kalevala landscape, surrounded by a lyrical chiaroscuro, has harmoniously joined with the massive devotion that lives round our old stone churches. (Sormunen 1942, 8)

All these points bring to mind the enthusiastic words of Satov in Dostoevsky's *The Possessed*, which have been analysed in an essay by the philosopher Georg Henrik von Wright (1989). Satov says that 'God is a synthetic personality of a nation. There has never been a God that has been common to all or several nations! Each has its own. When gods become common to several nations, it has always meant the disappearance of a nation.' Nietzsche (1989, 202) set forth in his *Gay Science* that the real invention of those who have established religions is in the first place that they prescribe a specific, common everyday life which holds the will in control (*disciplina voluntatis*) and simultaneously removes boredom. Secondly, they provide an *interpretation* for this life, which highlights the latter so that it becomes something good, something to fight for and something for which one's life can even be sacrificed. This constitutes the key to understanding the relation between religion and the state.

Paul Tillich (1973), a German-American theologian who has considered the role of boundaries from several perspectives, also ponders over modern nationalism as a condition in which space controls time. Tillich (1973) effectively employs metaphoric expressions and makes clear the ideological connections between nationalism and religion. He points out that in modern nationalism polytheism is an everyday element which makes the close connection between religion and nationalism apparent. He writes of the 'god of heathendom' and defines the latter as elevating some space or region as the ultimate value and the object of ultimate esteem. Finally, he points out that 'the god of one nation is fighting against the god of another nation, since every god is imperialistic due to his divine character'.

This basic idea of the enthnocentric exploitation of religion can be seen in hymns, as the following illustration shows: 'Oh Lord, Bless the Finnish people, give to it the abundance of your clemency, that it would be in all periods your own, the chosen people. Give us a faithful mind, success for the ways of the Finnish tribe' (Lutheran Hymnal 459, 2). The exploitation of religious ideas of a 'chosen people' is a much more common phenomenon on a world scale, and ideas of a 'nation closer to God than any other' or 'God's own country' have been typical in the construction of national social spaces. Typically 'we' are on top, our 'allies' come next, after that probably 'neutrals' and finally 'evil enemies' (cf. Harle 1990, 9).

This chapter can be concluded with a comment by David Lowenthal (1961, 249), which illustrates better the present secularized times in Western advanced societies but still indicates effectively enough the problems prevailing in international relations. It also refers to potential power relations and ideologies in shaping mindscapes:

Hell and the Garden of Eden may have vanished from most of our mental maps, but imagination, distortion and ignorance still embroider our private landscapes.

PART THREE

TOWARDS LOCAL EXPERIENCE

8

PLACE, BOUNDARY AND THE
CONSTRUCTION OF LOCAL EXPERIENCE

8.1 THE LOCAL SCALE

The construction and roles of territories and boundaries were in the previous chapters analysed in a context provided first by abstractions regarding the content and development of nationalism, nation building and inherent socio-spatial consciousness, and secondly by various aspects of the 'concrete', i.e. social practices that have 'put life' into these abstractions in the case of the Finnish state and nation and its relations with Russia. Thus the intention so far has been to trace the content and historical construction of the socio-spatial consciousness of the Finns by employing various materials to reveal more general social processes originating on the scales of state and world state systems. A further aim has been to discover representations which illustrate the content of socio-spatial consciousness.

It has been submitted in recent regional and urban studies, however, that without research into individual life-histories, talking of, listening to or reading about individuals' experiences, it is difficult to know what theories of social processes actually mean (Cooke 1987, 411). In fact, without these it is an arduous task to know what social and political processes in general mean, as opposed to mere theories.

The point made by Cooke is a crucial incentive for the latter part of this work, since it partly concretizes the fact that the production of social identity and the process of social reproduction are one and the same (cf. Abrams 1982). In order for any social formation to exist, individuals living within it must also have deeply internalized knowledge, so that they are able to construct a 'taken-for-granted world' simply on the basis of prevailing ideologies and cultural practices — on the basis of the narrative account that connects their personal histories with this collective history (cf. Carr 1986, Warf 1988). The aim of the following discussion is to analyse these questions both theoretically and empirically at the level of local everyday life.

To move from the analysis of socio-spatial consciousness manifesting itself on the national scale to the local scale and on to the everyday lives of the people is not merely to change the scale: it is also a move in understanding the premises regarding the constitution of socio-spatial life and its 'stretching' and sedimentation in time and history to cover various spatial scales. Furthermore, it is not, or at least it should not be, merely a move to a 'concrete' study in the sense that research at the local level should be somehow easier, purely empirical or anti-theoretical compared with more abstract questions of social theory (cf. Urry 1987).

On the contrary, what is needed are abstractions that contribute to conceptualizing the role of agents and individual and collective life-histories in the continual transformation of society and its regional structure. And an abstract social theory regarding the premises of human agency is not enough. Culture is not created by abstract actors — even if we may think so ontologically — but by living human beings, subjects. It is also impossible to understand human beings outside the context of daily social practice, nor can they be reduced to mere agents. Archer (1990) argues that the human being and the social agent are not identical, and that one sign of an adequate social theory is that it introduces one to the other punctiliously: to be human is to be social. But to extend the view as to what 'human' means, the argument put by Geertz (cited in Austin-Broos 1987, 141) is also challenging: 'to be human is not to be Everyman'.

During recent years research carried out by human geographers and other social scientists on the local scale has been much criticized because of the danger of empiricism. Similarly, studies concentrating on individual actors have been criticized. Harvey (1987, 371), for instance, states that 'individual biographies are valid data for all of us, but there is a long way from these to the understanding of social processes like inflation, deindustrialization, or even gentrification'. Hence, taken superficially, there seems to be an unbridgeable gap between 'micro' and 'macro' studies.

Nevertheless, some authors have been more optimistic and point out that individuals with their dynamic stocks of spatial experience are not to be regarded as manifestations of abstract individualism or voluntarism, rather 'as contextual modulations in a vast web of communicative interaction that is structured by determinate social relations acting through localized (but not necessarily local) social institutions in all kinds of ways' (Thrift 1987, 402). Hudson (1988) aptly points to the dual role of localities — and hence bridges the structural and more interpretative perspectives — so that, whereas various localities typically represent abstract space for capital, points which are evaluated solely in terms of their capacity to yield profits, the people who live in them evaluate them on a multidimensional basis. These localities are for them places, where they were born, went to school, have friends and pursue various social activities — simply contexts where they have been socialized as human beings and to which they often become deeply attached.

One significant solution for overcoming erroneous, abstract dichotomies between global/local or abstract/concrete phenomena lies in the conceptualization of scale. Smith (1992, 72–3) points out that much of the confusion in contemporary constructions of geographical space arises from an extensive silence on the question of scale: social life operates in and constructs some sort of nested hierachical space rather than a mosaic. Further, 'the making of geographical scale also results from and contributes to social struggles based on (and problematizing) class, gender, race and other social differences' (1992, 76).

Hence the perspective of everyday life practices, for instance, is essentially localized and the socialization of individuals always takes place in various local contexts. Eyles (1989, 109) remarks that a 'place' is not merely an arena for everyday life, or its spatial coordinates. Instead it provides *meaning* for that life. Everyday life does not consist only of local connotations but also of national socialization and the participation in a broader division of labour and a struggle over meanings through locally embedded institutional practices. It is through these practices that the forms and rules of territorial discourses are mediated and sedimented into the practical consciousness of individuals at the local level to become one part of their daily routines and the basis of the constitution of their social identities — a basis which does not necessarily have much 'discursive effect' on their everyday lives.

Local life is thus always bound to diverging institutional practices operating on larger scales, and in particular to social processes taking place on a regional, national or global scale. When deliberating over people's regional or local identity, it has been usual — particularly in humanistic geography — to concentrate on individual experiences of places and the meanings involved. The broader social contexts are not problematized or they are merely understood as being a frame in which the experience of place is realized. This easily psychologizes the problem of (social) consciousness. Nevertheless, the context of daily life is no longer bounded by proximity, but is a product of global rather than purely local processes (Pickles 1986).

Many forms of consciousness have for a long time been ignored in geographical research. As Cosgrove (1989, 120) writes, 'banished from geography are those awkward, sometimes frighteningly powerful motivating passions of human action, among them moral, patriotic, religious, sexual and political'. Cosgrove reminds us how fundamentally these motivations influence the daily behaviour of human beings and how much they inform our response to places and scenes. Hence culture is 'at once determined by and determinative of human consciousness and human practices' (1989, 123).

In this context ideologies, for instance, take a form of common sense and ideas are taken for granted. A good example would be the national stereotypes that people routinely reproduce in daily action, e.g. in the form of jokes, whether or not they really believe in these stereotypes (Jackson 1989, 51). In Finland, for instance, collections of stereotyped jokes about the inhabitants, cultures and political systems of all the neighbouring countries, including Russia and the Russians, have been published, and these doubtless constitute at the 'grass-roots' level one part of the Finnish identity in relation to other nation-states — 'our' relation to 'them'. As Klinge (1972) writes, anti-Russian feelings etc. are typically reproduced in civil society in the form of jokes. This leads to the point made by Cohen (1982, 13), that 'local experience mediates the national identity' and that it is impossible to understand the latter without knowledge of the former. Hence it is essential for a critical science to ask where various 'meanings' arise from, what kind of ideological constituents are involved in the construction of them and how they manifest themselves in social life.

8.1.1 Redefining place

Individual actors construct and symbolize boundaries continually as part of their individual and collective identities. This takes place in everyday life through participation in various social practices, e.g. ethnic, political, religious groups, communities and associations which exist and manifest themselves in different ways on various spatial and historical scales and in various social contexts. The boundaries of everyday life can be visible or invisible. Invisible boundaries in particular, having no symbolic objects on display, must be maintained through shared knowledge (Harré 1978, 150). In many cases these boundaries and the territorial and social units they demarcate are a ready-made context for social action, reflecting traditions and institutions sedimented in local practices. As Schwarz (cited in Norberg-Schulz 1971, 30) points out, 'the individual is born in the village which existed before him. But slowly this village becomes his homeland, a place lived in and full of memories. . . Paths and places become memories, time and space become the history of his life.' Also,

A human being identifies himself with places, where he has lived, has been happy and suffered. He is bound with matter, things and ground. Our places are our relation to the world. All places of our existence stay in us like the nails of a huge warehouse, where our memories are hanging symbolizing all the states of soul that we have experienced, the smallest shades of our feelings. (Tournier 1971, 13)

Each individual is a locus in which an incoherent and often contradictory plurality of relational determinations interact. Individual interpretations of socio-spatial structures, various things and episodes, are always personal and are bound to the experiences of one's life-world and *spatial life history*. The concept of 'place', one of the fundamental categories of geographical thought, is regarded here as an abstraction which refers to the spatial dimension of the personal histories and experiences of individuals. The intention is to broaden the traditional idea of place both in time and space by this means. In geographical discourse place is usually essentially bound up with some specific *location* or areal connection (Relph 1976, Entrikin 1991, Johnston 1991).

Place is thus not treated here as an objectified or experienced everyday environment of individuals, or as an administrative frame. It is treated as a unique web of social and material spatio-temporal life connections — bound to various localities — with associated meanings emerging on the basis of the life-world. Instead of bounding spatial experience with some specific locality, different localities and their structures of expectations are thus bound with the constitution of personal spatial practice and experience. Various regions and territorially bounded units may be sources of security and identity for individuals and groups of people, but their personal meanings are always structured in relation to an individual life-history. The following excerpt from de Certeau (1984, 108), although partly taken out of context, captures something of the idea of place employed in the present work:

Places are fragmentary and inward-turning histories, pasts that others are not allowed to read, accumulated times that can be unfolded but like stories held in reserve, remaining in an enigmatic state, symbolizations encysted in the pain or pleasure of the body.

However, there must certainly be some intersubjective basis in these localities, since individual histories are, above all, certain specifications of the collective history of an individual, group or class (Bourdieu 1977, 86). The connection with these collectives joins personal histories with those institutional practices — and discourses which are embedded within these practices — which constitute various regions or localities: villages, communes, provinces and nation-states. Relph (1981, 172) writes: 'The individual distinctiveness of a place [here a region or locality!] therefore lies not so much in its exact physical forms and arrangements as in the meanings accorded to it by a community of concerned people, and the continuity of these meanings from generation to generation'. Relph thus points to the local 'story' or structure of expectations that is connected to various territorial units and which connects individuals with the continuity that these units constitute.

By understanding the nature of place through personal histories, it is possible to understand more profoundly the nature of an individual's spatial experience, though not so much of the individual experience of some specific territorial unit in some specific time, which is what humanistic and behavioural geographers have typically tried to trace (cf. Paasi 1991a). This makes it easier to understand that for individuals participating in local discourses these spatial units do not necessarily form explicit closed or meaningful territories which they could shape discursively, but instead are embedded in their everyday practices and practical consciousness. These consist of those things, ideas and signs which actors know tacitly how to go on in the frame of social life, without being able to understand them reflexively and give them explicit 'discursive' expressions (cf. Giddens 1984, xxii–xxv). A point made by Berger and Luckmann (1976, 77–8) is illustrative of this:

The individual's biography is apprehended as an episode located within the objective history of the society. The institutions, as historical and objective facticities, confront the individual as undeniable facts. The institutions are *there*, external to him, persistent in their reality, whether he likes it or not. He cannot wish them away. They resist his attempts to change or evade them. They have coercive power over him, both in themselves, by the sheer force of their facticity, and through the control mechanisms that are usually attached to the most important of them. The objective reality of institutions is not diminished if the individual does not understand their purpose or their mode of operation.

'Place' is thus understood here as an abstraction referring to the cumulative archive of personal experiences and meanings which individuals gain from different locations and landscapes during their life-history. 'Place' is continually developed and modified in the course of the life-history of the individual, which literally takes place in various social spatio-temporally constituted situations. The dialectics of leaving and returning, characterizing the spatio-temporal organizations of human beings, are an essential constituent of places (cf. Karjalainen and Paasi 1994). Individuals and groups produce and reproduce through various institutional practices (politics, culture, economy or adminis-tration) regions, territorial units which have a relative independence in relation to the spatial history of individual actors, i.e. their 'place'. Through these practices individuals also produce and reproduce their personal and collective spatial identities. The episodes through which collective identities become part of the individual's life-history are time and space-specific (cf. Eyles 1989, 107).

Personal relations with specific localities manifest themselves in a 'sense of place', to employ a category put forward by humanistic geographers. This sense of place, Eyles (1985, 2) strongly argues, is derived from the totality of an individual's life, where life for its part is not merely equated with experience and feelings about a locality, because there are forces which affect and shape life (and sense of place) and which originate beyond immediate experience. Hence there is reason, once again, to emphasize that the practices that shape existence and experience are irreducible to the individual scale even though they manifest themselves at it. Personal experiences can be at times so deep that the actors cannot grow out of them even if the aim of the 'official' (geo)political socialization of the state, for instance, is the creation of new forms of socio-spatial consciousness. This could be seen very strikingly in the case of the social representations of the Finns — prevailing in the civil society — towards the Karelian areas ceded to the Soviet Union after World War II. Logically, it is easier to communicate new ideas on the relations of states, for instance, to new generations. It will be argued here that the category of generations is of crucial importance in understanding both local forms of spatial life and the reproduction of various forms of national culture (cf. also A. Smith 1979, 3).

Previous arguments make it clear why it is of great import to make a distinction between the identity of a region or territorial unit itself, i.e. between its physical, material and symbolic organization, and the regional identity of the human beings living there (their regional consciousness or place identity), i.e. the personal interpretations that people make of it, or their experience of place (Paasi 1986a, 1991a). Personal identity is organized through the life-world, through the interpretations that a human being makes of the world. The identity of a region (or nation-state), which is on every occasion a part of the continual transformation of larger regional structures, is composed of social practices which are of much longer duration than personal experiences and hence cannot be reduced to the 'place' experiences of individuals, which are self-centred in spite of the social 'frameworks' that characterize life in its relations to various social groupings, such as class or gender.

One problem in geographical studies of localities and regions has been the difficulty of relating individual life and experience to the territorial frames that the regions are seen to constitute. It may be argued, for instance, that people's time–space paths are differentially spread beyond the defined boundaries of the locality. Again, any one individual's social networks, cultural experiences and other aspects of his or her social life have highly complicated contextual geographies (Crang 1992). In daily life this can be seen in the construction of labour and housing markets which typically exceed administrative boundaries (cf. N. Smith 1990, 136).

The analytical distinction between regions and places helps us to overcome these problems. Regions and territorial units exist and their existence has a relatively independent logic. The constitution of individual places is dependent on these territorial units merely through institutional practices, in which individuals continually produce and reproduce territoriality, adopt, bear and produce new meanings.

The distinction between region and place is based on the fact that in the (post-)modern world the segment of the life-world actually inhabited by human beings consists of many small worlds, which are located both within the 'private' and 'institutional spheres' of existence (Luckmann 1978). A life-world should not be understood merely as an unreflected basis for individual action, for it is always grounded culturally and institutionally. Separate institutional spheres, for their part, have acquired, Luckmann (1978, 279) writes, a degree of 'autonomy' which allows them to develop their own rationally founded legitimations and to withdraw from a hierarchically interlocked system of representations of society as a whole. The division of labour and inherent role performances — in relation to individual geographies — are the reasons for the almost unique relations of individual places to regions.

It is thus important to note that the constitution and reproduction of the symbolic structures of reality, and thus also territorial communities, is a manifestation of the social division of labour and the power relations which emerge on this basis (Paasi 1991a). Social relations are thus between social groups, not between areas or regions, even if the idea of territoriality tends to reify power, identifying it with a place rather than social relations (cf. Johnston et al. 1988, 5). Accordingly, some people or groups (teachers, politicians, journalists etc.) who have 'authoritative power', are specialized in the production of narrative accounts, which bind people together and constitute otherness in various social practices.

On the grounds of the relation between the regional process and the personal interpretations of it, it is thus feasible to trace and comprehend the nature of social mechanisms through which individuals are social-ized as bearers of collective regional identities and spatial distinctions. It is obvious that the broader the territorial scale in question, the more invisible, abstract and less personal the constituents of an individual's regional identity. This does not, of course, mean that they are any less important than immediate personal experiences. Think, for instance, of the coercive power of the state to claim even the life of individuals in specific situations. Thus when speaking about 'places' as personal centres of meaning, it should be borne in mind that they are not spatially restricted only to specific localities but that we are dealing with a socio-spatial consciousness which, in spite of the fact that it is always located somewhere, consists spatially of larger forms of consciousness, such as nationalism. Similarly, this experience always reflects experiences of other localities. Diverging spatial scales of consciousness are sedimented within each other in the construction of individual and collective identities.

The key point can also be expressed as follows. The personal histories of individuals, the histories of certain regions which become institutionalized as parts of the continual regional transformation, the histories of nation-states and 'world history' constitute each other in specific situations. Individuals are always born into certain political-ideological world situations, which they can shape to a greater or lesser degree in institutional practices (economy, politics, administration, culture). But at the same time they always become members and reproducers of traditions and bearers of collective identities and memories (cf. Shils 1981, Coser 1992).

8.1.2   The interpretative approach in local studies

A geographer analysing local 'communities', 'place' and everyday life observes and strives to interpret and understand the particular ways in which space and time are represented as constituents of social action, how spatial experience and structures of meanings are constituted. Communities, for their part, are 'located' above all in our language, in our conceptualizations of various socio-spatial units. This holds good whether we are interested in social relations (or functional integration) or in the social identification (or moral integration) manifesting itself in these contexts, for the need to conceptualize the research subject is acute in both cases (cf. Paasi 1991c). One basic problem in traditional community studies has been to search for explicit empirical content for diffuse social constructions labelled as 'communities'. In fact, this problem is also common in the case of recent locality studies; where it has been typical to do research in which the spatiality of diverging territorial units has been more or less taken for granted or in which the 'spatial' has been understood as in the tradition of geography — as a mental vehicle of classification (cf. Urry 1987, Paasi 1991b).

What, then, is interpretation? Geertz (1973, 18) remarks that 'a good interpretation of anything — a poem, a person, a history, an institution, a society — takes us into the heart of that of which it is the interpretation'. In the case of local communities we can then, by and large, talk about the analysis of the historical construction of socio-spatial identities and how socio-spatial distinctions and connections are constructed — and how these are connected with material practices. In principle, interpretative geography searches for, and accepts as such, the definitions and meanings that human beings give to the social world in everyday life and practical consciousness, but of course it cannot remain at the level of these observations. The consciousness of practical everyday life is always to some extent 'thin' and not well articulated (cf. Eyles 1988, Alasuutari 1989a). In their routinized daily lives, people do not need to problematize deeper connections. Geographical research which aims at an interpretative or ethnographic approach strives for *thick description* — to employ the concept of Gilbert Ryle popularized by Geertz (1973) — in which the researcher has to query the 'self-evidence' and to put its content into a frame constituted by continually changing social, historical and spatial contexts. This also renders it possible to contextualize the results into broader frames (cf. Eyles 1988, 2–4).

Giddens (1987b, 62–5), putting stress on the contextual — temporal and spatial — constitution of agency, begins his evaluation of the basis of the constitution of meanings from a perspective in which practical consciousness is regarded as the central basis for the constitution and reconstitution of meanings in daily life. He points out that 'a good deal of what we do is organized knowledgeably in and through practical consciousness'. He remarks that the meanings given to words and actions do not originate merely in the differences created by sign codes or language, but from procedures which agents use in their practical action when they aim to reach interpretations and understanding of what they and others do. Knowledge connected with action has thus a 'methodological character' (Giddens 1987b, 63).

Alasuutari (1989a, 76) points out that whether the ethnographic method stands or collapses depends on one presupposition: that the subjects have something in common with each other (e.g. a shared world view, attitude towards life or ways of interpreting reality). The basic aim of the research should then be to construct a well-grounded theory of the structures of these 'common things', a theory which abstracts itself from the individual features of the subjects. A logical prerequisite for an interpretative study is thus that the people really have something in common on that specific level at which the researcher aims at constructing the structure he or she believes to be common for the members of the community or collective. If they did not, it is more a question of the imagination of the researcher. Alasuutari suggests that a serious problem in the case of a modern individualizing society is that it is more and more difficult to find clearly defined communities, or even informants who are able to tell researchers how 'they' live and think. Alasuutari puts forward the idea that modern individuals identify themselves with a multitude of groups where the 'world views' or codes of action required may be contradictory.

This view partly reflects the common myth of the disappearance of 'communities' in the modern world and is based on a vision of the traditional world as a continuum of distint communities, the modern world being without them. Nevertheless, communities as networks of social groupings — groups sharing common beliefs and aims — still exist on various spatial scales. But these communities are not necessarily strictly spatially bounded. Not all boundaries, and not all the components of boundaries, are objectively apparent. Cohen (1985), when discussing the *symbolic aspects* of community boundaries, points out that in so far as one aspires to understand the importance of the community in people's experience, this *symbolic* perspective is the most crucial.

Nevertheless, as Cohen (1982) argues, 'people become aware of their culture when they stand at its boundaries'. These boundaries are not natural or absolute phenomena but rather they are relational. Cohen remarks that they may be contrived and their existence is called into being partly by the specific goal on account of which one group of people distinguishes itself from another (1982, 3). Hence boundaries which give direction to existence, and in fact locate that existence, are the precondition of their own transcendence (Tester 1993, 8). This is because, as Tester remarks, 'without boundaries, without direction and location, social and cultural activity would itself be a simply pointless thrashing about in the world': boundaries create practices and forms, which for their part are the basis of meaning and interpretation. Boundaries are constructed on various spatial scales and in various social contexts, and vary in their personal and collective meanings. Further,

Borders and boundaries carry a certain mystery and fascination. They imply a transition between realms of experience, states of being; they draw an ineffable line between life as lived in one place and life as lived in another. (Ryden 1993, 1)

9

REGIONAL TRANSFORMATION ON THE LOCAL SCALE: THE INSTITUTIONALIZATION OF VÄRTSILÄ

The above discussion effectively indicates that the idea of a territorial identity is really complicated, and it becomes even more complicated when it is discussed at the local level in the Värtsilä area. Local life in the border area is constituted by socio-spatial boundaries and demarcations that are simultaneously both local and non-local and are to a greater or lesser degree social. Where important non-local socio-spatial forces or frameworks in the border area of Värtsilä are concerned, the location of the border between Finland and the Soviet Union/Russia — an entirely national and international question — comes first. But the role of provincialism, particularly the role of the province of Northern Karelia, also emerges, and similarly it is impossible to ignore the role of the national Karelian organizations that have continually aimed at creating abstract reference groups for the Karelian people on local, regional and national scales.

Hence local life is full of divergent spatial and social distinctions manifesting themselves in various discourses, through which people identify themselves with different groups at the same time. The groups to which human beings belong provide different contexts of justification and narrative explanation, different expectations about courses of action that individuals must follow and attitudes that they may adopt. Groups that vary in importance and membership in more than one way may also be a source of conflict for individuals (Carr 1986, 161–5, *passim*).

The boundaries of village, frontier zone or administrative commune are examples of territorial distinctions. The division of labour and division by gender roles that are sedimented in the history of the factory community are immediately mixed with these territorial identities. From the perspective of the continuity of the community the levels of historical understanding become crucial. Here it is necessary to make a distinction at least between personal historical continuity and the continuity of the community. Both are manifested, but in different forms and dimensions, and become comprehensible through the distinction between regions and places put forward above.

This section will briefly trace the process of institutionalization of the commune of Värtsilä in order to understand the historical connections and differences between successive generations which are part of the 'story' of the commune, but whose life-histories are bound up with the common story in different ways. The point of departure for the present analysis is a critical evaluation of my previous work. We will begin with the fact that spatiality (with history) is an essential constituent of people's world view, a structural feature that is crucial in the organization of the 'places' that emerge through practical and discursive consciousness, action and life history. These places emerge on the grounds of personal histories, experiences and meanings. Place is a continually produced and reproduced texture of spatially constituted meanings and experiences, which are connected through action to various real and utopian localities.

The analysis of the institutionalization of the Värtsilä locality will provide the historical context for an interpretative study of the radical changes in the boundary between Finland and Russia that occurred as a consequence of World War II (for details of the fieldwork in Värtsilä, see Appendix). The point of departure is that a historical understanding of the structures and meanings of the local community is necessary for appreciating the diverging meanings of the new boundary in the local landscape and in the everyday life of different generations living in the border area and elsewhere. The aim is hence to examine how the *places* of the individuals have been shaped on the local scale by the politico-ideological processes taking place on the scale of nation-states; how the local landscapes of war and peace have been shaped through economic, political and administrative practices operative on various spatial scales.

The details of the historical development of the community have sedimented in diverging ways in the memories of successive generations of individuals. Some people living in Värtsilä have had a deep interest in these developments and have bought books on the history of the area, for example, but most of the inhabitants are not much interested in the history of their community. The collective memory of the community is thus composed of heterogeneous elements. In order to provide a coherent picture of the background to this memory, it is important to outline the main features of the development of the community. Simultaneously these developments will be contextualized in their larger social and historical connections, in the changing spatial divisions of labour and in their societal backgrounds. An understanding of the constituents of the socio-spatial identities of the inhabitants of the region will thereby be facilitated.

Viewed in its entirety, the institutionalization of Värtsilä has been an extended, complicated process in which social practices taking place on different regional scales have been involved. The symbolic shaping of Värtsilä, however, has been a long but relatively straightforward process. Old Russian documents indicate that the name Värtsilä dates from the beginning of the sixteenth century (Kasanen 1981, 10). There was already an identifiable village at the site by 1631, for instance, consisting of 35 houses and forming part of the commune of Tohmajärvi. Värtsilä gradually developed into a distinct unit compared with other parts of Tohmajärvi which remained almost entirely agricultural, and by 1805 it had 28 farms and 317 inhabitants and was quite separate from the other parts of the commune (Rantanen 1950, Kasanen 1981, 26). Värtsilä was only established as an independent commune in 1920, although the name had been taken into administrative use some time earlier, a local church by that name having been founded around the factory in 1865 and an independent parish in 1909. The background to all these developments lay in the industrialization process beginning in Finland generally at that time.

9.1 RISING SPATIAL DIVISIONS OF LABOUR

Small and large regions pass through the same institutionalization process in the continual regional transformation as they become established units in the spatial system and regional consciousness. All this holds good in the case of Värtsilä, which developed rapidly — by Finnish standards — to become the location of a major industrial unit before World War II. Simultaneously the regional transformation took place on a provincial scale. The factory community was located in the administrative province of Kuopio in its early days, but from the end of the nineteenth century until World War II it became simultaneously a part of the emerging economic/cultural province of Northern Karelia. Before World War II it had a larger population than in Joensuu, the prospective capital of the province, and this fact was later to become a part of the myth of the commune's lost golden future.

The institutionalization of Värtsilä and the province of Northern Karelia took place in parallel with the institutionalization and nation building of the Finnish state — as part of the emerging 'modernity'. As was noted above, the activities that took place on the scale of the state during the second half of the nineteenth century, e.g. changes in legislation, in the dominant ideologies and in the construction of infrastructure (railroads, administration, education etc.), contributed to the rise of new industrial communities and provinces.

Hence the rise of the Värtsilä factory community was a typical example of emerging industrial capitalism in nineteenth-century Finland. To employ the categories of Taylor (1982, 1993b), it is easy to conclude that the origins of the constitution of the locality existed to a great extent on the scales of reality and ideology. One important reason for the rapid rise of the Värtsilä iron factory community in the mid-nineteenth century was international, i.e. it was connected with the emerging scale of reality, the heavy demand for iron in Russia for use in the Crimean War, while the economic and political conditions for these developments were created on the scale of ideology: they were part of the developing modernity in Finnish society. Nevertheless, the industrial history of the village of Värtsilä is longer. A sawmill had already been established there in 1834, the year in which the continuous history of the 'Wärtsilä company' is also deemed officially to have begun (Haavikko 1984, 7).

It has been argued that modern Finnish industry emerged very rapidly, in spite of its shallow roots (Hoffman 1982). The iron industry, which soon came to dominate the industrial landscape and life in Värtsilä, had a fairly long tradition in (Sweden-)Finland, originating from the sixteenth century, and it was thus the first to develop as a major industrial branch, giving employment to 600 persons in 1815, for example, and continuing to dominate the Finnish industrial economy until the late 1840s, the later years of the nineteenth century marking the end of this stage of development, at which there were only a couple of iron works in operation (Hoffman 1982). Although textiles totally dominated the Finnish industrial scene in the 1860s, Finland gradually emerged as a country known for its timber industries.

Industry had its own territorial features and spatial divisions of labour owing to the distribution of raw materials and energy resources. Such was the case with the iron industry. The ore resources of Finland have always been scarce, and in 1809, for instance, when Finland became a Grand Duchy of Russia, about 20 of its total of 28 iron-works were processing Swedish ore or bar iron as their raw material (Virrankoski 1975, 69, Härö 1991). This fact can be seen in the Finnish industrial landscape, in what Massey (1984) calls historically sedi-mented 'layers of investments', which reflect changes in investments and technological development, changes in the supply and demand for iron, and so on.

The first ironworks were established in the seventeenth century, mainly in the southern and southwestern areas of Finland (Laine 1907, 1, see Figure 9.1:I), developments that reflect the 'geography' of the industry, in that these areas were located near Sweden. The first period

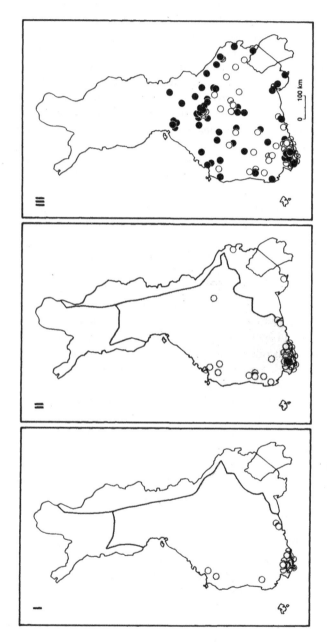

Figure 9.1 Ironworks opened (white circle) and closed (black circle) in Finland during periods I (from the sixteenth century to the Great Northern War 1713–21), II (from the Great Northern War to the end of the Swedish period 1809) and III (from 1809 to the end of the nineteenth century). Source: based on Moberg 1899

in the development of the Finnish iron industry, according to Moberg
(1899, 3), lasted from the seventeenth century to the period of the Great
Northern Wars (1713–21), while the second began about 1721 and
extended up to the separation of Finland from Sweden in 1809 (Figure
9.1:II). It should be pointed out that the role of Finnish iron production
after the Swedish period was very small as compared with Swedish
production, only 2% (Schybergson, cited in Kalpio 1988). When
Finland was a part of Sweden, most blast-furnaces employed Swedish
iron, but the declaration of peace in 1809 placed limits on this (Moberg
1899, 16). At the beginning of the third period, the nineteenth century,
the government was favourably disposed towards the iron industry in
accordance with the prevailing mercantilistic way of thinking. Accordingly the Finnish iron industry advanced rapidly.

This encouraging mercantilistic attitude, together with new political
ways of thinking, i.e. the liberation of Finnish economic and political
life, rapidly engendered a new spatial division of labour, and the role of
the iron industry became significant in this process, particularly in the
inland parts of Finland. These developments were mostly based on the
use of lake iron ore, although it was not very profitable as a raw
material. As a consequence, most of the new ironworks were closed
before the end of the nineteenth century (Solitander 1911, 34, Figure
9.1:III).

In the early stage of industrialization it was common for industrial
activities to be concentrated in certain centres. The iron industry had to
feel its way to the sources of raw materials, however, which were
generally located far away from the population centres. The locational
factors that were usually fundamental in the middle of the nineteenth
century were the amount and quality of ore available, the availability of
water power and charcoal, and the markets for the products (Lakio
1975, 167). In Western Finland the iron was mined (Figure 9.2), but
such activities decreased after the first half of the nineteenth century,
and the role of lake ore and bog iron increased up to the mid-1880s,
decreased after that but increased again in the 1890s, to collapse finally
in the early years of the 1900s. The iron industry in the inland parts of
Finland, and particularly in the east, differed markedly from that in
west Finland, using either ore taken up with hoop nets from the
bottoms of lakes or bog ore. This provided an important source of
income for the inhabitants of eastern Finland (1975, 128–9).

The iron industry, on which the development of the Värtsilä community was based, started in 1852 with the establishment of the
Wärtsilä iron foundry. The entrepreneur was N. L. Arppe, who had
purchased the sawmills in 1836. He was to become the major agent in
the economic development of the Värtsilä area and that of the economic
landscape of Northern Karelia. One of the basic prerequisites for the

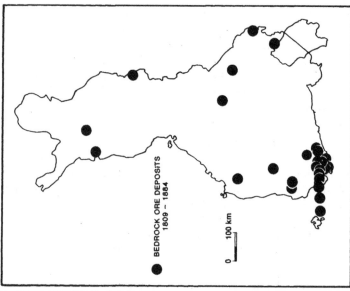

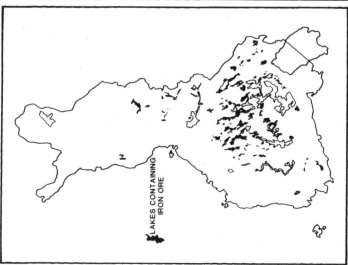

Figure 9.2   The physical basis for the rising spatial division of labour in the Finnish iron industry: the location of bedrock ore deposits and lakes containing iron ore in the nineteenth century. Sources: *Atlas of Finland* 1899, Laine 1950, 59, 353

capitalist mode of production is finance, which makes it possible to establish industries. Arppe had been involved in the timber industry earlier, but had had serious problems since the government discouraged this activity because of the threat of forest depletion (Kaukoranta 1935, 138, Laine 1948, 658–9). Arppe had begun to buy forests in the 1830s, and finally his huge properties proved an important source of capital, in that his factories were able to use wood taken from his own forests. These areas could also be sold when he needed capital for the investments. Eventually he owned almost 165 000 acres of land in Eastern Finland (Mustelin 1973, 135).

In 1853 Arppe bought the iron works at Möhkö, which was the most important competitor of the Wärtsilä factory at that time, and a new puddling and rolling plant employing steam power was completed in 1861. The fundamental factors in the establishment of the Wärtsilä plant were connected partly with the physical environment and partly with the rising capitalist mode of production.

The role of the physical environment is historically contingent in two specific senses. First, its role in the everyday lives of the inhabitants of a locality (i.e. a constituent of their place) is bound essentially to cultural and ideological considerations. The physical environment is a factor that cannot be understood ahistorically from the viewpoint of any specific locality. As regards the historical constitution of localities, nature is also historically contingent from the viewpoint of the accumulation process, and thus it is a critical resource in the shift from an extensive to intensive expansion in production.

The key prerequisite for the capitalist mode of production is nevertheless a market. As regards the constitution of Värtsilä, the demand for iron in Russia was considerable in the middle of the 1850s because of the Crimean War, and after the war Arppe had to reduce the production of pig iron and begin the puddling process (Laine 1948, 664–5).

The factory was dependent on world markets, and major economic recessions were soon reflected in the production of the Wärtsilä company. Four-fifths of the pig iron produced in Wärtsilä factory during the early 1870s and two-thirds even in the late 1870s was sold to Russia. Likewise, the recession in the Russian economy early in 1890 caused a decrease in production (Haavikko 1984).

Arppe had no professional experience in the iron industry, and he relied from the very beginning on the skills of foreign professionals to secure a high level in production technology, i.e. the development of the factory was dependent on the international scale. The company's first factories were planned by the Swedish-English mining mechanic Smith, and later another Englishman, Hill, continued the construction of the factory (Mustelin 1973, 176–7).

Haavikko (1984, 8) concludes that the Wärtsilä factories were confronted with the same difficulties and developments during the nineteenth century as other Finnish ironworks, conditioned by factors such as the unlimited markets in Russia, rapid changes in customs policy, the heavy demand for iron at times of war, the necessity to compete with old industrial countries and the ability to adapt production to changing economic conditions. Thus local activities were inextricably fused with the national and international scales.

As the analysis of the Finnish state and nation indicated, the state supported the liberation of Finnish economic and political life and the industrialization and capitalist mode of production that emerged in the 1860–70s. It is common to think that the Finnish society of the early nineteenth century was relatively stable, but Jutikkala (1976) argues that this is a myth. Such a view of the pre-industrial society is only true in a limited sense, holding good in the case of long-distance migration. This image is probably based on the fact that early nineteenth-century Finland was an almost totally agricultural state. Only 5% of Finns lived in towns in 1810, 6.3% in the 1860s and about 12% in 1900 (Rannikko 1980, 46). Furthermore, the idea of a stable society is probably based on the fact that there were many laws which prevented both social and spatial movement, and most people in any case did not have any special reason to move.

The liberation of legislation from the 1860s onwards which finally made movement possible is a good example of adaptation to the demands of emerging capitalism, which needed labour in the early stages of industrialization (Heikkinen and Hjerppe 1986). The role of the iron industry was unremarkable in nineteenth-century agrarian Finland, but as large industries began to develop, it became particularly significant for the landless people, who had become a major social problem as a consequence of the rapid increase in population in the countryside of eastern Finland during the first half of the nineteenth century. As Lakio (1975, 128) emphasizes, the iron industry became an important route for rising on the social scale at the same time as migration became more important.

In the case of Värtsilä the emerging iron industry was a significant employer for landless people and former crofters, and a new working class started to emerge in the 1860s. Figure 9.3 shows the migration field of the Värtsilä community from 1868–1900. At the first stage most migrants came from eastern Finland, where most landless dependent lodgers, 'redundant people', lived. Later the field expanded to include the areas of Southern Finland, typically other iron industry localities. Of the migrants presented in Figure 9.3, for instance, 17% were dependent lodgers coming in most cases from the province of Kuopio, where Värtsilä was located administratively at that time. Rasila (1982) writes that the emerging industry and marketing centres were important as

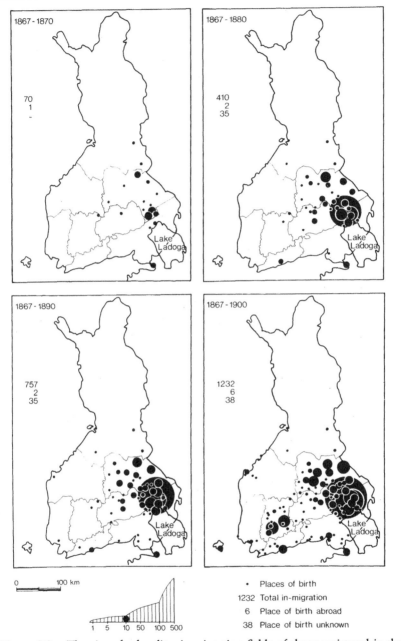

Figure 9.3. The rise of a locality: in-migration fields of those registered in the parish of Värtsilä in 1867–1900. Source: Värtsilä Parish Archives

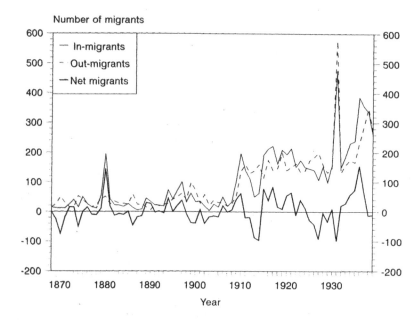

Figure 9.4   In-migration, out-migration and net migration centred on Värtsilä 1868–1939. Source: Värtsilä Parish Archives, Official Statistics of Finland, various years

sources of employment, and thus the labour market areas were large. The Finnish state effectively promoted the emerging capitalism, since the liberation of state policies helped the movement of workers by enabling former crofters and landless people to move freely and to sell their labour. The possibility of free movement can also be seen in the case of Värtsilä, in the considerable variation in net migration rates before World War II (see Figure 9.4).

The Finnish state supported industry, and especially the iron industry, because of the relatively high international demand for iron, especially in Russia. The construction of the state railways in eastern Finland, which helped the provision of traffic connections, was also very important. The Wärtsilä factory became a limited company in 1892 (Haavikko 1984, 8).

The number of inhabitants in the industrial community increased rapidly after the Wärtsilä factory was opened, and the establishment of the ecclesiastical parish of Värtsilä in 1909 broadened the area served by the church that began at the factory in 1865. This led to a drastic change in the social composition of the 'official' community. Now the most important occupations became agriculture and forestry (75%, vs. 2.2% in 1900) and the relative role of industry and construction

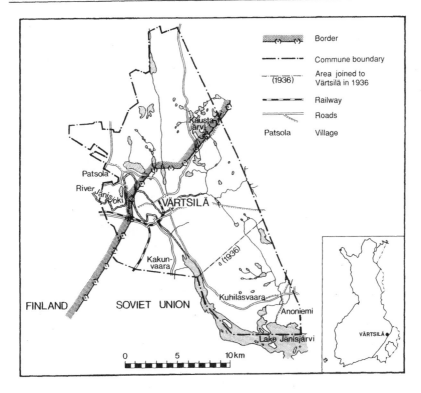

Figure 9.5   Location of the new border in Värtsilä in 1940

decreased markedly (19.7%, vs. 85.4% in 1900, see Official Statistics of Finland 1979). This was a 'statistical' effect, and it is extremely difficult to say what effects the extension of the parish had on the everyday lives of the inhabitants of Värtsilä.

In any case, the transformation of territorial activities was remarkable and the inhabitants who formerly lived outside the factory environment in Tohmajärvi commune were now separated from the larger Tohmajärvi parish. A more radical change in this respect was the establishment of the administrative commune of Värtsilä in 1920. This had direct effects on the organization of the administrative routines affecting the inhabitants, especially those who lived outside the factory community. The next stage in the development of the territorial shape of the locality took place in 1936, when four villages from the commune of Pälkjärvi were joined to Värtsilä (Rantanen 1950, Saloheimo 1963, Figure 9.5), increasing its population by more than 500 persons.

Initially, external communications from Värtsilä were directed towards Russia, where most of the iron produced by the factory was also transported. After Finland became independent in 1917 the effects were similar to those experienced in the whole of eastern Finland. As a consequence of political factors, occurring on the scale of the state (or of ideology), the boundary with the Soviet Union was almost totally closed. In the case of the Wärtsilä factory, production had to be directed principally to the domestic market and partly to some West European countries. Exports to the Soviet Union ceased entirely, although the factory had initially been established simply to export iron products to the East. The closing of the border was not so drastic as far as its market areas were concerned, however, since the railway connections with other parts of Finland constructed in 1894 made it possible for production to continue. Between the two World Wars the whole Finnish metal industry was concentrated more and more on the expanding home markets (Lakio 1975, 150).

When World War II began, the Wärtsilä company (and its locality) had the largest ironworks in Finland, with some 1000 industrial jobs. The cultural and economic life of the commune centre was lively, and its communications with larger centres were active because of the relatively good road and railway connections. Finally, as a consequence of the war, the commune was divided into two parts, one-third of its area remaining on the Finnish side of the border and two-thirds, together with all the factory, being ceded to the Soviet Union (Figure 9.5).

The latest change in the territorial shape of the commune took place in 1951 when the village of Patsola from Tohmajärvi was connected to Värtsilä, so that the commune received some 350 new inhabitants. There have been many discussions since World War II as to whether the whole commune should be merged with Tohmajärvi, which would mean that the locality would disappear in the regional transformation. Nevertheless the commune has remained independent — although its population is today less than 800, making it one of the smallest local government units in Finland.

9.2   THE FACTORY COMMUNITY AND ITS SOCIAL TEXTURE

The emergence of territorially bounded 'communities' is always a historically and spatially contingent process which is based on material practices as well as on the construction of a 'story' which binds the inhabitants together. By 'binding' together I do not refer uncritically to any assumption of a 'feeling of solidarity', since in practice the forms of integration and dis-integration can vary greatly from one community to another. What is important is not the feeling of togetherness or solidarity, but common social practices, which ultimately constitute the common 'story' with which people grow up and which help them to belong to a locality — whether they passionately identify themselves with that locality or not. In many cases they simply do not have any alternative. Hence, the community and individuals identify one another in a process of mutual scrutiny and recognition. Furthermore, since individuals constitute historical configurations, they are also historically constituted by them (Abrams 1982, 250–52).

As has been emphasized, a local community is ultimately not merely a local product, although it manifests itself in a time-specific context. As regards the old factory communities, the factory and its environment commonly form a 'texture'. A factory is not solely a workplace but rather an element which connects and often also controls all the essential dimensions of everyday life (cf. Hareven and Langenbach 1981). Furthermore, the director is not merely a manager but often also an indicator of the course to be taken by local forms of social policy and education, consequently taking care of the basis for the reproduction of labour (Koskinen 1987, Katajamäki 1988, 130). During the Grand Duchy period this type of 'patriarchal social policy' was very strong in Finland, and it also continued for a long time after independence (Paasivirta 1984, 211). When the factory takes care of the education system, it also attends to the nation building and shaping of territorial consciousness.

For the inhabitants of a small industrial community, the locality and its 'topography' — both physical and social — are the central constituents of their world. Any small community is a seat of face-to-face contacts where people know each other and in which their encounters are constituted and reconstituted in a rather restricted time–space context. These localities thus have a high presence availability and the basic institutional parameters of the social system are reproduced in daily practices (cf. Giddens 1984). The factory as an organization 'locates' every inhabitant in a specific position in the social field constituted by the community. People in the community are usually well informed of the structures of this social field, and knowledge of all changes soon spreads into the community. The social and habitual 'coordinates' and dispositions of the worker and his or her family — as also of the director and the community — become an essential part of the routines of daily life (cf. Talve 1983, 2).

Bourdieu's (1977) concept of *habitus* is useful for appreciating this. He defines habitus as 'systems of durable, transposable *dispositions*, structured structures predisposed to function as structuring structures, that is, as principles of the generation and structuring of practices and representations which can be objectively "regulated" and "regular" without in any way being the product of obedience to rules, objectively adapted to their goals without presupposing a conscious aiming at ends or an express mastery of the operations necessary to attain them and, being all this, collectively orchestrated without being the product of the

Figure 9.6 Workers at the Värtsilä factories in the 1890s. Source: National Museum of Finland. Reproduced by permission

orchestrating action of a conductor' (1977, 72). Hence the 'collective consciousness' is based on practice which emerges from the relationships between context and the actor's habitus. Practices emerge neither from the immediate objective conditions under which they occur nor from the historical conditions which produced the habitus. In fact, habitus mediates between these two and structures present in terms of a logic derived from the past experience itself structured by the habitus (Miller and Branson 1987, 218).

The factory community of Värtsilä was the most modern in Finland during the mid-1800s, and the directors of the factory systematically organized many activities that were of crucial importance for the rise, social organization and control of the community — and of course for its reproduction. A 'factory school' was started for the children of workers in the late 1850s, six years before the general primary school statute was passed in Finland, and the history and geography of the fatherland were among the major subjects on the curriculum (Juvonen 1991, 523–4). A factory church with its own priest was set up in 1865, and the representatives of the factory soon aimed at directing the free-time activities of the workers by establishing an association which met in the evenings 'since there are a lot of people in the factory who spend their free evening moments in useless amusements' (1991, 334, and Figures 9.6–9.9).

Figure 9.7 The Värtsilä factories in 1909. Source: National Museum of Finland. Reproduced by permission

Before World War II, the Wärtsilä company intervened in many of the reproductive functions of the administrative commune, e.g. taking care of fire precautions, allowing the factory physician to work as a communal health officer, providing cheap electricity for the commune, and maintaining sports facilities and children's playgrounds (see Kasanen 1981, 34). Arppe even had his own banknotes for workers (Mustelin 1973, 221). In the case of a delimited factory community it is tempting to discuss the local version of the idea of 'collective consumption', which has an essential role in the reproduction of a labour force. Collective means of consumption usually refer to the totality of educational, cultural, medical, sports facilities etc. The 'local version' in the case of Värtsilä did not involve state intervention — the sense in which the term collective consumption is usually employed — but company intervention in the local state and the everyday lives of the people. For the company, this was important both for the reproduction of the labour force and for the social control of the workers. One of the former inhabitants recalls these relations as they were in 1915 as follows (cited in Kasanen 1981, 55):

When I arrived at Värtsilä, the locality was, except for the factory area, an imperceptible, tiny country village with a few farms and cottages. Only some of the farms were big enough to maintain people, and all the others were partly or completely dependent on the factory, since if the factory was not in operation —

Figure 9.8   A view of the old market place of Värtsilä in 1915 and the Wärtsilä factories in 1920s. Source: private collection

Figure 9.9   View of the village of Värtsilä in the 1930s. Source: National
Museum of Finland. Reproduced by permission

because of lack of water or for some other reason — the whole locality was
worried about the situation. Similarly the commune was completely dependent
on the factory and the work it provided, even though the expenses of the
commune authorities at that time were minimal.

The basis for the emergence of the community of Värtsilä was the
rapid increase in the numbers of workers and their families in the area.
The population is a good indication of the institutionalization of a
locality, and especially of its territorial shaping. The rapid growth of the
factory and the increase in the number of workers (18 in 1853, 54 in
1860, 157 in 1865, 363 in 1890, 402 in 1900 and 779 in 1920, see
Lakio 1975, 149, Juvonen 1991, 331) and their families did not take
place without outer or inner conflicts. The factory had much power in
the political life of the administrative commune, Tohmajärvi, because of
the forests that it owned, and Sahama (1985, 14) writes that conflicts
between the factory and the farm owners of the western part of the
parish were a typical feature of the political life of Tohmajärvi during
the nineteenth century.

A particularly important figure in the 'grand narrative' that emerges
from the written histories of Värtsilä (e.g. Kaukoranta 1935, Kasanen
1981) and the Wärtsilä Company (Haavikko 1984) appears to be
Wilhelm Wahlfors, who became director of the Wärtsilä factory in
1926, at a time when the company was nearly bankrupt and its shares
were in practice valueless. He lived in Värtsilä until 1937, when he

Figure 9.10   The director, Wilhelm Wahlfors (front left), and a group of his old workers at the 75th anniversary of the ironworks in Värtsilä in 1926. Source: private collection

moved to Helsinki to become the Director General of the whole Wärtsilä Company and an extremely influential figure in Finnish economic life — and to some extent also in political life.

During his time in Värtsilä he was very active in the life of the community (see Figure 9.10). He became a key figure in the commune, and was a member of the local council. This probably provided him with the opportunity for initiating many projects in communal life that had an effect on the activities of the company. He was also a member of the Civil Guard, and contributed to the activities of the workers in many ways, e.g. through the sports club that the factory organized in the early 1930s (cf. Zilliacus 1984, 131–3, Hako et al. 1975). The members of the Civil Guard were usually directors and foremen from the factory, only in a few cases workers.

Wahlfors was very deeply involved with the national defence organizations, and he was sent by Field Marshal Mannerheim to the USA during the Winter War to try to buy arms and ammunition for the Finnish army (Zilliacus 1984, 153–7). After World War II he was a member of the Finnish delegation to the peace negotiations in Paris, and tried to influence the final location of the boundary line, proposing a line which would leave the mid-Karelian and Northern Karelian areas to Finland on the grounds that the major Finnish factories located there would be necessary when paying the war indemnities to the Soviet Union. His efforts were in vain, as we saw in earlier chapters. After the war he was active for several years in discussions aimed at recovering the ceded areas, thereby arousing the fury of leading politicians, who were inclined to be very cautious in their new neutral political relations with the Soviet leaders (Zilliacus 1984, 312–16).

Since the war indemnities were paid in the form of products of the metal industry, Wahlfors became a key figure in Finnish–Soviet trade relations after the war and built up very confidential relations with the Soviet authorities. Legation Counsellor Poselyanov labelled him a 'progressive capitalist'. Minister Mikoyan, for his part, suggested that 'If we had had such directors in Russia, we would not have needed the revolution' (Zilliacus 1984, 294–5). All these tasks that Walhfors undertook point to the fact that 'local' life, a huge abstraction, is not inevitably local as far as its social, political and economic consequences and contexts are concerned.

Wahlfors soon reorganized the activities of the company with a firm hand during his years in Värtsilä. In the case of strikes, for instance, he evicted the strikers from the company's houses (Zilliacus 1984, 107). These episodes are colourfully described in the collection of workers' autobiographies edited by Hako et al. (1975). In the early 1930s, during the international economic recession, the workers accepted a 25% reduction in salaries. For the generations that experienced the factory environment before the war, Wahlfors became even more important than N. L. Arppe, who was a 'ghost' of history for them just as Wahlfors is for post-war generations.

It can be seen from the autobiographical memoirs and interviews that Wahlfors is presented almost unanimously as a benefactor among the workers who lived in the community before World War II. It is interesting in general terms that an essential part of the ideal picture of the community that most of the interviewees carry in their memories concerns certain persons such as Wahlfors or personifications of holders of prominent positions in the community, such as the people who took care of law and order (the policemen and police chief), the coordinators of the moral landscape (above all the priests) or those who took care of health and welfare (the doctors). Finally, some representatives of the ordinary working-class people became important symbols in the community, usually people who had challenged the absolute authority of the masters and 'gentlefolk' through particular actions. It is typical for the interviewees to associate a kind of heroic quality with the representatives of these categories, and such myths about community founders and leaders, together with other actors, contributed to the collective definition of the locality. These became part of the account constituting the symbolic community and were exploited together with other symbols to serve the integration of local life and to overcome political and economic conflicts (cf. Hummon 1990, 27).

Even though many of the activities in the factory were organized before the war along the dimension workers–owners, this fact was not emphasized when important participants in the community were discussed, neither did it figure strongly in the autobiographies. The workers who were interviewed did not, for instance, identify Wilhelm Wahlfors as a bearer of social relations who was antagonistic to the interests of the workers. On the contrary, almost all of them emphasized that his characteristically severe and straightforward actions were in line with the common interests of the community, so that both the workers and the company as a whole benefited from them. According to the interviews, the memories of the factory workers seem to 'stop' at the manager, and it is highly exceptional for interviewees to discuss the spatial and social scales and logic of the factory's operations, e.g. the structures of ownership, shaping of the markets, etc.

Another interesting point is the fact that people did not usually talk about the 'immoral landscape', i.e. the drinking of alcohol, prostitution etc., which according to police records was an essential element in community life. It appears that all these attitudes give clear support to the idea of 'deference' which characterized social integration in the paternalistic community at the moral level (cf. Joyce 1980).

Local associations were doubtless important vehicles of socialization into 'membership' of the locality, inasmuch as they generally created a network between the actors and consequently also shaped their local consciousness. In particular the choice of a name for a given association is an expression of territorial identification (cf. Ryden 1993, 79). From the register that was begun in 1919 it can be seen that 34 associations using the name 'Värtsilä' in their titles were established in the commune before World War II, while only 13 were given some other title. Those established during the 1920s were in most cases political associations or branches of larger organizations such as national labour organizations.

9.2.1 Community and class

It has been pointed out that individual biographies can tell us about the local manifestations of industrial restructuring, the role of gender in the developing labour markets, etc. (Pickles 1986). There are many ways of approaching these matters through biographies, and for this reason an attempt will be made to relate the biographies of the people who lived in the old Värtsilä factory community to their broader social categories, particularly class. The research material does not render possible any profound analysis of gender relations in the community, and these will be discussed merely on the basis of the interviews with workers.

Traditional factories represent a typical expression of the relationship between capital and the worker, i.e. the worker sells his or her labour to the capitalist for a salary, which partly covers the labour sold by this act. To keep the system going, capital and the state have to guarantee the reproduction of labour. This is done in many ways.

There are vast differences within the working class as regards the roles of skilled and unskilled workers, but an attempt is made here simply to outline the expressions of community solidarity among the workers and to find some explanation for it. Similarly, the role of gender in the construction of daily routines and social reproduction will be characterized briefly.

When information regarding the construction of the Värtsilä community was collected by interview, it was typical of the results obtained in the first interviews in 1987 for both people who had been workers in the factory and people who had lived and worked elsewhere in the community to view the community in highly romanticized terms as a centre of harmony, an ideal body in which social relations between various groups and classes were good and the director of the factory was a righteous man. Social conflicts thus seemed to have been almost non-existent in Värtsilä.

Does this superficial observation support the extreme critics of the oral history method, for example those who claim it to be 'marginal', 'suspect' and 'trivial': 'marginal' because it is always limited to the distinctly 'modern' sphere and accessible people, 'suspect' because old folks' tales are likely to say more about the present than about the past, and 'trivial' because it indiscriminately amasses mountains of data that will 'never be of use to anyone' (Plummer 1983, 26)? There exists one significant theme that has to be discussed before this issue can be resolved.

It must be remembered that a factory community has a social texture and much of the activity necessary for the reproduction of society is provided by the factory itself, where the relationship between the workers and the employer is strong, even emotional. This points to the existence of a certain 'deference'. Social historians have defined this as the social relationship that converts power relations into moral ones and thus ensures the stability of a hierarchy threatened by a less efficient, potentially unstable, coercive relationship (see Joyce 1980, 92). Whether or not we call this 'theoretically' a false ideology which is sedimented in the hegemonic structures of the community and the state, Joyce seems to be on the right track when he talks about social or cultural hegemony rather than simply ideological hegemony.

As far as post-1850 English factory society was concerned, the family at work and its reflection in the structure of communal status and authority outside work were instrumental in creating a sense of place, of belonging, a source of integration. Paternalism had to deliver the economic goods, too, and it has been pointed out that failure to do this could rapidly result in the disintegration of deference (see Joyce 1980, 93). In essence, the idea of deference works on the same principles as identification with the nation, which is organized by the state for delivering collective 'goods' to reproduce the state and its ideological landscape. 'Responsibility' and 'security' are probably the keywords that characterize this situation of exchange. Both of these factors are essential constituents of 'ontological security', i.e. existential parameters of the self and social identity (see Giddens 1984).

The moral codes included in paternalism can in some cases penetrate deeply into the relations between the worker and the company, as can be seen from the following excerpt from a written autobiography of a worker from the Wärtsilä factory, which is characterized by a deep internalization of the Protestant work ethic:

I went to the factory office on 29th November 1922 to ask for a job. I was immediately asked to come next morning at six o'clock to the hoop iron rolling mill to begin working as an apprentice. Thus a completely new stage started in my life. I had found my position in society — which I did not then realize — I had entered new circumstances, a new environment, with machines, in the service of industry... I got used to the fact that my salary depended on the results I achieved... It was not long before I started to understand the machine, my fellow worker, who worked for me and did not look askance at me. I have said that I was used to working hard to earn more money... The words of engineer Enbom have been engraved permanently in my mind: 'We are all members of the same family. Machines are means, which serve human beings. If we achieve better results from machines with the help of workers, it is to the common advantage of all of us. A better result achieved as a consequence of co-operation between machine and worker will produce higher profits for the employer who owns the machines. It will also produce better results for the employees in the form of higher salaries — this is true.

When discussing later nineteenth-century industrial society, Joyce (1980, 110) writes that the emphasis on territory and the development of a local or parochial consciousness were to be sources of employer control and influence. The social situations in which deference worked were nevertheless invariably full of tensions and ambiguities, and the realities of poverty, dependence and authority were never far from the foreground in such situations (1980, 95). This held good in the case of Värtsilä, too. But before it is possible to understand this phenomenon, it is necessary to refer to the complicated role of history that was discussed when considering the aims of this research. The personal histories of the interviewees are located at various 'points' in the larger-scale histories of local institutions, which are for their part connected with the history of social formation and the rise of the Finnish state and nation and their final independence.

It would thus be an exaggeration and a mistake to look at Värtsilä — or any other industrial community — simply as an ideal community without any social or political conflicts. This also became obvious in later, more specific interviews and in various documents and publications containing stories of the lives of the workers written by themselves (Hako et al. 1975). These stories record diverging forms of conflict that took place in various industrial environments. Particularly effective memories are associated with the Civil War in spring 1918.

Just before the Civil War there were 432 Finnish communes that had a Civil Guard and 375 with a Red Guard (Ylikangas 1993a, 16) and, as in numerous Finnish localities, radical class-based action also emerged in Värtsilä in 1918, as it was among the first places in the province of Northern Karelia where Red activities occurred (Sahama 1985, 14–16). Admittedly the group of Red activists decided not to start armed violence, but when they would not give up their weapons to the representatives of the Civil Guard, a conflict ensued in which several people were shot. Finally, it was claimed that all the Red activists (about 100 men) should be shot, but eventually about 20 people were executed (Paavolainen 1967, Soikkanen 1970). The Civil War divided the Finnish nation for a long time, and these events — local manifestions of more general social processes — prevailed for a long time in the collective memory of the community and its activities. Some of the older interviewees still remembered personally this depressing episode in Finnish history.

People were in many ways dependent on their relation to the Värtsilä factory. The most fundamental fact from the viewpoint of social reproduction was that they usually lived in houses owned by the factory. The position of the workers was also manifested in their children's educational possibilities, so that usually only one child in a working-class family was able to go to secondary school or to continue beyond the compulsory years of primary school.

Even though the factory participated in the reproduction of the labour force and the families of the workers in many ways, social tension between the workers and the 'better folk' was apparent in the whole factory community in a variety of forms. In a way it was constitutive of the practices of the entire way of life. Värtsilä was a typical working-class area, and it was also among the first places where workers' cooperative shops were established after the law which enabled this was accepted in 1901. The first shop was founded there in 1902, and according to the Social Democrat newspaper *Pohjois-Karjala* (Northern Karelia, 21 November 1972) the reason for this was the 'despotism of the private shopkeepers', which made the workers almost 'slaves to debt'. More than 200 people joined immediately, which the newspaper looked on as an expression of the political consciousness of the workers, i.e. the *politicization* of everyday life (cf. Eyles 1989, 107).

The conflict was also present in daily shopping behaviour. Many interviewees pointed out that it was felt to be one's duty to support the workers' 'own' cooperative shop, and some people still remembered how their father often reminded them of this duty. The workers had also organized their free-time activities communally in many ways. They had their own sports club, for instance, the activities of which were disturbed by the club organized by the director, Walhfors. According to an autobiography, one reporter once asked Wahlfors if he believed he was some kind of Governor General when he had asked the workers to join the factory's club. 'No, I do not believe I am a Governor General,' he had answered, 'but I do not accept the activities of sport clubs other than Wärtsilän Teräs', i.e. the factory's club (Hako et al. 1975, 279–80).

One of the most visible activities of the small community was this interest in sports. The workers' own club was part of the nationwide workers' sports movement, and 'Wärtsilän Teräs' was established to attract the workers, and as one worker writes, to oust its predecessor (Hako et al. 1975, 279). The workers were usually active in their own association but the sympathies of the managers were on the side of the factory's association. The members of the latter were allowed to train during working hours and were given better jobs in the factory.

The factory also stamped its image on the political landscape of the commune. Hence support for the left-wing parties in Parliamentary elections was over 50% until 1958, and as much as 71% in 1945. During the 1920s, when the administrative commune was established as an independent unit and Wahlfors arrived at the factory, the left-wing share of the vote in all communal elections was more than 60%.

According to the interviews the division of labour between men and women was clear. Most women contributed to the reproduction of the family by working at home. As the Wärtsilä Company was an ironworks, only a minority of the workers were women, mostly welders in the radiator factory, cleaners in various departments and employees in some lighter jobs in other departments (Figure 9.11). Women also worked in the large farmhouse owned by the company. The women who worked at the factory were usually responsible for their home and family as well, since the division of labour was deeply embedded in the community. It was typical, according to the interviews, that if a woman lost her husband, she was usually given a job at the factory so that she could maintain her family.

Figure 9.11 View of the radiator factory, where most of the workers were women. Source: private collection

9.3 EVACUATION AND CEDING OF THE AREA

Before the Winter War, the Finnish–Soviet boundary was located far away, more than 100 km. east of Värtsilä. It was *there*, not here, it was not part of the immediate surroundings and everyday landscape of the people living in the commune. When Finland gained its independence the role of the boundary naturally became directly apparent for the people of Värtsilä in that traffic routes to the east were severed and the economic orientation of the company had to be turned to the domestic scene, but the boundary itself was still located 'somewhere out there', and was probably more evident in the general socio-spatial consciousness and forms of national socialization than as an explicit line. As was noted in earlier chapters, the boundary in the period between the World Wars was a concrete manifestation of the territoriality which was inherent in nation building and which showed itself very deeply in the socio-spatial consciousness of the Finns, as a national socialization aimed at filling the national space both physically (settlement of the border areas) and symbolically (the creation of nationalism).

Between the wars the right-wing movements put forward claims for a Greater Finland and the existence of the Civil Guard also reminded the people living in Värtsilä of the persistent division of the nation into two separate groups. The activities of the extreme left, as it were, were forbidden from the early 1930s, and the children of the community learned from their school (geography) textbooks how the boundary between Finland and its huge neighbour was an artificial one and the social system of the Soviet Union an alien one. Its inhabitants were depicted in highly critical terms, for they constituted the hereditary enemy of the Finns. In a way, all these tensions as well as the immediate political conflicts in internal and foreign policy sedimented into the practices of everyday life. This was strengthened by the fact that, according to the interviewees, the radio increasingly became part of Finnish life during the 1930s. Towards the end of 1939 Finland's position on the world political map had become more and more difficult.

This was the situation when everyday life in Värtsilä was turned upside down and the abstract question of Finland's eastern boundary became a very concrete one, as the following extract from a written autobiography indicates.

The Winter War began on 30th November 1939. A lot of evacuated people were brought into our village — located 200 km from the boundary — to rest. They were given food and clothes before they were sent on... On 8th December the first bombing raid came in Värtsilä. One of the buildings of the factory was damaged, and the body of a woman worker was removed from the ruins. Our Värtsilä was now part of the theatre of operations... Heavy bombers flew over our village on every beautiful, clear day, and they often dropped bombs. Even bright moonlight was dangerous. The longer the winter went, the more severe became the tempo. After mid-winter had passed the time between alarms was regularly only 4 hours, the time the enemy needed to fetch a new load. One day before peace, information had leaked from somewhere that an evacuation would be ahead. The next day, 13th March, peace came. It was greeted everywhere with joy, but the conditions, the sacrifices that were made for it, can be understood only by people from the evacuated area. Värtsilä belonged to the evacuated areas. Some people cried and were howling, others cursed, were bitter, accusing, some were quiet, broken. It was a small consolation that there were half a million people who were doomed to be homeless. A sorrow is easier to bear when you do not need to bear it alone. There were companions in misfortune, with whom one could cry and groan. At one time half a million Karelians became beggars, when they had to leave everything and take to the road to beg from the state and private persons. In the forced peace of the Winter War most of Värtsilä was among the areas that had to be ceded, only the fringes of it remained on the Finnish side. Our destination was Pielavesi. We had 13 days before our village was to be ceded to its new owners... On 26th March, the Annunciation, the Russians appeared, and after a couple of days our main room was full of soldiers. Without a word they stayed, living with us in our main room. We understood immediately that we had to clear out. That night we slept on our trunks and 20 soldiers on the floor.

During the Winter War Värtsilä was an important crossing point for many activities. Between 3 and 8 December 1939 a total of 10 000 evacuees moved through the area. Large military operations, e.g. the movements of soldiers, also took place through there, the fallen were brought there and the wounded passed through on their way to hospitals in other parts of Finland. Prisoners of war were also brought to Värtsilä and, as mentioned above, Wilhelm Wahlfors was sent by the Commander-in-Chief, Mannerheim, to the USA to try to buy war materials. Meanwhile, about 70 members of the factory community died at the front or as a consequence of the bombings (Kasanen 1981, 43–5), together with almost 24 000 other Finns.

Finally, when the Winter War ended three months later, people were not at all certain on which side of the new boundary the community of Värtsilä would finally be. The new boundary line was not marked on the land until 8 August, and people immediately restarted the routines of everyday life under the restrictions that prevailed. The civilians living in the border areas had many difficulties in adapting to the new regulations. Travelling was forbidden over large areas, and a security zone was established in which no one was allowed to live and in which working was allowed only in specific circumstances. The breadth of this zone was 400–1000 metres on the southeastern boundary and even more on the eastern boundary. The aim of these acts was to prevent possible conflicts in the border areas, but the Soviet border guards took prisoner dozens of Finnish soldiers during the summer of 1940 (Paasikivi 1986, 32–3, cf. Kosonen and Pohjonen 1994). In some cases it was also possible for civilians to work in the zone.

We went hay-making at Savikko, and immediately after that life began again. It was possible to live as near as 1 km to the boundary and make hay 100 metres from it. The boundary line had now been drawn, 10 metres broad, edged with barbed wire and cleared through the forests. (From an autobiography)

The commune of Värtsilä as a whole was damaged badly during the war. The factory and the homes of the workers and clerks were now on the Soviet side of the border, and the machines that the company had evacuated had to be returned to the Soviet authorities. Nevertheless, the aim of the company was to provide jobs for its workers at its other factories in Southern Finland (Zilliacus 1984, 158).

Since both the boundary drawn after the Winter War and the final boundary ran exactly down the middle of the commune of Värtsilä (see Figures 9.5, 9.12 and 9.13), the inhabitants had to be evacuated altogether three times during the war years: first after the Winter War, secondly when the Continuation War began and finally after it ended. The strange bulge in the boundary line was obviously caused by the location of the Wärtsilä factory, for if the boundary had been a straight line, it would have been exactly on it!

Figure 9.12 The Soviet suggestion for a new border in Värtsilä, on a map with Russian place names, April–May 1940 (compare Figure 9.5). Source: Archives of the Ministry of Foreign Affairs, Map Collections, Map No 26, Helsinki. Reproduced by permission

Thus people literally moved between landscapes of war and peace, building their houses under severe conditions. For some inhabitants the evacuation after the Winter War was the last:

Värtsilä had to be ceded on 27.3 and I left the parish with the fire brigade on the afternoon of 26.3. On this beautiful winter's day I looked at my home area with deep emotion and with an oppressive feeling in my heart that I would never return here. When I mounted the fire engine at 3 p.m. and we drove towards Joensuu I cast my last look at the factory through a hole in a river bank and I will never forget that. I had finally left my home area and since then I have not been there, not even after the occupation by the Finnish troops. I remember it such as it was when I left, and the village such as it was before the war. (From an autobiography)

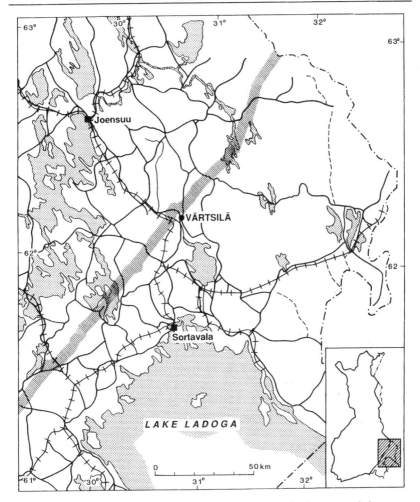

Figure 9.13   Transformation of the functional space: the location of the new Finnish–Russian boundary in relation to main roads and railway connections

Others moved back to Värtsilä after the boundary determined by the Winter War had been established, and they had to leave again in June 1941, when the Finnish troops began their attack to re-occupy the ceded areas.

It was Midsummer Day. I was working the dough when my sister-in-law came in timidly and said, 'Now it has come, at three p.m. we have to go along the street with our livestock and the train for evacuees will leave Kaurila station at 5 p.m. if anybody is going in it'. . . . A 20 km broad empty frontier was created. All the inhabitants were evacuated, but in spite of the order some old people stayed in their forest cottages to take care of their livestock. And nothing happened to them. They milked their cows and sold their milk to the soldiers to the last drop. . . . When we were approaching the town of Joensuu, we were just crossing the last field, when a large squadron of massive bombers flew above us. At first we thought they were ours, as we were not in war. Then a rain of bombs came down on the town. We were cursing, since we were in an open field without any shelter. The town in front of us started to burn. On June 28th we heard that Finland's Field Marshal Mannerheim had declared that hostilities with Russia had begun, because the Russians had crossed the border in many places and bombed several Finnish towns. . . . . In one day a gunner came and warned us that we should not be frightened if they start to shoot. It was July 10th, 1941 when it started. The firing lasted about two hours and it was heavy. When the firing ceased, the horizon turned red. They said, 'Now the Russians are burning Värtsilä down as they leave.' It was terrible to look at the sky above Värtsilä shining a fiery red. Later heavy rain and a thunderstorm arose which put out most of the fire and prevented it from spreading. I heard that our house was still standing. Later in the day I went with my husband to look at our home. We cycled and stayed on the border line. We were looking at the boundary marks, white on the Finnish side, and red on the Russian side. 'The boundary opens up like a crack', as it were. Forming a steep bend, it turned up the hill. Most of the buildings located in the open boundary area had been re-built, and the boundary was fenced with many layers of barbed wire on the Russian side. (From an autobiography)

When the Finnish troops crossed the boundary after the Continuation War had begun, fighting in Värtsilä was severe. Wilhelm Walfors inspected the area one week after the Finnish troops had re-occupied it and met with almost complete destruction. Värtsilä was regarded as one of the most severely ravaged communities in Finland (see Figure 9.14). Some of the factories had already been taken into use again by September, however (Zilliacus 1984, 159–61).

Three years later the next evacuation of machines took place, with their surrender to the Soviet Union. And finally — for the inhabitants — the boundary was closed for almost the next 50 years.

The people evacuated from Värtsilä were resettled in neighbouring communes, not elsewhere in Finland as in the case of most other evacuees (Paavolainen 1958, 1960), and in many cases they wanted to settle down in the immediate vicinity of the localities where they had lived before. In this situation it was difficult for some evacuees to accept the permanence of what had happened. In 1987 one old man remembered that 'they couldn't believe it. There were some who would not unpack their vans. They were just waiting to get back. Now they are beginning to understand that there is no return.'

As many of the people evacuated from Värtsilä as possible stayed in the commune. Only a small minority went (this time) to our relocation commune Karstula. Then all the old people from Värtsilä would often come to visit the hill of Patsola and cry. A stream of evacuated people came to Patsola, but not all could get a house, so they had to return to their relocation commune. The talkative Karelians did not feel at home in the inner parts of Finland. There was a longing which drew them to the border. (From an autobiography)

Figure 9.14   The landscape of war and peace. Finland's President Svinhufvud and the troops of the Värtsilä Civil Guard at the centenary celebration of the Wärtsilä Company in 1934 (above) and the same place in summer 1941, after the Finns had recaptured the area from the Soviet troops. Source: Wärtsilä Company/INDAV, Helsinki. Reproduced by permission

This longing is easy to understand on the basis of the existential experience of space, since when evacuees were relocated in distant areas, adaptation to the economic and cultural environment was always more difficult. It disturbed the rhythm and the knowledge with which their daily life 'goes on' and the practices through which this usually took place; in brief, their ontological security (Giddens 1984, 22–3). Karjalainen (1988, 98) remarks that the 'boundaries' of the life-world are boundaries of common experience, and that familiarity, similarity and normality constitute the 'border signs'. A radical, violent change in these basic parameters challenges the whole identity of the actors (cf. Eyles 1989).

The people still living in the border areas were now inhabitants of the borderland between East and West and had to meet with its new symbols and severe restrictions on action in their daily lives, both in local activities and in broader social consciousness. This caused also problems in adaptation, even if the locality was partly familiar.

Now we had again to adapt to the conditions and shrug off the last remnants of wishful thinking. The electricity was the first of the good things of Patsola. It was almost free. As the electricity plant of the Wärtsilä Company exploited the river of the village, it had to give free electricity to the village. The electricity was now used without any pity. Even the outdoor lighting was used every-where. In the dark, Patsola was like a small town. Suddenly we received a severe reprimand: the lights should not be seen beyond the border. They said that those on the other side of the border had ordered that they should not see any light from the Finnish side. Outdoor lighting now ceased at the houses located near the boundary, and even the windows facing the border had to be covered. All shooting was prohibited in border areas. Even rifles had to be covered and could not be brought on one's back nearer than 15 km from the boundary. (From an autobiography)

After the two wars the factory and the industrial workplaces were left on the Soviet side — to become numbers in the long chain of ceded factories in Karelia. Similarly the old commune centre was ceded, together with all the activities and symbols connected with the varying spheres of social action and the everyday life of the inhabitants. These included administration, basic services, the Friday market, and import-ant symbols such as the church, the graveyard etc. These were soon to become the essential symbols of the lost area in the collective memory of the inhabitants, symbols that almost all the people interviewed always returned to in the course of our discussions. In a way, the essential coordinates of the old community were broken.

After losing all the industrial workplaces in the factory, the commune became a peripheral border area with agriculture and forestry as the main forms of economic activity. A small foundry had been founded at Peijonniemi in Tohmajärvi at the beginning of the war — now named

Uusi-Värtsilä (New Värtsilä). This used electric power produced by the two hydro-electric power stations owned by the Wärtsilä Company and gave work to a couple of hundred workers. It has been said that Uusi-Värtsilä was located at Peijonniemi because the owner of the factory, and particularly its director, wanted to continue industrial activities in Karelia.

Nevertheless, the location of Uusi-Värtsilä was seemingly decided on rational economic grounds, i.e. minimum distances from power stations and railway connections, and on the possibility of obtaining cooling water (Haavikko 1984, 82). The opening of the Uusi-Värtsilä foundry was not a serious substitute for the lost workplaces, however, but merely a solution to make use of the energy produced by the power stations. Its activities began to decline at the end of the 1960s and the Wärtsilä Company sold it off. Like many other firms based merely on one economic branch, the firm collapsed. Simultaneously the local people lost their last direct contact with the symbols of the Wärtsilä Company — all of which were now beyond the boundary. They also lost the practices that shaped their everyday lives and their associated meanings.

The final ceding of the area completely destroyed the infrastructure and the population base. Just before the war the commune had some 6000 inhabitants, whereas after the war its population was less than 3000, and since it could not provide jobs for its inhabitants, out-migration continued at a rapid rate. The population in 1950 was 2000, that in 1960 about 1700 — including those from the Patsola area, which was incorporated into Värtsilä in 1951 — and that in 1980 922. Today there are less than 800 inhabitants.

The Wärtsilä Company, the original constituent of the locality, also ceased all industrial activities in the commune. Wahlfors had already begun to expand the firm on a national scale in 1936–39, and several old enterprises were fused with it to concentrate and rationalize Finnish industry. Zilliacus (1984, 151) writes that there are many ironic features in the relations between the Wärtsilä Company and World War II. The war crushed the 'mother factory', burned down, destroyed and robbed it, and altogether almost 200 men of the community were killed on the front, a fact which stamped itself on the collective memory of the inhabitants, just as the more than 85 000 dead did in the case of the Finnish collective memory (cf. Kasanen 1981, 48). But, after the Winter War the company received compensation for the lost machines from the Finnish government, and its new docks, located on the coast, received orders from the previous enemy and the company expanded (Zilliacus 1984, 159). After the Continuation War the machines again had to be surrendered to the Soviet Union.

In the late 1930s just before World War II, the deterioration in international relations provided a new field of operations for the

Figure 9.15   Two key symbols expressing the social and historical basis and territorial-administrative aspects of the identity of Värtsilä: the shield of the Wärtsilä Company (left), and the coat of arms of the commune of Värtsilä which was adopted in 1956 (right). Source: Published by permission of Metra Corporation and the Local Council of Värtsilä

Wärtsilä Company — the armament industry. During the years of the war and afterwards, when Finland had to pay indemnities to the Soviet Union, the company still expanded. Wilhelm Wahlfors was a member of the delegation which made the decisions regarding the nature and price of the commodities that were produced as war indemnities, as well as being the managing director of one key company producing these commodities. Zilliacus (1984, 151) writes that now 'Wärtsilä found out that it had changed to a modern, effective export industry plant, to which the whole world opened its markets'.

When we think of the schema of three spatial scales put forward by Taylor (1982, 1993b), the operations of the company became directed at the scale of 'reality': towards global markets. Up to the mid-1980s its activities expanded worldwide. Simultaneously the locality whose name the company still carried had lost its most essential constituent. The company still regarded Värtsilä as its home and paid some taxes there, and it still carried in its trademark the armoured hands and sabres of the Karelian coat of arms — the symbol that had once so annoyed the Soviet ambassador (Zilliacus 1984, 152, see Figures 9.15 and 9.16).

After the Continuation War the workers had to be placed at the company's other factories, while for those who stayed in Värtsilä this meant separation from the company and the jobs it provided. For them Värtsilä, or what was left of it, was still part of their place, but for the capital of the Wärtsilä Company the village was now merely a point in space and its new location was no more effective in the space of investments (cf. Hudson 1988). After the war the company still did some charity work in the commune and parish of Värtsilä, besides having two hydro-electric power stations there.

The name 'Wärtsilä' has recently lost much of its historical

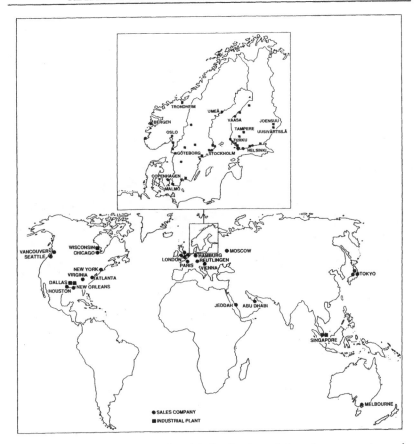

Figure 9.16 Capital in space. Spatial distribution of the sales companies and industrial plants of the Wärtsilä Company in Scandinavia and on a world scale in the mid-1980s. Source: Based on Haavikko 1984, 169, 172–3. Reproduced by permission

connotation since the concern has fused with other Finnish companies under the general heading of 'Metra', and only one part of it, Wärtsilä Diesel, still carries the name of the original firm. This concentration of industrial capital may thus lose its last symbolic contact with the locality it once created. The commune of Värtsilä, for its part, will become one of the new tourist access routes to Russia, and is looking forward to a new, prosperous future.

9.3.1   The establishment of the new boundary after World War II

After the war the Soviet Union received the main part of the home commune of the company, and almost as if mocking the Finns, the old

Figure 9.17 The landscape of loss and dreams for old inhabitants of Värtsilä. The view from the hill of Patsola (1992), which became an important sightseeing place and place of memories after the war. The white buildings in the middle of the picture are part of the Russian settlement of Värtsilä

centre of Värtsilä with its factory buildings and chimneys was left clearly visible but nevertheless just a couple of kilometres beyond the new boundary. This fact was significant for the formation of the 'identity and memory rites' of the local inhabitants and others coming to Värtsilä. Visiting the hill of Patsola and looking 'illegally' from there with binoculars at the houses, factory and chimneys beyond the border became an indispensable rite for those visiting Värtsilä, as became obvious in interviews in 1987 (Figures 9.17 and 9.18).

At least old friends who were born in Värtsilä can come here and enjoy going to the hill, where they can see it and refresh their memory. (Woman, 80 years)

I wish one could go there. People would like to look at it. One man who visited us once — he was one of the old workers of the factory — said on the hill that when he comes next time, he will lie here for a whole day with binoculars and look at it. But he died before that. (Man, 73 years)

The boundary, it's the boundary that interests people. They come here with binoculars, and during the summer they — the tourists — come by the busload and look at it with binoculars. They do not have permission to take photographs of the Soviet Union, but they do take them. But the Soviet Union is not allowed to be seen in a picture. (Man, 70 years)

As Figure 9.19 illustrates, the new border and the frontier zone established in 1947 were deeply canonized in the everyday routines of the inhabitants. Many of the houses were now located in the frontier

Figure 9.18 One of the many. Director Wilhelm Wahlfors looking at his old empire from beyond the new boundary in 1950. Source: Zilliacus 1984, 317

zone, where the strict rules of behaviour were something totally new for inhabitants of the region.

The new border between Finland and the Soviet Union also had many drastic effects on the everyday lives of individuals and families. First, the war cut off all connections with the East. Thus the essential constituents of the community and of the everyday practices of the inhabitants were lost, i.e. the ironworks and about 1000 industrial jobs. More than 20% of the economically active population worked in industry in 1940, but less than 6% in 1950 and only 2% in 1960. Thus an industrial community of almost 6000 inhabitants reverted radically

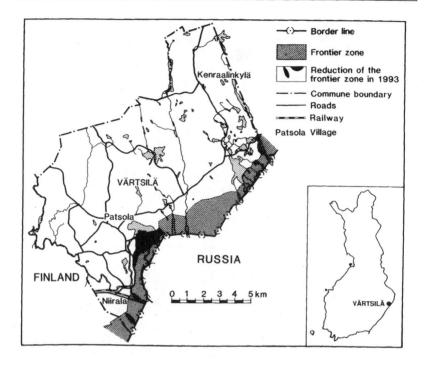

Figure 9.19   The area of the commune of Värtsilä today

to an agricultural community of only about 2000 inhabitants. It also lost most of its infrastructure (office buildings, roads etc.) and all private services.

The loss of the factory community was as radical an upheaval for the individuals as for the whole community. The new border caused problems for some small farms and forestry villages and for the everyday lives of the inhabitants. The livelihoods of many families were severely disturbed, because the ironworks had been the most important source of wages. Also the road network had been cut off, and this made movement very difficult.

The inhabitants who had had to move from the present Russian side of the border and who still stayed in Värtsilä were forced to begin their new lives in quite a peculiar situation. They were virtually continuously faced with the old environment with its factory and houses, but it was impossible to visit it. Most of the older people interviewed during the summer of 1987 said that this new geopolitical situation was initially felt to be a strange one to a greater or lesser extent. The old border had been about 100 km. to the east, and the new one was located right in front of their eyes. The image was without doubt exaggerated by the content of national socialization, e.g. education and 'public opinion' in general, in the first decades of the twentieth century which had created a very deprecatory and frightening picture of the aims of the Soviet Union. Furthermore, people were forced in some cases to watch the parameters of their former ontological security literally being torn down in front of their eyes.

We saw our home beyond the border in front of our eyes, on a high hill. It was the birthplace of my husband. He had lived there for the whole of his life, worked for it, and sacrificed his strength for it. There was nothing left but the beads of perspiration as rewards for his trouble. We couldn't even cry any longer. Those who came from further away couldn't do anything but cry. (From an autobiography)

My father's brother, who had just built a new house on the other side of the river, died behind his house in the Continuation War. The house remained there and they did not burn it down. Every summer we looked to see if anything had happened to it. Then they broke it down timber by timber and took it away. (Woman, 47 years)

A rather human psychological reaction after the creation of a border following a war is fear among those living in the border area, particularly on the side of those who have lost the war (cf. Muir 1975). This fear is particularly easy to understand in the case of ideological boundaries. The new boundary at Värtsilä naturally aroused fears among the inhabitants after the war, particularly among those who lived close to it. Because the landscape opened up directly onto old Värtsilä and some people could watch for years what was happening to their old houses and fields, this also aroused bitter feelings among them. Their nervousness was further strengthened by the fact that for some reason the new inhabitants on the Soviet side made a lot of noise and people often heard shooting in the middle of the night, for instance.

Well, at the beginning I felt gloomy and horrified when the boundary was there and I had to move near to it. I was somehow afraid, since the former rowdy element were there and I wondered how we could manage together. Now I have got used to it. At first it was simply horrible. (Man, 62 years)

Personally I felt nervous — maybe a reaction after the war — since our neighbours on the other side had a lot of shooting practice. When I woke up in the middle of the night, I wondered if peace has not come after all. (Man, 62 years)

9.4   THE TERRITORIAL IDENTIFICATION OF THE INHABITANTS

Since people live their everyday lives in a jungle of regions, localities and inherent social distinctions, social and spatial frontiers and boundaries are essential for organizing their routines. Whatever human beings do is usually motivated by what they have experienced in their 'own' domain(s) and nowhere else. The boundaries of these domains are shaped in daily life, and the effective radius of their activities does not extend beyond these physical or symbolic boundaries (Heller 1984, 238). The spatial extent and capacities of social power vary markedly with the social and spatial division of labour.

Personal territorial identification is also a complicated hierarchical phenomenon, as is the collective regional consciousness and regional identity expressed in the structures of expectations of territorial units. It should be noted that people do not necessarily have such categories as 'identity' or 'regional identity' in their practical consciousness. These things are thus lived out in daily practice rather than being thought of reflectively. People can express their territorial identification in everyday practice in many ways. They can participate in various activities which aim at promoting the development of the region, they can combine their territorial interest with political interests to secure the existence of their community, they can subscribe to newspapers published in 'their' community or region, and they can form territorially bounded associations.

People are seldom able to label these activies as explicit expressions of their territorial identification, for instance, and this probably holds good as regards the whole question of expressing territorial identities. This arises partly from the fact, pointed out by Marcus (1992, 315), that the identity of any person or group is produced simultaneously in many different locales of activity by different agents for different purposes. Also, 'one's identity where one lives, among one's neighbours, friends, relatives or co-strangers, is only one social context, and perhaps not the most important one in which it is shaped' (Marcus 1992, 315). Hence, the social and the spatial become in a complex way interwoven in the construction of identities.

Discussions with people, asking more or less explicit questions regarding their identification with various spatial units, can thus be at worst a rather violent intervention in their socio-spatial representations and obviously a typical expression of scientific colonialism or communicative hegemony, i.e. imposing 'geographical' communicative norms and categories on those consulted, for instance (cf. Briggs 1986, 121). At best it can of course be a truly dialectic discourse, in which both the researcher and the subject(s) change categories and mentalities and some kind of learning process can be presumed to take place (Marcus 1992, cf. Clifford 1988, 34).

All this harks back to the difficulty of matching everyday language with the language of science. Schutz (1971, 119–20) discusses this problem, and begins with a comment that what the sociologist calls a system, role, status or role expectation, for instance, is experienced by the individual actor on the social scene in completely different terms. This probably also holds good for the geographical categories such as regions, places etc. Schutz points out that for actors all the factors denoted by these concepts are elements of 'a network of typifications' — typifications of human individuals, of their patterns of action, of their motives and goals, or of the socio-cultural products which originate from their actions. Typifications on the common-sense level — in contradistinction to typifications made by social scientists — Schutz writes, 'emerge in the everyday experience of the world as being taken for granted without any formulation of judgments or of neat propositions with logical subjects and predicates'.

The complicated problem of territorial identification will be approached in the following discussion by using several sources and by setting them into conceptual frames. The basic interpretations are based on the interviews carried out in 1987 and 1988, where several open-ended and closed questions were used in an attempt to illuminate this problem. In the first place, the interviewees were directly asked to which 'region' they felt they belonged and on what grounds. Then they were asked whether they were able to name their home area. Finally, they were asked to choose from several alternatives the spatial unit with which they identified most strongly. They were also asked which newspapers they subscribed to and whether they participated in the activities of any associations operating in the locality. Apart from the interview material, the conclusions are based on previous field studies and on various documents which illustrate this problem and the relations of territorial identities to other social identities.

In principle, there are four categories of people identifying themselves with the idea of 'Värtsilä' as a consequence of the war. First, those who identify themselves with the lost and present Värtsilä area, secondly those identifying themselves with the present Värtsilä, thirdly those who now live elsewhere and carry with them memories of Värtsilä, and finally those who identify themselves vaguely with the Wärtsilä Company. Most of the interviews were conducted with people who still live in Värtsilä, but four former employees of the Wärtsilä Company were interviewed in Joensuu in 1989.

In the light of earlier research into the territorial identification of the Finnish people, those who were born and still live in Värtsilä have in principle four territorial units which can be of some importance in this respect. The local context is their own village inside the commune, and the present area of Värtsilä still contains several historical villages which people distinguish in the territorial structure, notably Niirala, Patsola, Kaustajärvi and Kenraalinkylä (see Figure 9.19). The commune is the lowest 'official', administrative spatial unit in Finland, i.e. the local state, and in this case we are concerned with that of Värtsilä, nowadays one of the smallest of the 19 communes in the province of Northern Karelia. The latter, for its part, is one of the administrative provinces of Eastern Finland. Finally, Finland itself is a central territorial identification unit. For most people who were interviewed in the course of the project the question of their 'home area' was rather clear. Värtsilä is the area they belong to more often than belonging to some other territorial unit.

The role of Värtsilä in the interviewees' regional consciousness also became apparent in the fact that, with only a few exceptions, they also strongly supported the independence of their small commune. There have been several discussions since World War II as to whether it should be merged with the neighbouring commune of Tohmajärvi to make the population base larger, but the inhabitants of Värtsilä have resisted this, partly for emotional reasons and partly for practical ones. First, some of the inhabitants wanted to maintain the territorial unit which is essentially associated with their life-history, and secondly, the great majority thought that union with Tohmajärvi would lead to the peripheralization of their daily and communal services, or to a rise in taxes.

I believe that we would have played second fiddle here and that they would merely have developed the centre of Tohmajärvi. Nevertheless there are a lot of buildings and other things that we can obtain with our own funds. (Man, 52 years)

Värtsilä was also the most popular unit of identification among the ready-made alternatives given in the interviews. When asked which of the following alternatives they feel that they identify with most — their own village, the commune of Värtsilä, the province of Northern Karelia, or Finland — only a few people said they felt that they belonged to units other than Värtsilä, generally to Finland.

Earlier results regarding the territorial identification of Finns indicate that people identify most commonly with the administrative commune in which they live, which represents the local state in the Finnish regional system. This is based on the fact that this unit is close to people's everyday lives, and people have an explicit administrative, political and economic relation to it. They pay taxes to the commune, have political representatives in its administration, and most collective services are provided by it. The provinces, for their part, are units that are conceived of in political and cultural rhetoric as emerging from 'below', but people do not identify with them very often. This is based on the fact that they do not have any explicit relation to these territorial units. Nevertheless, there are major differences between the Finnish provinces as far as their shaping of regional consciousness is concerned (Paasi 1986b).

Provincial identification among the interviewees became apparent in connection with the newpapers to which the households subscribed. With only a few exceptions they took the major provincial newspaper, *Karjalainen* (The Karelian), a supporter of the Coalition Party. Provincial newspapers are very strong in the Finnish regional system, and it is not unusual for more than 80% of the households in the area to subscribe to the main one for their province even if it is the organ of a party that the people do not support in elections. Newspapers have also been essential economic and cultural forces in the creation of several new provinces in Finland from the nineteenth century onwards, and they also have very sharp boundaries between their distribution areas, so that a strong regional point of view is one means of sales promotion. All these features contribute to the fact that newspapers are important media for creating provincial identities in Finland (Paasi 1986b).

Karjalainen, for instance, exploits the idea of provincialism effectively — as do other provincial newspapers. It presents itself as a Northern Karelian newspaper, fairly often writes of the inhabitants as Northern Karelians or even simply Karelians, and is markedly regionalistic in its attitudes. Newspapers have in principle two functions as regards the role of regional identity. First, regionalistic and provincial articles are indicative of the aims of certain individuals and organizations to create identity and solidarity among the readers and to provide material for these purposes. Secondly, articles are an expression of the existence of this identity among some actors. Provincial identity in Northern Karelia is fairly strong on account of this strong regional press and the visible activities of regional activists.

9.4.1 Where do I belong?

Nevertheless, as far as the open-ended questions were concerned, some interviewees simply seemed not to have the words to reply to the scientific categories or even 'territorialize' their identification. For instance, when the interviewees were asked to which region they felt they belonged, several of them simply could not answer the question by expressing their identification in these spatial terms. This is a fitting illustration of the character of practical consciousness (and of scientific imperialism!).

How aware are people of their local forms of life, spatiality or culture? What is it of which they are aware? Cohen (1982, 5) states in the case of culture that it is not usually experienced as a set of coherent systems of ideas, as in the idealized, abstract and therefore somewhat unrealistic accounts which researchers typically believe and present. Cohen indicates rather that people know *their* way of doing things, and thus the customary mode of thought and performance.

It is in any case obvious that people identify themselves with their home region, but many of them simply do not reflect these facts in spatial terms in their daily lives. Rather than 'knowledge', identification with the community represents 'feeling' (cf. Hummon 1990, 17). The following extract from a transcript of one interview illustrates this:

Q: Can you say what is the region to which you first and foremost feel that you belong?
A: This is it, I have lived here the whole of my life.
Q: Right and this (what?) is your homeplace?
A: This is our homeplace, or the one beside this.
Q: The one beside?
A: Yes.
Q: Can you give some name to your home place?
A: Uh, how to name it? I can't say.

It is immediately clear in this specific case that the person interviewed simply does not think in the geographical or spatial categories that were the basis for the interviews. This citation is not the only one. As many as one-fifth of the people born before World War II and 40% of those born after the war simply did not 'regionalize' their identification. Some interviewees identified themselves with a 'forest area', a 'little village in the countryside', a 'home area', etc. Several people also had difficulties in naming their region. This makes one more and more convinced of the basic premise of the institutionalization theory: that 'regions' become constituents of one's place, i.e. a personal spatial history is formed simply in various institutional practices, in action. For many people Värtsilä is not a 'region' but simply their home, the lived context of everyday life, where most of their social interaction is located and where most of their meaningful daily routines take place. Thus the meanings of the locality have more to do with everyday living and doing rather than thinking. As Stewart (1990, 74) states, regional consciousness is probably less a matter of 'geography' than it is a state of mind; it bears no necessary relation to artificial administrative lines created by governments or administrators.

This returns us to the conceptual theme discussed particularly by humanistic geographers. Relph (1981, 171), for instance, points out that the words 'place' and 'landscape' are misleading abstractions and have meaning merely to the extent that we can relate them to specific scenes and situations. Buttimer (1978), for her part, points out that 'even to *discuss* place we have to freeze the dynamic process at an imaginary moment in order to take a still picture'.

It has been said that names of various 'places' or spatial entities are among the things that link human beings most intimately with their territory and that the naming of regions and other territorial units is indeed a part of a fundamental structuring of existential space. Localities and their names are sources of identity and security (Relph 1976, 16, Ryden 1993, 79). Karjalainen (1988) illuminates this problem effectively in his discussion of the experiences of place among evacuated Karelians, talking about objective and subjective places, by which he refers to objective factual locations and 'lived experience'. He concretizes his discussion by referring to a novel by Unto Seppänen, a well-known Karelian novelist, who describes the feelings of the evacuated people after they had been assigned a commune to which to go (cited in Karjalainen 1988, 97):

Risto then wrote on the cover of his diary, for safety's sake, the names that had been given to his fellow traveller, which one could follow to reach the destination: Ilmee, Savitaipale, Tuohikotti in Valkeala, Selänpää in Valkeala, Iitti, Urajärvi in Asikkala, Kurkila in Asikkala, the parish of Lammi, Tuulos, Vanaja, Renko, Tammela, Jokioinen, Humppila. . . After leaving the office, men were leaning on Risto's load and staring for a moment at the cover of the diary, slowly spelling out this long main street of names. Some of those names had surely flitted by in newspapers in various connections, as small, unattainable, faraway places. But now, when they had to drive through them, they turned out to be huge and frightening: in fact they felt that they had to drive into those names, into their established course of life, as small beggars in unknown landscapes.

The interviews conducted in Värtsilä also support the argument that localities, and above all their names, can be understood as signs, the message of which is not the same for everyone who 'reads' them. As Keith (1988, 42) remarks, different regions and localities signify both collective and individual experiences. The memories bound to a locality may be the property of one, two or a handful of people. Locality formation always contains an inherent element of power and manipulation, which can manifest itself in the construction of the meanings attached to space. This has been typical of the construction and marketing of the new, symbolic Karelia on the Finnish side of the boundary since the 1970s.

9.4.2   The sedimentation of historical experience

In order to understand the nature of territorial identifications, it is sensible to point again to the personal histories and their sedimentation and to the histories of families and the community as a whole. As was argued above, the factory community in Värtsilä constituted a texture in which the activities of the factory were crucial for the production and reproduction of daily life. This texture created in many cases a historical continuity in which a person's own history together with those of the other family members was connected with the history of the community, which created a feeling of security (cf. Eyles 1989, 105). This continuity often stretched through several generations, as the following excerpt from an autobiography, published in Kasanen (1981), shows:

My family has had foundry workers in it for three or four generations — the work has passed from father to son and I am the last. I remember and know my grandfather best. He worked both as a foundry worker and as a miller in the mill which was next to the foundry. I started working in the foundry with my father when I was less than fourteen years old. . . In the evenings, when the working-day was over, we would sit on the stairs, often in large groups, and discuss. The atmosphere was good. To my mind, this is a typical picture of the sense of belonging together that the people of the factory had. We would talk about work and common affairs. Everybody earned his living at the factory, and there were hardly any other possibilities. The children grew up and the parents knew that the factory would give them work when they came to that age. Then the economic situation of the family would become better. The children went to work and lived together with their families until they established a family of their own. In those days people did not get married when they were very young.

Probably one of the longest connections with the company is that of the family of Helge Ratilainen, born in 1918. His grandfather Juho, born in 1830, came into the service of the company in 1869 and worked there until he was 92 years old. Both Juho's son Jooseppi and his son's wife Olga worked in the company and their son Helge came to the factory at the age of 15 with false identification papers, because he was underage. He worked at the factory up to the war, moved to other works owned by the company in Southern Finland for a couple of years after the factory was ceded to the Soviet Union, and came back to Northern Karelia in 1949 to work at its plants in Uusi-Värtsilä and Joensuu until his retirement. His wife also had a long career in the service of the company, and their son and his wife also worked for it for a couple of years (see Figure 9.20). Their family history and the history of various generations is hence totally sedimented into the fortunes and misfortunes of the company, so that Värtsilä merely as a specific territorial unit is not their Värtsilä. For them, Värtsilä is the company and vice versa.

This very concrete illustration makes it easier to understand what we mean when we argue that if we want to understand the personal meanings and the historical character of spatial experience, the idea of place cannot be reduced merely to some specific location. The

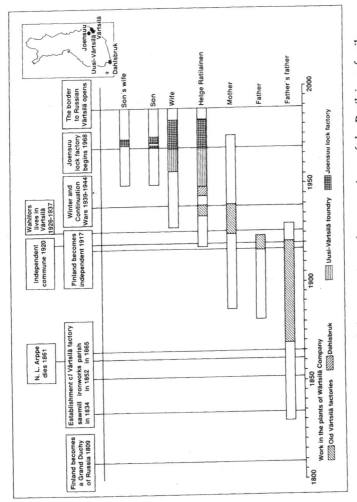

Figure 9.20 Time, space and the generations: the connections of the Ratilainen family with the Wärtsilä Company and the commune since the early nineteenth century. Source: Published by permission of Helge Ratilainen, Joensuu

territorially and historically bounded commune of Värtsilä and the practices and common memories are merely one constituent of the spatial identity of those who have lived there or still live there. Värtsilä for Mr Ratilainen and that for his son, born in 1950, are completely different things, even if Värtsilä as such has some common meanings for both. For Mr Ratilainen, as well as for others who lived in old Värtsilä and worked in the factory, the ceding of the area meant a deep sense of loss. The structures with which they had identified in institutional practices during the important years of their lives were, to employ the expression of Hareven and Langenbach (1981, 109), ruthlessly and unceremoniously swept out from beneath them. In 1990 there were still 114 people in the commune of Värtsilä who had been born there before World War II — only 13% of the whole population, although altogether 46% of the 853 inhabitants in 1990 had been born there.

As a conclusion regarding the identification of the inhabitants with various territorial units, it is obvious that their identities are much more fragmented and 'taken for granted' than the analytic, academic categories usually require. Also, it is important to note that various territorial identities are constituted inseparably from each other and other social identities (e.g. gender and generation, status and even class identities). The most essential distinction in Värtsilä appears to be the struggle between local political groups, which is reflected in the everyday discourse of ordinary people. During the interviews people were asked whether the commune constitutes a solid 'whole' or whether it consists of smaller groups who are pursuing their own interests. People were almost unanimously convinced that the situation had been better in former times, when people were actively thinking in terms of common interests. Surprisingly, a common scapegoat for these problems was also identified, the new administrator of the commune, who had recently entered the community from outside and was regarded as being too 'political'.

The social identities of the people in Värtsilä hence seem to be rather fragmented. Whereas the factory provided a common denominator before the war for almost all possible activities and also the social conflicts that took place in the commune, no establishments creating a common identity existed any longer. The increased social and spatial division of labour in the commune has brought new residents into it and created new social identities. The number of elderly people in the population is large, more than 25% being over 60 years.

All in all, various social and spatial identities are part of the naïvely given world. They are expressions of something which can perhaps be labelled as the 'hegemony of the given'. Identities reflect above all the sedimentation of collective memories in individual life-histories and present-day practices. 'Sedimentation' may therefore be phenomenologically understood as a process by which various elements of knowledge and their personal interpretations and implications are integrated into the layers of a person's previous knowledge. The sedimented 'items' are then mixed with already existing typifications to constitute the core of new, emerging typifications (Schutz 1971, 322). By this means they gradually become an individual's habitual possessions and part of the world he takes for granted.

Here we come back to the idea that 'in each of us, in varying proportions, there is part of yesterday's man; it is yesterday's man who inevitably predominates in us, since the present amounts to little compared with the long past in the course of which we formed and from which we result. Yet we do not sense this man of the past, because he is inveterate in us; he makes the unconscious part of ourselves. Consequently we are led to take no account of him, any more than we take account of his legitimate demands. Conversely, we are very much aware of the most recent attainments of civilization, because, being recent, they have not yet had time to settle into our consciousness' (Durkheim, cited in Bourdieu 1977, 79). Now the object is to search for the keys to understanding how this takes place.

9.5 TERRITORIAL TRANSFORMATION AND HUMAN GENERATIONS

The social sciences commonly emphasize class position, gender, education etc. as fundamental vehicles of social distinction and the construction of social identities. These are doubtless very important categories in the construction of daily life and of the various social identifications — as has also been seen in the case of Värtsilä. Nevertheless, encounters between individuals in space and time through specific episodes can create collective experiences which are unique in their collectiveness and which cannot be understood in the life-world of other individuals or through purely 'external' social categories such as those above. These episodes are not necessarily local. A fitting illustration of this is the role of wars in shaping the consciousness of individuals.

At the local level and in the context of everyday life, the dimensions of 'history' become complicated and new concepts are needed to depict the variety of local consciousness and its constitution. Among the most important 'mediating' social categories as far as local life is concerned is *generation*. It is a category providing people's spatial or local consciousness with common cultural elements, identity and thus frames for interpreting new experiences (Paasi 1986a). As to the role of individual history, generation becomes perhaps the most important vehicle of social and cultural distinction. There are also other bases for distinction, such as occupation, social class etc., but in the constitution of a local community generations bind people with the episodes of the history of the locality and the whole society.

It can be argued that local identities form a continuum, since memory is furnished not only from the recollections of events which an individual has experienced but from memories of older associates. 'Layers' of personal history always include elements of the history of the 'larger self', family, neighbourhood, locality and nationality (cf. Shils 1981, 51), which unites the individual as a part of the histories of these entities. Nevertheless, since personal histories are necessarily bound to specific localities, to centres of meaning, and to the developmental features of the society through institutional practices, the time–space-specific, socially contingent encounter of personal histories engenders different forms of action and thought. A personally experienced past validates one's present attitudes and actions by affirming their resemblance to former ones, writes Lowenthal (1985, 40).

Here it may also be argued — on the basis of previous discussions — that in the case of Värtsilä its name makes a big difference between the generations that have lived in the commune. This became apparent after the interviews in 1987, the first year of this research project. It is common to argue that the succession of the generations, together with religion, class consciousness, economic mode, internationalism and war, is one of the most important mechanisms of socio-cultural change. Generations, sometimes regarded as independent of other social forces, produce pressures for change. Lambert (1972) writes that this occurs through (1) death and the resultant social forgetting, and (2) birth and the resultant fresh contact with the socio-cultural system and necessity for socialization, a process which is never perfectly reproductive of tradition and modal personality. Further, Lambert argues, politico-cultural consciousness normally develops within an approximate age range of 18–26 years, so that there is a most probable or 'best' time at which events have the most impact on the development of consciousness (Mannheim 1952, 300, Lambert 1972, 24–5, Schuman and Scott 1989).

One problem in using the idea of generations in social and cultural analysis is, as Williams (1976, 142) remarks, that in a period of rapid change the 'period' involved is likely to shorten and to fall well below that of the biological generation.

Generation is nevertheless a useful concept since it permits an understanding of the relations between life-histories and larger-scale institutional and social histories. Also, although it lies somewhat aside from geographical discourse, the concept also has spatial implications — or perhaps it is more useful to label them as geo-historical. As was argued above, places, as centres of meanings, are constituted through personal histories, and are bound to the history of society and various regions through institutional practices. The time–space-specific, socially contingent encounter of personal histories engenders collective forms of action and thought, and the concept of the structure of feeling proposed by Williams (1961, 1979) is useful for appreciating these forms. Mannheim (1952, 303) wrote about *generations as actuality*, which involved the co-presence of persons in a historical and social region, and the participation of this historical and social unit in the common destiny. He went on to define the parameters of generations and develop an idea of *generation units*, groups within the same actual generations which continually work up the material of their common experiences in different ways (1952, 304).

The fieldwork in the Värtsilä border area suggests that various generations can have radically different regional identities which develop out of their personal experiences. The main watershed in the case of Värtsilä appears to be World War II. Age, both biological and social, is thus a constraint in this respect. For the elderly, their present locations are often the foci of the constrained social lives of their reminiscences and often of their nostalgia for past places (cf. Eyles 1989, 109). Regions and states are a contingency, not a universal necessity — and it is even more of a contingency how individual life-histories become bound up with the logic of the emergence, transformation and disappearance of these territorial units, the institutionalization of specific regions. But individual (and group) interpretations of these processes are something that cannot be reduced to the institutionalization processes themselves, processes which usually take place on a time-scale that exceeds individual life-histories.

Roos (1987), having studied Finnish life on the basis of autobiographies, makes a distinction between four generations that emerge from life-stories written in the early 1980s. These generations have common experiences, but also differences in their lives. The first, the generation of war and shortage, was born in 1900–1912. For them war has been a crucial element in the construction of their life-histories and in many cases the war meant the loss of their 'best years'. The second generation was born in the 1920s and 1930s, and has experiences in common with the older one, particularly of war and shortage, and Roos labels it the generation of post-war restoration and affluence. The third generation was born during the war years or immediately after and became active in the 1960s, at the time when the shortages were over and the great period of transition proceeding in both the Finnish economy and the external cultural influences acting on it (cf. the post-war generational conflicts). The youngest generation in Roos' typology was born in the 1950s, and now, in the 1990s, the generation born in the 1970s is entering the typology.

In practice the generations distinguished by Roos can also be found in the case of Värtsilä. The proportion of the economically inactive population, i.e. mostly elderly people, increased rapidly after the closure of the factory (1940 — 2.7%, 1950 — 5.2%, 1960 — 10.6%, 1970 — 22.1%) and the increased out-migration was selective. As mentioned above, 114 inhabitants, 13% of the whole population, were born in Värtsilä before the war, and 36 of them before 1920, so that they must have been at least 20 years old before the war began, i.e. they experienced the critical age emphasized by Lambert (1972) and others regarding the crucial forms of socialization.

War is a much more significant stage of life-history for these generations — as for all those generations who had to move away from the ceded areas — than for those who did not lose their homes and localities, the practices of which became the constituents of their everyday life. Almost half a century of peace has meant that two generations of adults have grown up whose spatial identity and ties with the local environment, as well as their attitudes towards the border areas, differ from those of the generations who personally experienced the years of war, the moving of the border, the ceding of the old areas of Karelia and in many cases the loss of their place of work, home and social community. These are all events that have impinged deeply on the personal experience of the inhabitants. This spatial effect is only one part of the formation of the way of life for different generations.

For those born in the old, larger Värtsilä, who experienced the factory community and perhaps worked in the factory, identification with it is constituted on the basis of a continuum between the symbolic and the material. The factory, employment and daily life in the locality are all connected with the idea of Värtsilä that people have in their mind. The factory was the locality, and the locality was the factory. Identification with the symbiosis of the historical development of the locality and the company is clear for most of the elderly people, who point to the role of the company in the history of the locality:

Q: Do you know the past history of Värtsilä? What things are the most important and spring to mind?

A: Well, one famous name from Värtsilä, Wilhelm Wahlfors comes to my mind. He was the developer of Värtsilä. And in fact the Wärtsilä Company — this name will never disappear from the map, it is so famous. There it is, within sight, the home of the Wärtsilä Company [points with his hand]. There you can see the chimneys of the factory, almost through our window. (Man, 56 years)

The local 'communities' with which successive generations identify themselves may differ. This is particularly obvious in the case of Värtsilä, where the changes in the community have been so fundamental. The community with which the older generations identify contains elements of utopian thought. Some older inhabitants still retain the image of the old idealized factory environment and remember every detail of it. The old community (and the period of prosperity it experienced) are features that older people continually want to recall. Younger generations receive hints of this community through the stories of the elderly population, through literature and so on. Most of the younger inhabitants are to some extent interested in the history of the Old Värtsilä community. Nevertheless, only a few have read anything of its history, and even fewer know the important places or any of the key persons that still exist in the discourse of the elderly inhabitants. Very likely these memories will become less and less important to rising generations, who will no longer have any personal contact with people who knew the locality in pre-war days.

When part of the administrative commune and the factory — the central constituent of the commune — were ceded to the Soviet Union, the 'glue' that fixed the post-war generations to Värtsilä was only a small region, its name and various institutional practices. The larger community was present only in the narratives of past times. In other words, the constituents of the life-world are completely different for the younger generations than for the older ones. They have no golden store of memories in their minds, since they have always lived in the border area. For them the state border is a geopolitical fact, and it has always been located in the same place. Most young people identify themselves with the present community of Värtsilä.

For some inhabitants the name of the company is a sign representing the continuity of their own experienced local existence and identity, while for most Finnish people the name 'Värtsilä' denotes the late Wärtsilä Company without any explicit spatial content.

9.6 LIVING IN THE LANDSCAPE OF WAR AND PEACE

It has become obvious that boundaries are not merely neutral elements or lines in a cultural landscape, as traditional political geography has tended to represent them (cf. Prescott 1965, 26–7). In the case of all state border areas, particularly in the case of the border between the capitalist and socialist worlds, the local landscape and environment have of necessity carried strong ideological connotations. As Meinig (1979b) writes, an ideological landscape is commonly understood as a symbol of the ideals and values of a specific culture. A border landscape can also be understood as constituting and reproducing a particular social order and set of power relations. It is simultaneously a crucially important manifestation of the distinction between us and them, as well as of the state's power and authority that extends directly to all individuals within its boundaries (cf. Nisbet 1990, 93). In the case of the border between the Western and Eastern worlds, the ideological dimension turns out to be particularly significant.

Apart from these general observations, every aspect of the landscape, and particularly its interpretation, is loaded with meaning, meanings of home, of history, also of scientific interpretation and of personal histories (cf. Jeans 1988). For local people, the landscape and its symbols are first and foremost displays of (local) history and of the events associated with it — even though local symbolism often reflects simultaneously more general social and ideological processes, particularly in the case of the symbolism used to mark boundary lines.

The landscape of a specific local community is also a cumulative archive of the work of human actors and their relation to nature. Landscape, in this specific sense, always manifests itself as a specific locality, a unique focus of felt values and personal experience. Personal meanings, for their part, are created in the course of one's personal history in relation to broader social and spatial structures.

As far as the experiences of the boundary in 1987–88 are concerned, the 'generation gap' between those who have experienced the old environment and those who have not becomes immediately clear. The younger generations have been completely socialized to the existence of the boundary. It is part of the routine of everyday activities and part of the security that springs from the routinization of action. In fact, this feeling of security is in some cases emphasized to the extent that it may be asked whether this may not be a matter of cognitive dissonance, i.e. whether in this curious situation the expectations and beliefs of those coming from elsewhere are transformed into a kind of 'over-emphasized identity'.

It is much safer to live here, with all the border guards who are always watching the boundary. Nowhere else is so secure. (Woman, 23 years)

You know, the border is not frightening for us. At least I have grown up here. The border is there and it does not matter at all. . . When you have always been here, you don't put any special emphasis on it. The border guards, of course, are seen moving about here more frequently, but there is nothing more to it than that. (Woman, 28 years)

In fact, some people interviewed were rather amused by the reactions of those living on the other side of Finland, and particularly foreigners, who entertain mystically laden representations of the role of the boundary. This points to the fact that for most Finns — as opposed to foreigners — the boundary is an invisible representation of socio-spatial consciousness and a 'mental political landscape' which is totally different from the concrete, daily landscape that constitutes the context of life for the inhabitants of the border areas.

They ask fearfully how we can live so near. Aren't we frightened? What should we be afraid of? It is so safe here. It has not changed at all. (Woman, 29 years)

The generations who have not experienced the war live in a socio-spatial context which is characterized and limited by certain geopolitical facts. For them the boundary has always been where it is now, and they simply have no experience of any other situation. Some representatives of the younger generations, contrary to the older generations, are also able to analyse this situation discursively:

The situation is now what it is: the boundary is there and we have had to get used to it. There have not been any problems with this, since it has always been there — for this generation. But the previous generations have a completely different attitude towards it and the most extreme opinions are expressed by those who had to leave all they had and in practice had to begin again from scratch. Isn't this quite a unique situation, that the inhabitants escaped the occupation? This does not occur everywhere in the world, that people will leave their home region. . . The attitude towards the boundary today is ambiguous. The previous generation, who have personally experienced the situation and know the effects of the new boundary, think consciously that the older days were golden times for Värtsilä, and they very often return to those days. Our generation takes a slightly different attitude. We have the present Värtsilä and we have to struggle on this basis — our views of the world vary somewhat. (Man, 34 years)

As far as the experiences of the older generations are concerned, when the new boundary was established after the war it was a shock for those who still lived in the area. Those people living in the peripheral villages of the Värtsilä area suffered greatly for several years after the war with regard to the organization of everyday life.

During the war, when we did not have any vehicles, we had to walk to Värtsilä to bring all the necessary food, leaving in the morning and getting back late in the evening. When one had a burden more than 10 kilograms on one's back one could not run. After the war the traffic was lively here, several buses a day, and post buses, too. And now no buses come here, no post van, no buses. (Woman, 72 years)

The roads were cut off and we had to bring food from Värtsilä — as the journey to Kutsu was 15 km. I had to bring things, too. I took 15 kg of food in my bag and a can of oil in my hand. Then I walked this 15 km in both directions, as there were no cycles, and when my husband was at the front, our horse was there, too, so that we did not even have a horse at home. (Woman, 85 years)

9.7   PRISONERS OF SPACE? LIVING IN THE FRONTIER ZONE

The ideological nature of the landscape along the border between Finland and the Soviet Union rested the whole time on people moving or living in the border regions. The ideological role of the border had — and in fact still has — a concrete manifestation in the lives of Finns and foreigners. This manifestation is the *frontier zone* that was established after World War II (Figure 9.19). Another strong element in the ideological landscape is provided by the watchtowers of both countries, which continually remind people, particularly those coming from other parts of the country, of the peculiar nature of the local environment. In the Värtsilä area the border landscape communicates these things very effectively, not merely because of the signposts that inform one about the border zone but also because of all the visible signs of the character of the border. The frontier zone is fairly narrow, having varied from 500 metres to 2.5 kilometres up to spring 1993, which makes the ideological expression of the landscape easier for outsiders to sense.

In his book on elderly people, Rowles (1978) speaks of them metaphorically as prisoners of space — that is their abilities to move in space are restricted. This metaphorical expression also holds good to some extent in the case of the people living along the border area, where space really makes a difference. As we saw, the boundary zone was established in 1947 in a special statute, the basic idea being to prevent the worsening of relations between the two states as a consequence of frontier violations (Kirjavainen 1969, 342).

The existence of the frontier zone has been problematical, particularly for those people who have been living in the zone, but also for farmers living outside it who have their fields within it. They do not have any personal problems about moving into the zone because they usually have a permit that is valid for several years, but the real problems are in contacts with other people, e.g. firms which have business with those in the zone. The situation was particularly difficult when one had to pay to obtain permission to go into the frontier zone, i.e. if you wanted to visit friends living in the border area you had to pay 10 Finnish marks, more than 2 dollars, at the beginning of the 1980s. Similarly, if somebody needed a mechanic to repair something in their home they had to pay the extra costs.

Even though people in general have accepted the rather normative rules that have been established in the zone and have accepted them as being caused by the curiosity and blundering of tourists, some have been rather frustrated by them. This became apparent in the interviews conducted in the zone, and also in some cases where people who lived outside the zone had their fields etc. in it. Some of the most active people have also contacted the Ministry of Internal Affairs and

Members of Parliament, but no major changes took place until spring 1993. In some minor cases the outer boundary of the zone on the Finnish side has been changed to leave new houses outside it. The interviewees also generally emphasized that they have no problems with the guards themselves. Their protest was directed towards the faceless bureaucracy and nameless authority which they could not concretize or identify in their daily lives.

In a newspaper interview in 1981 some very frustrated people living in the zone along the southern part of the Finnish–Russian border argued that 'whereas prisoners can leave prison after completing their sentence, we have a life-long sentence' (*Uusi Suomi* 3 February 1981). Hence this problem was probably much more important. The following excerpts from the interviews characterize typical problems.

Life in the border zone certainly differs from life outside it, since we are in a sense prisoners. We need a permit to move about or stay there, and so do all our guests — otherwise we will be fined heavily. In a way our freedom has been taken away, so that we are under continual control, and there is no complete freedom. . . I think this is an expression of the rigidity of the Finnish authorities that they guard the boundary to the extent that people cannot come here freely. Movement in the border areas should be freer in a free world and state. (Man, 62 years)

Yes, we have problems. The commune organized an information meeting in the sports building — the decision-makers of the village versus the village. So he (a man) complained that it is curious that people should be living as though they were in a zoo, that their own relatives have to pay for permission to visit them. At that time a fee was charged for the permit. So he got the zone sign removed to behind his house. This complaint caused the sign to be moved. It was moved behind the house so that the door of the monkey cage was moved behind his cottage. (Woman, 47 years)

The repressive nature of the zone has manifested itself particularly in the fact that people coming into it from the outside have needed permission to visit inhabitants living in it. In the Värtsilä area, for instance, more than 10 families lived in the zone in 1987. Several quite specific restrictions also apply to the zone. It is forbidden to take photographs of the border or the land on the other side, to shout across the border, to discuss with the border authorities or civilians on the other side, to throw or in any other way send a newspaper or document etc. over the border, to light up the border or land beyond it, and so on. Rules like these have been difficult to understand for foreigners coming from Western Europe, as seen from the scornful evaluation of the British journalist Hamilton (1990). Thus the basic premises of Finnish–Soviet relations in foreign policy remained somewhat mystical to foreigners and have doubtless contributed to their understanding of the nature of Finlandization.

As we have seen, it has been common for political geographers to think that 'frontiers' are politico-geographical areas which lie beyond the integrated region of the political unit (Glassner and de Blij 1980, 82, Pounds 1972, 67). This is essentially more a question of spatial scale than of absolute fact. There is certainly a boundary between the states of Finland and Russia, but when we come to the level of everyday life in the border areas, the idea of the frontier zone as an interface between two social systems, as a relatively closed zone of contact, becomes clear. It contains features from the traditional idea of boundaries, i.e. as symbolized lines between two social systems, but it similarly contains some elements of a frontier, i.e. although the frontier zone is inhabited, legislation has a 'sedative' effect on movement within it (cf. House 1968).

According to the statutes, persons of more than 15 years of age living outside the frontier zone are forbidden to move about or stay within it without a special permit from the appropriate police authorities. The rules were still more rigid for foreign citizens up to 1993, as they required a permit issued by the Finnish security police. It is also still prohibited to take photographs in the zone. These examples are illustrative of the *taboo* character of the boundary, and also of the 'invisible' side of state boundaries. As Glassner and de Blij note, a boundary in fact is not a line but a vertical plane that cuts through the airspace, the soil, and the subsoil of adjacent states (Glassner and de Blij 1980, 84).

Life in the boundary zone has differed from life outside, and in some cases the zone has made everyday life rather complicated. People therefore speak of the 'curses of the boundary'. This refers to the fact that all visitors, even friends and those performing various services, have needed a permit to enter the zone. If people living outside the territory do not own land in it, they must have somebody who they are going to visit in the zone before they can obtain permission.

People simply cannot go into the zone, e.g. to look around without any objective. This lends a certain mystery to it, and many of the people interviewed expressed amusement at visitors entering the zone as if it were a secret area with a view to telling their friends at home about this exotic experience.

People coming from southern Finland and elsewhere, those who have not visited here before, wonder how they dare to come here. Once, even if they had permission they did not dare to come, and we had to fetch them here. (Man, 66 years)

According to the newest Basic Map of Finland (1:20 000) there are more than 20 houses that have been in use in the present frontier zone in the commune of Värtsilä. Nevertheless, only some 12 households actually lived in the zone in 1987 and 1988, when the interviews were conducted there. The number of inhabitants has decreased, and by now most of the houses located in the zone will have been 'liberated', since the authorities are planning to move its western border.

The new relations with Russia have probably led to this more open attitude of the Finnish authorities towards the frontier zone. In June 1993 the border legislation was changed so that in the case of Värtsilä the western edge has been adjusted so that no households live in the zone any longer and there are only two or three volunteers who wanted to keep their summer cottages within it (Figure 9.19). The border legislation regarding foreigners has also become less rigid, and permission to visit people with cottages in the zone is now issued by the Headquarters of the Border Patrol detachment for Northern Karelia instead of the security police. Hence the prerequisites for social communication have become much better. One key distinction that has characterized the internal organization of the social community of Värtsilä since World War II has now been removed, and the regional transformation of social and administrative practices is continuing.

9.8   BETWEEN THE EASTERN AND WESTERN WORLDS?

9.8.1   The role of the boundary in everyday life

A basic idea when undertaking the interviews was that the unusual spatial context and the presence of the boundary would characterize the whole way of life of the people living in Värtsilä. One way to map the importance of these questions in their daily routines was to ask what they would like to show tourists visiting Värtsilä. Surprisingly, a typical object for people was not the border but Lake Sääperi, which is famous for its numerous species of birds, and the physical environment of the commune in general. These answers probably reflect partly the relative unimportance of the border in the routines of daily life, and partly its 'secret' nature, which means that even though the border characterizes the whole landscape very clearly, the frontier particularly is hidden in the sense that its signs cannot be photographed and it cannot be visited 'just for fun', and so on.

During the existence of the Soviet Union, the Finnish–Soviet boundary was typically represented as an ideological boundary between West and East, the capitalist and socialist worlds. The boundary has also been represented to tourists as a cultural boundary between West and East. It has acquired different meanings for different generations, for groups coming from other parts of Finland and for visitors from other countries.

As regards the foreigners and people from western or southern Finland, the boundary, and particularly the frontier zone, was a mysterious place throughout the existence of the Soviet Union. A couple of the people interviewed described this in the following way:

Especially for tourists, the boundary raises their curiosity. This is clear, for as you know this tourist route is the Road of the Bard and Boundary. During summer-time many tourists and foreigners stay here and take photographs of the prohibitory sign. So they are interested in the boundary, but very few of them go into the zone without permission. (Man, 70 years)

They cannot understand how someone can live so near the boundary, a couple of hundred metres from the settlements of the superpower, the Soviet Union. These people are not necessarily foreigners. My friend from military service, for instance, visited us once and when we were walking beside the barbed wire fence, he wanted to go away from it because it sent a cold shiver down his spine. (Man, 24 years)

The division between East and West was (and still is) much more invisible among the local people, whether they belong to an older or younger generation. In fact the inhabitants have many jokes to tell about the curious context of their lives. This is owing to the fact that for them the boundary is part of the structuring of everyday life and not merely a representation reflecting the content of a wider socio-spatial consciousness:

No, we live in 'Western Russia'. I always say this, when people ask where I come from. So I introduce myself as a western Russian. . . People here say many times that when we live on the Finnish side of the border, we live in Western Russia. He [a man in southern Finland] really believed that this boundary is on the Russian side. (Man, 24 years)

I do not regard this boundary as a boundary between east and west. Of course it is a fact that there is a break between the two economic systems within 2 kilometres, a socialist system and a capitalist one. It is a kind of boundary, but I have never seen any other boundary. (Man, 30 years)

Some younger people suspect that the whole idea of the boundary between East and West has been created by businesspeople and is of importance only for tourists. The older generations are also accustomed to the boundary but, as the quotation above indicates, many had been bitter and afraid of it at first. The ideological border between East and West was literally in the backyard of some farmers, who in many cases had lost some of their land, which they could still see on the other side of the river which was now the new boundary.

The people living near the border varied in their attitudes to the

Soviet Union. Space-preference research carried out among Finnish university students in the early 1980s indicates that the Soviet Union was the least preferred country for Finnish students, while England was the most preferred (Paasi 1984b). According to student opinion, the most negative feature of Soviet society was its socialist economic system. The interviews carried out among the people living in the border region, however, showed that the Soviet Union was commonly regarded as a very closed system. This was a typical opinion for people who had some experience and memories of life beyond the new border, but it is difficult to say to what extent their attitudes were coloured by the above mentioned socialization and the establishment of the view of the Other — which was clearly coloured in a negative manner during the years preceding the war.

The Soviet people as individuals, on the other hand, were relatively highly respected among the people interviewed, even though some bitter feelings still existed. Several of the older inhabitants living in the border zone had become personally acquainted with them during World War II, since Soviet prisoners had been working on their farms. During the war years Värtsilä was a crossing-point through which prisoners were brought, and many people emphasized that they had often overstepped their authority in the treatment of prisoners, regarding them in many households as ordinary family members, and this had raised the suspicions of some authorities.

All in all, the interviewees in 1987–88 had a rather neutral attitude towards the Soviet Union, a neighbour with which they had lived for almost 50 years.

So, I do not have anything against the people, but I cannot accept the system that prevails there. An iron curtain is an iron curtain. I am sure that the people there are friendly; it is just that they have been the objects of such a harsh discipline. (Man, 28 years)

I am now 47 years old. After the war everything was so negative. A hatred of Russians prevailed when I was a child. All the men who had been in the war were full of bitterness, and they discharged this bitterness on us children. Today I am studying the Russian language and I am very favourably disposed about everything. I even go there. (Woman, 47 years)

10

BACK TO KARELIA

10.1 THE FORMS OF UTOPIA

There are certainly many other categories of social distinction, but the role of the generations appears to be of critical importance in the creation of a collectively experienced past in a local community such as Värtsilä, and this also appears to hold good generally among the evacuated Karelians. It is easy to conclude from the personal interviews that the people who live in Värtsilä today have quite a strong *ideal* regional and social identity of living there, based on a more or less vague general knowledge about the history of Värtsilä as (1) an administrative unit, a commune, (2) a social community with an industrial past, and (3) personal involvement with the history of the locality.

The matter of the generations is of particular importance as regards the elements of the 'Värtsilä utopia'. It is the utopia of a people who have experienced the existence of the factory community in their personal history, and equally the utopia of the older generations who have experienced personally the loss of the old community with all its essential symbols and activities (Paasi 1988). These older generations are in a completely different situation from the younger generations who were born after the new border with its symbols and restrictions became part of the everyday routines of the locality.

Utopias can be regarded as one of the distinctive characteristics of human beings: a form of thought which is often connected explicitly with time and space. Utopias may have only an ideal or desired content for something better, without any explicit relation to concrete social reality. Nevertheless, they may also have a content which is connected with the expected developmental tendencies and regularities of a local community or an entire society. Utopia is typically a social manifestation of hope, but it is also an essential potential for the life-history of an individual or group: 'an ability to have day-dreams is a condition for every human step of progress', as Ernst Bloch wrote (cited in Rahkonen 1985, 18).

In utopian thought it is possible to exceed the boundaries of time and space. As Porter and Lukermann (1976) point out, the dream of the utopian is to annihilate space as well as time. Much utopian literature reflects a deep sense of loss: we have been driven from the garden of Eden and long to return. Utopias can thus also be manifestations of an idealized past, a projection of the lost onto the present time and perhaps also onto the future. The perceived identity of each scene and object, writes Lowenthal (1985, 39), stems from past acts and expectations, from a history of involvements. In this specific sense the content of utopia recalls the connotations of the concept of 'community': the social ideal of a 'good life and place to live'.

The unique embroidery of personal experiences makes the individual and the spatiality of his personal history unique. This is true even though individual experiences are not simply the property of certain individuals but a part of the collective memory of the community. This holds good in the case of utopias, too. Porter and Lukermann (1976, 206) suggest that a utopia often rises out of the ashes of a terrible war. This is exactly the case with Värtsilä, and in principle with the whole of the ceded area of Karelia, where the new boundary permanently separated more than 420 000 Finns from their homes and broke the spatial bond of their social identity.

In Värtsilä, a new socio-spatial setting gave rise to utopian thinking among the inhabitants of the area. The utopia is that of yearning for the lost home. In the memories of the evacuated people the old factory community, with its workaday sounds, the lively bustle in the market and the many free-time activities, turned into a golden land of dreams in the course of the years.

Oh, how beautiful was the village of Värtsilä. It was pretty and had a factory and all those things. There were two shops at Kaustajärvi. I hope they will some day give Värtsilä back to us. (Woman, 75 years)

From all perspectives, Värtsilä was such a good place that there cannot be a better one. (Woman, 63 years)

For almost 50 years this community was partly real and present on the Finnish side of the border and partly utopian, unattainable, on the Soviet side. The old industrial community and its territorial elements — both functional and emotional — are deep and real in the memories of older people.

Of course, those old things, you can't forget them. I would remember almost all the places in Värtsilä. The railway station is 7 km away from here. (Man, 73 years)

You see the road. All the curves in the road are fresh in my memory. I remember every curve and every hill. (Woman, 76 years)

The utopian community has been strengthened by the fact that the ironworks have been in operation on the Soviet side of the border since the war, reminding the older inhabitants and factory workers from day to day of their existence. Also, the works could be seen across the new boundary for a long time, and can still be seen from some places on the Finnish side.

The manifestions of the utopianism of the Karelian refugees have been divided between two spatial scales, and have been directed towards both the past and the future. The loss of the Karelian areas has activated local and national organizations within civil society which have aimed at the continual production and reproduction of the collective identity of the resettled Karelians. This identity is partly maintained by organizing large annual meetings of Karelians and their descendants, for the latter are important for the reproduction of this identity. It is equally maintained by local associations, which serve to strengthen the collective memories of Karelia and to socialize the new generations into adopting this identity. The evacuated Karelians have created their own 'sub-culture' with annual general and parish-based meetings, with their own press, reports and books of collected memories. Their identity is also based on habits and narrative accounts that have been maintained between friends and relatives (Lehto and Timonen 1993, 88). The instilling of traditions into younger generations was found to be a problem by some of the old people interviewed.

Old Karelia is so beautiful. There are many places, many sights for old Karelians who have lived there and who know and feel the places. The present youngsters do not know them. It is all totally strange for them. How can we make the offspring of Karelians interested in looking at the homes of their fathers? Karelia was such a beautiful area. There are certainly things there to look at. (Man, 62 years)

There is also the 'written Karelian identity', the representation of the identity as it is produced and maintained by the media at the national level, and more especially in Eastern Finland (cf. Paasi 1988). This collective identity sustains the utopian concept of a community which is lost but nevertheless present in the collective action and meetings of its people and in their dreams.

Whereas the context for the operation of local associations is always restricted and the activities take place in specific localities, the nation-wide organizations of evacuated Karelians have, for their part, been active in constructing the mythical Karelian image of the border areas. This has taken place partly through active individuals, on the basis of their cultural and economic motives. In this respect they recall the 'ethnic activists', discussed by Allardt and Starck (1981), but they nevertheless have a more explicit territorial or regionalistic tune. A new symbolic space for Karelia is 'located' in Eastern Finland, or even Northern Karelia, in their discourse. Here we cannot neglect the marketing of tourism by the provincial associations and local authorities, as discussed above, for the inhabitants of Northern Karelia themselves do not often identify themselves as Karelians (Paasi 1986b). In the interviews conducted for this project, for example, the question of Karelian identity was, with a few exceptions, not especially evident among the people interviewed.

The boundary was loaded with strong ideological and mythical connotations after the war, these emerging partly from the strict legislation regarding it and partly from the active marketing of its heritage of mystical Eastern qualities for the purposes of tourism. Today various media play an essential ideological role in the social and cultural production of the meanings of various localities. Marketing myths are often more significant than the 'reality' itself, particularly in the simple visible facts (cf. Burgess and Wood 1988).

Several symbols have been exploited to depict and symbolize 'Easternness'. Among the most significant is doubtless the 'Road of the Bard and Boundary', along which the fruits of the Eastern cultural heritage and the boundary of the two worlds meet each other in a number of features. These include wooden houses, which are a curious synthesis of iconic and indexical signs in symbolization and which market Karelianism and its history. Iconic signs refer to visual representations which have a connection with the concept of the referent, e.g. with the idea of a Karelian house. Indexical signs, for their part, refer explicitly to the referent, e.g. to a certain Karelian house (on iconic and indexical signs see Burgess and Wood 1988, 108). The meanings associated with the Korpiselkä House (Figure 5.11) as an object of tourism are firmly connected with the idea of the boundary and cultural Easternness.

The official tourist information provided by the commune of Värtsilä does not emphasize Karelianism, which is easy to understand inasmuch as there do not exist any Karelian symbols in the commune as there are in many other locations along the 'Road of the Bard and Boundary'. Värtsilä is marketed as a focus of trade across the boundary and of tourist routes — as a 'Gateway to the East'. It is possible to argue that the boundary characterizes the past and the present of Värtsilä perhaps more than is the case in any other Finnish commune.

Another local form of utopia has been typical of the people living in Värtsilä, where part of the old community is on the other side of the border. This is a yearning based ultimately on the idealization of past social life and social relations associated with the lost environment (Kanter 1972, 32–57). In the case of Värtsilä it has also been based on dreams and memories of the old factory community.

One specific form of local utopian thinking in Värtsilä is directed towards the future. It is the idea of a *lost future,* and mirrors the lost ideals, beliefs and hopes of the inhabitants, the lost 'great future' of the Värtsilä area. Most old inhabitants are of the opinion that *if* Värtsilä had remained on the Finnish side of the border after World War II, it would be now a prosperous town, an obvious centre for Eastern Finland, and the interviews suggest that these feelings have partly been transferred to the younger generations. These ideas are not analytical. Accordingly, such a utopia is not contextualized at all within the framework of the development of the Finnish iron industry, the problems that industrial communities have confronted since the war or the premises of regional development. This would be an excessive claim from the viewpoint of local daily life. An academic discursive critique of these views would be absurd and external and would not have much to do with the practical world view that has arisen on the basis of common experiences and interpretations.

Even though the inhabitants of old Värtsilä have not had the possibility of visiting their old home area — except during the last few years — local experience, identification with it, the violent divorce from it and the ceding of Old Värtsilä are essential features of the place that has arisen in the course of the life-histories of those who were born before World War II. The desire to visit old Värtsilä once again was an essential part of the life-history of the people interviewed in 1987. This is to be understood in the light of the observation of Tournier (1971, 26) that to lose one's place implies a serious moral trauma: a human being who has lost his or her place can no longer be taken into account.

God damn, I want to go to Värtsilä! (Man, 64 years)
It is like a curtain, you can't go there however much you want to. (Woman, 63 years)

Here is a point to support the argument that both history and space make a difference. These people who have experienced the years of hard work, the years of war and the loss of their factory environment have experienced a radical change in some of the most important parameters of their everyday lives. Above all, the loss implies a change in familiarity, which usually constitutes the self-evident, unquestioned aspect of everyday life (cf. Eyles 1989). As these phenomena retreat into the distance, the lost area seems to become a romantic land of dreams. This is obvious, to employ the expression of Taylor (1993b), on the scale of experience in Värtsilä, since very few of the people interviewed stressed the social conflicts between the classes, poverty, unemployment or prostitution which were an essential part of the everyday life in the community. This is typical of 'root-searching': the past is usually contemplated as a golden age (cf. Wilson 1993, 192).

10.2 THE FULFILMENT AND DISAPPEARANCE OF THE UTOPIA

With the opening of the border to tourism as a consequence of *perestroika*, it has become possible for the older people to bring their 50-year-old utopian dream alive and visit their old home areas in the ceded parts of Karelia. There are still about 180 000 people alive who were born there, and the opening up of possibilities to visit the old home areas immediately gave rise to a boom in travelogues which were published in journals and magazines intended for the Karelians. Systematic collections of travelogues have also been compiled since 1990. Lehto and Timonen (1993) have typologized the materials of these collections, and point out that even though all the narratives are unique, there are many similarities between them. The culmination of most accounts is the search for and discovery of the old home, or at least its location. The stories transform and develop one and the same theme of the realization of the utopia: the story of visiting their home.

This stereotypic presentation of experiences partly reflects the fact that most visitors have already become acquainted with the code of the stories that have been published earlier. It is probable that the 'grand narrative of visiting Karelia' that has been established during the last few years is partly reflected in the stories told by individual visitors: what is seen, how things manifest themselves, and so on. But as Alasuutari (1989b, 82) stresses, modern individuals commonly shape their past in the form of stories, as a whole with a plot, not in the form of fragmented or disconnected episodes.

Lehto and Timonen (1993) distinguish between three generations of visitors to Karelia. The first is the generation who were born in Karelia and lived and worked there. Their perspective is not restricted merely to the home but the larger context provided by their own village also comes out in the stories they have written. For this generation a visit to Karelia is as much a visit in time as it is in space, a visit into the arms of the past. The second generation comprises those who were born in Karelia or during the evacuation, so that their visit is above all a journey back to their childhood, in many cases a discovery of their Karelian identity. For the youngest generation Karelia is a place that is familiar from the accounts of the older generations, and their stories,

Lehto and Timonen (1993, 93) claim, can be arranged on a scale from 'indifference' to almost 'ecstasy' at finding a familiar identity.

Soon after the opening of the border it became obvious that a visit to Karelia actually destroyed many people's utopian dreams. This was documented on television and in the other media immediately after it became possible to visit the region. For the media, this also provided an occasion to evaluate maliciously the results of the Soviet system.

When the border was opened in 1989 and the first greetings had been exchanged with neighbours, the shock was great, since the situation on the other side was very different from what had been believed. The Finns were crying over the dirt, the collapse, the confusion and the shortages prevailing among their neighbours, and the Russians were crying out for tidiness, progress, order and consumer goods (Väistö 1993).

This also comes out in the narratives that Lehto and Timonen (1993) exploit in their account and in the interviews for this research, where a typical comment was that the Russian Värtsilä is no longer their Värtsilä. Many visitors do not want to go there again, even though it is now possible.

10.2.1 The opening of the Russian Värtsilä and the beginnings of tourism

At the beginning of this project it was almost impossible to imagine a visit to the Russian side of Värtsilä. The commune administration had some official connections with it, and there had been discussions about the possibility of opening the border to tourism since the 1960s, but in practice it was *perestroika* and finally the disintegration of the Soviet Union that made tourism possible. Even though there were plans for Värtsilä to be a crossing point for visits to other parts of Karelia, visits to the industrial community of Värtsilä itself did not seem probable in 1987. For this reason the original draft of the questionnaire did not contain any explicit questions regarding a possible forthcoming visit to old Värtsilä. This theme was nevertheless discussed with most of the interviewees, and many of them, particularly those who had been born before the war, expressed a hope that they would be able to visit their home area some day.

The Russian Värtsilä area is nowadays a part of the administrative region of Sortavala and has some 3000 inhabitants (as of 1989, see Klementjev 1991, 68). The factories are still in operation and are the most important employers in the village. Inasmuch as the infrastructure built by the Finns has to a great extent been ruined or is not in use and the Russians have built new departments at the factory, it is rather difficult to estimate the number of workers. Local guides estimate that to be some 800, which is thus smaller than at the time when the works were on the Finnish side. The factory has had problems following the dispersal of the Soviet Union, as its raw materials were previously brought from the Ukraine and the independence of the former Soviet republics has changed this situation dramatically. The Russian part of Värtsilä is looking forward to a brighter future, since new industrial development will take place in the area, funded by Russian and Finnish capital. This development is one part of the increasing Russian–Finnish cooperation that is taking place in the border areas, areas which have at times been labelled the 'Rio Grande of the North'. Along with this cooperation the idea of the old industrial community of Värtsilä will be revived. This does not mean, however, that the old territorial identity would be established again. Social and economic practices are not inevitably in congruence with collective and personal memories and identities. The latter point will be considered in the last section of this chapter.

The inhabitants of Värtsilä on the Russian side have been recruited from various parts of the Soviet Union, and the workers reported that the immigration had taken place on the basis of personal contacts, i.e. those who came to Värtsilä may have asked their friends to follow them and the final acceptance will have taken place on a 'higher level'. For this reason most of the inhabitants are Russian-speaking and those who speak either Finnish or the Karelian language are very rare (Klementjev 1991, 68). This is also owing to the settlement policy of Soviet authorities after World War II (Paasi 1995). Only 4.2% of the population of Russian Värtsilä were Karelian-speaking in 1989, a situation similar to that prevailing in all the areas located beside the Finnish border, where the inhabitants are typically Belorussians and Ukrainians (Figure 10.1).

It has been estimated that about 1.26 million Finns crossed the boundary at the seven existing checkpoints during 1991–2 and that many of them were visiting their old home areas (Lehto and Timonen 1993). The opening of the border has also been very significant for the small commune of Värtsilä, since some of the traffic to Russia goes through Niirala, a village in Värtsilä (Figure 9.19). During the last few years the number crossing the border has grown greatly at most checkpoints, and this has also been the case in Värtsilä, as the statistical information from the Finnish Border Guard detachment in Table 10.1 clearly indicates.

Some groups were already allowed to visit Värtsilä in 1990, and the Värtsilä associations were active in this respect at first. When the boundary was opened up the numbers of members in these associations almost doubled, i.e. the opening up of the possibility to visit one's old

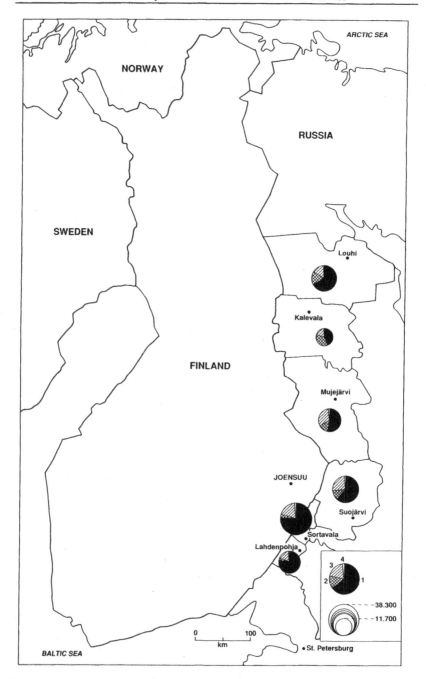

Figure 10.1   Proportions of (1) Russians, (2) Karelians, (3) Belorussians/
Ukrainians/Vepsians and (4) Finnish-speaking people in the administrative
districts of Russian Karelia in 1989. Source: based on data in Varis 1993

Table 10.1 Border crossings

Year	Numbers of persons
1989	25 943
1990	75 338
1991	105 244
1992	138 808
1993	191 158
1994	202 287

home area probably activated or re-awakened the regional identities of many people. Altogether more than 100 persons participated in the visits organized by the association in the town of Joensuu in 1990 and about 80 people visited Värtsilä with the association the following year. Since 1991 a travel agency in Tohmajärvi has been organizing visits to old Värtsilä, and about 170 travellers visited the area during the first year, in groups of approximately 20 persons. In 1992, however, there were only 54 visitors, which suggests that saturation point has already been reached.

The travellers have come from various parts of Finland, particularly from other iron industry communities where people moved from Värtsilä after the war. Some visitors have been immigrants or their descendants from abroad, e.g. from Sweden, USA and Germany, who have come to search for their roots. The representatives of the travel agency inform us that tourism in Värtsilä has been characterized by a specific 'group spirit', i.e. the travellers have been 'chained', so that at least one member of the group had lived in Karelia before war. Those born after the war generally participate in these visits only when they have some older person with them. There are two typical groups among those visiting Russian Värtsilä: those born in the 1910–20s who once worked in the factory, and those born in the 1930s who spent their childhood in the area.

During the previous visits it had already become apparent that the utopian dreams of the beauty and harmony of the community did not exist in Värtsilä: the Russian side was no longer the locality that had been an essential constituent of their imagined place since the war. An almost unanimous comment was that old memories were now shattered, even destroyed, as was the case with the old 'landmarks' such as houses, roads, features of the landscape and the old factory, which were in bad condition (Figure 10.2). In Värtsilä, as in other places, the Russian material culture had come as a shock for the old images and likewise the bad condition of the environment and the buildings, i.e. the physical and cultural landscape of the area.

Figure 10.2   The broken memories of place. Some of the ruined buildings of the old Värtsilä factories in 1992 (photograph by the author)

10.2.2   Visiting Russian Värtsilä

The aim of this section is to document a one-day tourist visit to old Värtsilä in terms of the results of participant observations and personal and group interviews carried out during the visit. Through an analysis of the relations between human beings and a historically constructed environment, an attempt will be made to synthesize the perspectives of participant observation, oral history, and the knowledge gained in previous research.

Värtsilä and the memories connected with it constitute a unique set of experiences for all its inhabitants, but this five-year investigation has also revealed many common elements, so that it is reasonable to talk about a common collective horizon of experience. This is to a great extent sedimented on mutual knowledge (Giddens 1987b, 65), i.e. knowledge of the convention which actors must possess in common in order to make sense of what they — as well as other actors — do in the course of their day-to-day lives. Meanings are produced, Giddens continues, via the practical application and continued reformulation in practice of 'what everyone knows'. Thus mutual knowledge consists of the ability to continue the routines of social life.

Old Värtsilä is undoubtedly lost in the physical sense. It belongs to the past. But it lives on and has effects on present-day practices, and will perhaps also do so in the future through the memories and stories that people maintain as part of their personal and collective identity. A good illustration of the existence of a horizon of such collective knowledge and experience in the case of Värtsilä is the fact that activists

have founded two associations in Finland for people who were born in
Värtsilä, the first in Helsinki in 1968 (now having 200 members), and
the second in Joensuu in 1982 (with some 300 members). The members
of the associations are typically old people and, as in the case of
Karelian associations in general, their activities have not attracted the
younger generations.

The visits to old Värtsilä themselves are also an indicator of collective
memory. All the people visiting there have simply and voluntarily made
a decision to do so, taking care of all the necessary formalities. As
Giddens (1987b, 66) writes, the knowledge of what agents do is not just
incidental but is constitutively involved in that doing.

The observations and experiences will be contextualized in this
chapter in relation to earlier results and interpretations achieved on the
basis of interviews, life-histories and photographs. The role of earlier
results becomes crucial inasmuch as one traditional claim of participant
observation — an observation period that is 'long enough' — is not
satisfied here, simply because the group was observed only for one day.
For this reason particular stress will be laid on two other basic ideas of
participant observation, the documentation of the process in which the
research material is collected and recorded, and the construction of a
theoretical and conceptual framework that directs the actions of the
researcher. This framework was crucial when making the observations
and interviews. Participant observation is a sensitive method. Smith
(1988, 21) refers to it as an art which is necessary for the development
of human geography into a social science. She points out that no
knowledge, particularly knowledge of the social world, arises apart
from the creative imagination of the observer and the observed.

In June 1992 I took part in a visit to old Värtsilä together with seven
tourists. The intention was to share the visit as one tourist among
others. The other visitors were all about 70–80 years old. Four of them
were women and three men — and all of them had lived in Värtsilä
before the war. Two of them now lived in Central Finland, and the
others in Northern Karelia. Three of them had already visited old
Värtsilä, two of them three times. I did not ask them their education
level or former profession but I tried to draw conclusions by listening to
their discussions. Since all of them had some connections with the
factory or other working life in Värtsilä (maids etc.), I concluded that
all of them were of the working class or lower middle class, with a
minimum of education.

When I was planning the visit, it was clear that even if I minimized
my participation and maximized the observation (cf. Smith 1988), I
could not pursue my research by a non-reactive method, i.e. so that the
group would not at all know that I was trying to collect information
about them and their experiences. I was an outsider, not a member of

the 'we' that these people formed. It was obvious that my aims would become clear to them, since my object was to observe their movements in Old Värtsilä, to ask direct questions about their feelings and to document their discussions. In addition, I aimed to record the localities that we visited.

On the whole, my intention was to listen to their comments as we travelled on a journey which for many of them was probably a fulfilment of long dreams and hopes, but also a somewhat traumatic experience. I had also reserved some questions which I would ask the members of the group and the guide on appropriate occasions. The themes of these questions were where the members of the group (and visitors to the ceded Karelian areas in general) have come from, whether they had been born or lived in Värtsilä, whether this was their first visit to Värtsilä/to the ceded parts of Karelia, why they were taking part in the visit, whether they wanted to see the past or the present of the area, what they hoped to find out from the ceded area, what were the things and places with which their memories were most commonly connected, what were the pleasant and unpleasant experiences during the visit, and whether they would visit Karelia/Värtsilä again later.

It was natural to suggest that most of the travellers had actively planned their visit over the border, and hence probably had a good idea of the conditions prevailing in the ceded Karelian areas, as the Finnish television and other media had on many occasions made it clear that old images regarding the lost home area of almost 50 years ago were no longer valid. This fact was especially obvious for those who had already visited Värtsilä.

Since the group consisted of only eight people, we travelled by minibus and mostly walked around the Russian village of Värtsilä. This also made it easier to hold group interviews. I wrote notes continually, but I never made it explicit during the visit that I was a researcher who was trying to find out through observation and interview what features made the visit meaningful for the members of the group. It seemed that they regarded me as an eager home area activist, and hence an interpretative approach became a self-evident point of departure for this part of the project. Together with unstructured thematic interviews, free discussions with the people and my previous materials, these observations became the keys for the interpretations.

A particular comment is needed on the use of photographs. While geographical studies generally use photographs as a means of illustrating the objects of research, and while photographs are invariably of minor importance compared with maps, ethnographic investigations and especially visual anthropological ones have for a long time exploited photographic material both as an important source and as a medium of ethnographic description. Both old photographs taken before World War II and new ones taken during the project, and in particular during the visit to Russian Värtsilä, are employed here, the object being to employ them for both comparative and documentary purposes, so that it is the situations and objects that the pictures illustrate rather than the pictures themselves that are of consequence. In this way an attempt is made to illustrate the features to which people attach meanings in Russian Värtsilä.

10.3 BACK TO KARELIA: TRAVELLING INTO THE PAST OR INTO THE PRESENT?

Do the old inhabitants visit the Karelian areas because they want to revisit the past golden times that came to an end with the war, or do they want to see something of the present Russian village of Värtsilä? Even though this distinction is an analytical and somewhat absurd one, many observations indicate that the answer is essentially the former, or at least that this is a visit in time as much as in space (cf. Lehto and Timonen 1993, 97). People generally want to visit their past history and the physical locality with which it is connected, in which they try to find symbols and elements that they can relate to their own personal history there. This has been a typical evaluation made by experts in the tourist industry as well as by investigators of the narratives that the visitors have written. The visitors are typically old people who simply wish to visit the areas where they have lived before. Younger people are not usually interested in these visits and if they go there, it is mainly on account of the wishes of their older relatives and parents. This does not prevent them from having a deep experience of the Karelian past, however, even if they have had no personal contact with the area.

The same conclusion can be drawn at the local level from the interviews carried out in Värtsilä in 1987–91. As we saw above, the generations live in different worlds as far as their territorial boundaries and identity are concerned, and the observations and interviews made during the visit to the Russian side confirm this. The visitors assured me, one after the other, how Värtsilä would have been a flourishing city in Eastern Finland nowadays if the area had remained part of Finland after the war. When I asked if they had often thought about the old times in Värtsilä, one of them said, with emphatic support from the others, that 'You cannot forget your home area. Värtsilä is always in my mind. My father and mother always hoped that they might be able to visit it again.' It was obvious that the members of the group were visiting the area because they wanted to see what kinds of memories from the old days they could still revive.

Hence people did not come to see what Värtsilä could provide for a

Figure 10.3 The church of Värtsilä was one of the key symbols in the mindscapes of Old Karelians. The church before World War II and its ruins in 1992. Source: above: private collection, below: photograph by the author

visitor today. A clear symptom of this was the fact that one visitor who had a camera with her took only a couple of pictures during the whole day, one of the steps of the ruined church and the other of the old market square. According to previous interviews both had been essential symbols of Old Värtsilä (see Figures 10.3 and 10.4). A typical

Figure 10.4 Numerous interviewees constantly returned to the lively bustle of the market square. Värtsilä market square before World War II and in 1992. Source: above, National Museum of Finland. Reproduced by permission; below, photograph by the author

reaction was that all the visitors described to each other enthusiastically the atmosphere that prevailed in the market square on Fridays in the pre-war years. The reaction was similar in the case of other objects that the group wanted to see.

One member of the group also carried with her old pictures of the

Figure 10.5 Searching for a lost place: an elderly evacuated Karelian woman trying to find concrete evidence for her broken roots and memories of home when visiting the Russian village of Värtsilä in 1992 (photograph by the author)

house where she had worked before the war, and was constantly showing the other members of the group these pictures and explaining their contents. She also wanted to pay a visit to the place where this house had stood. She could not find any sign of it, but she did find a tree which she identified and then told the other visitors about its role in her memories. Another old woman wanted to see the ruins of her old home near the Russian frontier zone, but she could not find the place. The general feeling of the group was not bitter but nostalgic (Figure 10.5).

One part of the journey was a short visit to a Russian family. In a way this pleasant and intimate coffee break was a concrete illumination of the new era in Finnish–Russian relations at the level of civil society. At the same time, it was also a culmination, providing as it did concrete evidence for the visitors that Värtsilä was now something different from that prevailing in their memories. This small episode together with Russian-speaking people, a different culture, a different landscape served to make this quite clear. The members of the Russian family, ordinary people, were no longer the strange Other for these old Finnish people who had experienced the loss of their homes and who had studied the character of Russians and the Soviet Union from their geography textbooks before the war.

This new situation is equally well illustrated in the results of Lehto and Timonen (1993), who found that in some cases the evacuated people have begun to pay visits to their old home areas in Russia as a hobby every summer. In some cases they have even become friends with the present Russian inhabitants. The latter have in many cases lived the whole of their lives in the ceded areas and identify themselves as much with these areas as do the old Finnish inhabitants.

10.3.1   How was Värtsilä perceived by those who returned?

When, in 1991, I interviewed seven workers from the old Värtsilä factory who now live in the Finnish commune of Värtsilä, I received an impression of deep disappointment. The conflict between the old images and the present reality had been too great. Of the people participating in the visit to Värtsilä in 1992, three had already had previous experience of the changes that had taken place on the other side of the border, and one of these had seen videotapes of the Russian village. In any case, all the visitors had a clear idea about what to expect. They knew that the population was mainly Russian, and that the cultural landscape is totally different from the pre-war landscape. Similarly, they knew that the local population are in the habit of letting their livestock wander freely around the streets of the village. Those who had been working in the factory knew that a large number of the old factory buildings had been removed or were in ruins and that others had been reconstructed with a new outward appearance. The shock had been much more acute for those who had visited the factory and the village immediately after the opening of the border.

Once they had crossed the boundary and the customs formalities had been completed, the visitors clearly became excited. Some of them were visiting their old home for the first time in 50 years. Owing to the previous information that they had, the atmosphere was unconstrained. The centre of the old commune is 3 km. from the railway station, which is almost immediately on the Russian side of the border. Straight away the members of the group began to comment on the differences in the landscape compared with former times. Everything had been much better in the olden days than it was now: the physical and cultural landscape, the building, the state of tidiness, etc.

It seemed to be rather difficult for them to define the area, and it was the dominant lines in the landscape, e.g. the roads, landmarks and some familiar symbolically laden objects such as old workers' houses, that were noticed most clearly. Above all, the old main roads were easy to recognize, but the new roads and buildings that had been constructed since the war made identification of the area more difficult. At this stage in the visit I asked the members of the group what were the things that they most often thought about when they harked back to old times in Värtsilä. They mentioned the market square, Lake Jänisjärvi, the

Figure 10.6   Two key objects in the memories of old Värtsilä people: the rebuilt Kisapirtti house and the shore of Lake Jänisjärvi in 1992. A pioneer camp for children from Russian Värtsilä has been located on the shore until recent times (photographs by the author)

Kisapirtti house, built by the company, where the people used to watch films and have parties. These objects were the same as had emerged in the previous interviews (Figure 10.6).

Landscape is a common symbol or metaphor for home in the Finnish culture (cf. Lehto and Timonen 1993, 96), and the physical and social topology of the landscape is usually deeply sedimented in the memory. When the parameters of this topology are destroyed, the identity of the individual is shaken. In the Värtsilä area several of the most important symbols of identification that people had mentioned during earlier interviews had permanently disappeared from the landscape, some had been shaped into a new form, and some were still almost the same as when they had been left by the Finns almost 50 years ago. According to the old inhabitants, though, there were less than 10 buildings in the area that still reminded them of the Finnish period. As an explanation for the poor condition of the environment, the Russian authorities have told visitors that Moscow would not assign the community adequate resources to prepare or maintain its infrastructure — the school and hospital apart (Figure 10.7).

Statues of prominent figures in socialist ideology and other symbols were an essential part of the political landscape and iconography of the former Soviet Union (see Pickles 1992), and this socialist ideological landscape was still visible in Värtsilä in summer 1992, in the form of a statue of V. I. Lenin in front of the main gate of the ironworks. None of the members of our group paid any attention to it (Figure 10.8).

It was pointed out above how nationalism and religious rhetoric have always been closely connected. Lehto and Timonen (1993) state in their analysis of accounts written by visitors to Karelia that religious and mythical terms are very common when people recount their visit to their old home area. A 'holy land' or 'promised land' are metaphors applied to the ceded territory. They refer to their 'pilgrimage' and to themselves as 'pilgrims'. According to the authors, this is not merely an accidental terminological resemblance between religious pilgrimages and more secularized visits, but rather they are inclined to find structures common to both: an orientation towards an important centre of experience, a strong communality among people belonging to different generations, and the integration of various temporal scales. Furthermore, Lehto and Timonen (1993, 102) write that a sense of humour and of celebration is an important part of the visit, as is a certain lack of restraint.

On the occasion of the 1992 visit perhaps the most surprising thing was that the church was not mentioned — even though it was among the most important symbols to emerge from the interviews conducted in 1987–91. The members of the group knew that the church had been destroyed during the war years and the only thing left was the stone steps, but the new religious symbol was the cemetery, where some of the visitors had relatives buried. Even though I was a complete outsider, it was clear to me that the visit to the cemetery was a deeply moving experience for some old visitors, and particularly for a woman who wanted to see her child's grave for the first time since the war but could not find it. The visitors were shocked that all the old Finnish graves had been destroyed and the tombstones had fallen down (see Figure 10.9).

Cemeteries are commonly regarded as the territories of culture,

Figure 10.7 Views of the present-day village of Värtsilä on the Russian side (photographs by the author)

heritage and memories. Norbert Elias (1993, 32–3), a cultural theorist, writes that 'imperishable tombstones contain a deaf message from the dead to all living people'. Hence they are a concrete illustration of the continuity of human, individual existence in communities, in the same way as certain other material and spiritual cultural features are an expression of the collective heritage of the community. For the people visiting the Värtsilä cemetery, the fate of the old Finnish tombstones was a concrete message of an interruption in the individual and collective cultural continuity of their old community.

Figure 10.8 Rudiments of the Soviet ideological landscape. V. I. Lenin still dominated the landscape in front of the main gate of the Värtsilä works in Russia in 1992 (photograph by the author)

Figure 10.9 The cemetery in the Russian village of Värtsilä in 1992. In the foreground is an old Finnish grave (photograph by the author)

Andersson (1980) and Nuolijärvi (1986), having studied the relations between migration and spatial identities, both point out that after people have migrated from a locality, their memories and connections with the social networks and persons in that locality will lose their meaning in the course of time, particularly after new social relations have become established in new localities. Simultaneously the role of more permanent factors will increase as the basis for identification with the old locality. Nature, the forests, lakes, buildings and landscapes, will become important as a source of identification.

The search for roots in the old Karelian areas is a fitting illustration of this change. Innumerable old people wandering around the villages or standing in the forests and fields of Karelia have been searching desperately for signs, small things that would change their memories of the past into concrete experiences here and now. It has been typical for returning visitors to bring back with them soil, flowers, water, and other such things as will remind them of their old home. Lehto and Timonen (1993, 99) write that people 'take their home with them' in pieces, in memories which they can smell and touch and which become concrete physical manifestations of their previous home and its landscapes. In some cases this earth is brought to the graves of their parents, or the water often used in the christening of grandchildren, and so on.

In the case of the Russian Värtsilä these tendencies are natural, since the social content of the commune became void when all its inhabitants had to leave after the war. None of the visitors in our group wanted to bring anything back with them to Finland, however. Even though the landscape and physical features had become significant in their memories in the course of time, this does not mean that social factors were no longer significant in the utopian elements that were part and parcel of their spatial and social identity.

Ryden (1993, 81) writes that the folkloric sense of locality is usually closely related to local community history and particularly as it circulates in oral tradition. As we observed above, in addition to the memories of the landscape, culture and social activities in Old Värtsilä, certain social figures also emerge as important features in the narrative accounts of old inhabitants, the best-known being Wilhelm Wahlfors. These figures are typically positive constituents of the socio-spatial identity in the stories of the old inhabitants rather than mere bearers of social relations. Through them the personal identity of each individual becomes a part of the social and collective continuity that lives in the memories and practices of old people. It is typical for these characters to be retained in the memory through general stories rather than personal experiences. They are part of the collective heritage of the old Värtsilä community, which the new boundary once divided.

The old community still lives in the memories of the older generations, and it has now become possible to compare these memories with the present community. The younger generations are growing up in a different direction, without having any personal connections with the old community or the local geohistory. They have been socialized into territorial identities (national, regional and local) and structures of expectations that are changing relatively slowly. Old Värtsilä does not belong to the constituents of their place. How long this collective heritage will live on, how long the memories of the old, larger Värtsilä will remain, will depend above all on the interest of the younger generations in the history of the community and their will to maintain the heritage of the older generations. It is clear that if the younger generations do not keep up the tradition, we can conclude this analysis of territorial transformation on a local scale with the words of Norbert Elias (1993, 33):

When the chain of the memory is broken, when the continuity of a community or of a whole human society comes to its end, then the meaning of human achievements that have lasted thousands of years will also disappear, and the meaning of everything that has once been regarded as important.

11

EPILOGUE: TOWARDS A GLOBAL
SENSE OF PLACE

Johnston (1989) opens his review on the role of the state in geographical research by noting that a major component of the spatial
organization of the Earth's surface is its division into 150 or so
sovereign states — today almost 200. Each of these states has a well-
defined, though in some cases contested, territorial extent. Also, the
state differs from all other institutions in that it is necessarily a
territorial body associated with a clearly defined area. Even though the
'state' as such is the arena for a multitude of conflicts between classes,
ideologies and institutions, the key element in the reproduction of the
territoriality and sovereignty of a state in the continual spatial
transformation is its boundaries — whether open or closed, political,
cultural, economic or ideological.

Boundaries have served as fixed points in the international relations
between states. They have been and still are the manifestations of state
control over movement and interaction, closure and distinction in the
world system. Boundaries have also been the key element in the
changing world political map, and in the landscapes of peace and war
— both in concrete meanings and symbolically. People have killed each
other because of boundaries, and are still doing so today. People
negotiate and sign agreements because of boundaries.

Boundaries and frontiers have always been a major field of discourse
in the tradition of political geography, and hence state boundaries, for
instance, are regarded as manifestations of state sovereignty, specific
types of landscapes and so on. The basic aim of the present work,
inspired by recent trends in cultural and social theory, has been to
broaden the traditional view of boundaries provided by geographers in
a more theoretically, methodologically and historically sensitive
direction. Thus boundaries are not understood here merely as concrete
lines or visible landscapes located between socio-spatial entities, but

above all as phenomena that are 'located' in the socio-spatial consciousness and collective memory of people living in territorially constructed units on various spatial scales, whether communes, provinces or states. Boundaries are understood as structures that are produced, reproduced and contested in and between territorially bounded groupings of people.

Our use of the abstraction 'socio-spatial consciousness' does not mean that a society can be reduced to the ideas that its members may develop of it. Social consciousness does not float freely and ahistorically above the heads of people as a causal power controlling their actions, but is deeply sedimented in various historically constructed social and material practices and relations, e.g. within politics, administration, economy, education and culture. Through these it is reproduced and shaped by individuals, groups and classes who perpetually aim at maintaining these practices. The production and reproduction of practices takes place both discursively and through practical consciousness in the daily struggle over material conditions and representations. In these practices the discourses will be created through which distinctions between us and the Other are produced and reproduced. In the case of nation-states, for instance, the narratives of national identity are constructed in this way, but this basic framework also holds good on other territorial scales. Nevertheless, it is the state that controls the internal and external relations of nations, and accordingly it is the state that usually defines the criteria for belonging to a nation, whether people identify with it or not.

Both territoriality and social consciousness are hence deeply contested categories, where power — again sedimented in diverging social practices (politics, the economy and administration) — is an essential constituent as far as the organization of various practices and forms of consciousness are concerned. This points in particular to the political dimension in the construction of space and its representations, as well as in representing otherness, a crucial element in the construction of territorial identities and narrative accounts of us and them. The contents of these accounts — as represented in media, or national socialization (e.g. school geography and history), for example — are always expressions, even avowals of the results of these contests. Social consciousness is produced and reproduced by individuals and groups in various practices — individuals usually enter into these 'already given', historically created practices, which they can shape and modify according to the power they have obtained in the division of labour.

It is common to think that in the present-day world the roles of state boundaries will be eroded as interaction between political, economic and cultural institutions increases on various spatial scales. This is probably true as far as the movement of capital, goods and people is concerned, and this will inevitably lead in the long run to lower boundaries between the nation-state and the Other — and to a new regionalization of diverging social and cultural practices on various spatial scales.

Nevertheless, the constant transformation taking place in the global system of states, of which the most recent and closest example for Europeans is the rapid process of change manifesting itself in the collapse of the Socialist states of Eastern Europe, the emergence of new states and the occurrence of terrible civil wars, is a serious indication that geography and territoriality, and nationalism as their concrete manifestation, have not lost their significance in the present-day world even though the Cold War belongs to history. The changing economic, political and cultural geographies, and even physical geographies, witness to the continual struggle between forms of action which are primarily global, national or local.

Spatial consciousness and the material-economic, political-administrative and cultural interests that are connected with it are constantly shaping each other via either peaceful or violent processes. The spiral of violence triggered by the transformation in Eastern Europe, again a reflection of nationalism, is communicated to us globally through the mass media, but there is still a fundamental process of nation building and nationalism proceeding all the time inside all states — a striving towards the production and reproduction of a congruence between a political and a national unit, even though it cannot always be explicitly observed.

As a concrete illustration of the construction of socio-spatial consciousness and territoriality, the nation building process as it applied to Finland and the institutionalization of the commune of Värtsilä as a local manifestation of the former was analysed here, and the changing representations of the Finnish–Soviet/Russian boundary and the material practices connected with these changes were examined as manifestations of this process. It became obvious that the representations continually reflected the changing political relations between the two states, these relations being part of the international geopolitical landscape.

It was emphasized throughout that territorial units and regions, states and nations — and their representations — are in a continual state of flux, rising and disappearing in a perpetual regional transformation. Since national identities and other socio-spatial representations are contested categories which are continually being redefined as expressions of the power relations in society, the contents of the national identity narratives also change over time.

Hustich (1972) realized this basic idea well, and was already asking rhetorically two decades ago whether Finland was merely at a temporary stage (in territorial transformation). His answer was in the affirmative. The problem for Hustich was whether Finland would become a part or a sub-state of some larger Nordic or European whole during the next generation or whether this would take place only after several generations.

A territorial transformation process and inherent self-analysis is evident nowadays in Finland, particularly in the form of the discourse regarding the European Union and its effects on state sovereignty, national identity and economic power. Similarly, nationalistic arguments are part of the debate on the possible social, political and military effects of the opening of the Finnish–Russian boundary. It has been argued in the most extreme opinions that the decision to enter the EU will invalidate the whole foundation of Finland's independence, whatever this means literally in the (post-)modern world with its increasingly global 'space of flows' in capital accumulation and cultural heterogeneity. It is also forgotten that in fact the content of national culture is constantly being redefined. As remarked by Löfgren (1989), every new generation produces its own national sharing and national frames of reference. It is not the nation that is falling apart but rather older versions of the national ideal, old narratives of the content of national identity. Löfgren points out that new, rising generations then select items from the 'symbolic estate' of the earlier generations (1989, 21).

What is important in this selection is, of course, the fact that most people have nothing to do with these definitions, but rather they are expressions of the hegemonic structures of the state which manifest themselves in various institutional practices. Owing to the division of labour in society, it is specific positions and power groups in the society who define and redefine the content of this abstract identity, who create the fundamental vocabularies and grammars for the languages of integration and difference. But this also holds good on other spatial scales and in other territorial identifications. In the realm of everyday life, spatiality and identities — no matter what scales they are bound to — are lived and experienced, not usually reflected on.

To some extent ironically, the discourses of nation building appear to be empty words in Finland just now, inasmuch as the collapse of the country's economic conditions — connected essentially with changing international economic relations — together with the apparent gradual demolition of the welfare state, is dividing the nation between those who have a job and those who have not, now 20% of the labour force. This basic division is more and more often taking place between the generations, beside the traditional division between social classes. All these trends — probably for the first time since the Winter War or the Civil War — will probably put strong pressure on the roles of the 'civil religion' in the future, i.e. on legitimation and the moral landscape in social integration, a key function of the state (cf. Johnston 1989). One critical question will be the identity of the younger generations who grow up in a Finland of unemployment and hopelessness. Will it be based on extreme nationalistic tunes from the 1930s? Or will it still be based on the globalizing perspectives that have prevailed since the 1960s?

All these trends are indicative of many faces and layers of territoriality. As Taylor (1994) points out, the state as a power container tends to preserve its existing boundaries, but as a wealth container it usually tends towards larger territories. Finally, as a cultural container it tends towards smaller territories.

The present findings indicate, however, that the wide-ranging discourse on the effects of EU membership on Finnish identity is by no means a unique situation in the history of Finnish society and its inherent ideology of nation building. It has always been typical to adapt the major processes and mechanisms involved in the constitution of Finnish nation building and nationalism to 'external' pressures, which has also enhanced the role of 'geography'. This examination of Finnish nation building has made it clear that geography is crucial to any understanding of the state just as much as the state is crucial to the understanding of geography (cf. Johnston 1989, 292).

'Geography' has been comprehended here in three ways, (1) as the geographical construction of the world, (2) as an academic field created by government action to introduce territorial organization into this geographical diversity and to educate geography teachers, and (3) as a theoretical and conceptual form of discourse. All these meanings have been crucial as far as the construction of Finland and Finnishness is concerned. Geography as an academic discipline was established to create national identity and geographers prostituted their conceptual categories and empirical effects during World War II in service to the leading politicians, to support their efforts towards a 'Greater Finland'. After the war the role of 'geography' became the key argument in the rhetoric of internal and foreign policy. Thus, in the Finnish case, geography can be said to have played a role of major importance in the exercise of power.

Particular emphasis has been placed on the significance of the education system for the production and reproduction of language, which is the key discursive instrument of national identity. In the same way, school geography, among all the applications of geography in Finland, has always been a crucial instrument for the production and reproduction of the language of territoriality and social distinctions — and hence socio-spatial consciousness. As far as national integration is concerned, the ideological cornerstones of this language have been its normative character and its collective nature. Both have promoted the creation of a relatively firm classificatory scheme to establish a fixed picture of the Finnish cultural 'semiosphere', its symbols and boundaries as the visions of the Other. The semiosphere is essential to the maintenance of territorial identity in the regional transformation. The Finnish language as such is already one cornerstone in the national identity, since it is a very closed and unique instrument of communication.

For Ilmari Hustich (1972), Finland's eastern boundary seemed to constitute one basic fact which defined the country's (geopolitical) location and economic and political relations, but the boundary has been under continual transformation. It has changed many times through the years, and accordingly the content of the socio-spatial consciousness of the Finnish people has also changed dramatically. Both time and space make a difference in the formation of the content of this consciousness, since episodes taking place in various locales constitute the fundamental frame in which the production and reproduction of social consciousness are realized. The content of this process is not confined merely to the local level since territorial identity is a hierarchical phenomenon into which various scales are structured.

As to the formation of the representations of the Finnish–Soviet boundary and its role, the indisputable watershed is World War II, which makes a distinction in socio-spatial consciousness and divides Finnish generations with different spatial experiences. Those born before the war have experienced a time when the relations between the two states were commonly organized on the basis of an 'enemy image', whereas the generations born after the war have been socialized into a much more favourable outlook towards the Soviet Union. Today increasing interaction between the Finnish and Russian communes and enterprises in the border area appears to be reducing the traditional meanings of the boundary as a line between the Eastern and Western worlds.

As a consequence of the opening of the border, it has become possible during the last few years for both Finns and foreigners to visit Russian Karelia, and this has partly removed the old trauma. It has also provided an opportunity to establish cooperation in both directions. The Karelian authorities aim to develop the area by means of tourism and increasing economic interaction. All these activities are gradually changing the nature of the boundary and rendering it more open. It is no more the longest boundary between a Western capitalist state and the leading socialist state.

But the symbols and practices that have prevailed in the border area since World War II are still the same in spite of the changes in the world geopolitical order, and under the recent unstable conditions the situation probably will remain the same. As far as the cultural consequences are concerned, the opening of the boundary may cause a diminishing of the role of Karelianism in the Finnish areas in the future, since it is now possible to visit the 'original' Karelia instead of a simulated one constructed after the war on the Finnish side of the border.

The more relaxed geopolitical atmosphere has encouraged some Finnish elements to claim that the ceded areas should be returned to Finland. On the other hand, there exist some hints of suspicion as far as the aims of Finnish cooperation in Karelia or the debate on the ceded Karelian areas are concerned. In Eastern Karelia, the region that the Finnish troops occupied during the Continuation War, some debate has arisen as to whether the aim of the Finns is 'neo-colonization' of the Karelian areas in an economic sense and the exploitation of their natural resources (see Sykiäinen 1993). Nevertheless, it is very unlikely that these areas will become a seed of new territorial conflicts. The Finns have learned that geography is still important in their external relations.

But Finland and the 'grand narrative' of its existence as a sovereign state cannot remain unchanged in the changing world system of states and ideologies. On the global scale, there are no longer any grounds for postures of hostility between a capitalist West and a socialist East. According to NATO, Russia is not a significant threat. This, of course, makes it more difficult to define the content of the 'we-ness' of those living along this old ideological boundary — as has been the case in the Finnish state. Joenniemi (1993) has noted that the effort to achieve security has by tradition bound the Finnish state and nation together, while threats have traditionally rendered possible the creation of social distinctions and the construction of boundaries — strengthening the distinctions and reproducing a strong identity for 'us'.

This became clear in the present study of the Finnish territory-building process and analysis of the changing forms of national socialization, which gave concrete meaning to the words of Paavo Haavikko: 'The state, economy, culture and nation are a whole — created by the threat of Russia — which has prevailed for the last hundred years' (cited in Joenniemi 1993, 23). Joenniemi points out that, even though the changes in the world political and ideological map in recent years have been remarkable, the narrative of the threat of Russia has again been given publicity, principally to maintain the idea of 'us', and simultaneously to reproduce certain features of Finland as an 'ontological, state-centred project'.

Hence some actors are attempting to articulate new narratives which set forth a new representation of the threat posed by Russia. The image of Russia emerging in this discourse is not a modern post-Soviet Russia but a historical, old Russia, of which Finland was a part up to 1917. The field of rational argumentation has been narrowed in recent debates, Joenniemi (1993) writes, and generally places these debates in the realm of defence policy, which appears to have remained traditional in the changing world. Hence discourse is still based on the ideas of repelling a danger, possessing distinct boundaries, defending these boundaries, maintaining an alliance between the state and the nation, and so on. There still exist attempts to define Finland's eastern boundary as the eastern boundary of Europe, and Joenniemi notes that this is doubtless grounded in intentions to maintain the old state-centred national identity. According to his forecast, the position of Finland in the ongoing European integration will become more open, more internationally based, and hence the encounter between Finland and Russia will be based on more interactive ways of acting. These will permit Finland to renew and maintain its national project more in the spirit of the 'Kantele', the traditional Finnish harp, than in that of the 'war trumpet'.

The results of the 1993 election in Russia and the rise of extreme nationalistic movements have of course become a new worry for the Finns as well, and have given rise to debates on the question of national security. One winner in the Russian elections, Vladimir Zhirinovsky, has on several occasions alluded to a future 'Greater Russia', which would include the independent Baltic States and Finland. It remains to be seen whether this will provide a new enemy for NATO or not, but it has in any case provoked a new geopolitical self-analysis in the states surrounding Russia.

One future theme in Finnish relations with the Other will be the question of refugees. The Finnish state policy on this matter has been the strictest in Northern Europe, reflecting the common exclusive nature of Finnish nation and identity building. This policy becomes apparent first and foremost in the almost neurotic interest in what foreigners think of Finland and the Finns, and secondly — which will be more important in the future — it manifests itself in a critical attitude towards 'the Other here', particularly in the case of refugees. Refugees are commonly represented in Finnish publicity and debates as aliens who pose a threat to the traditional Finnish national identity — whatever that is in practice. And the old vision of the Other coming from the Third World is now being matched by a vision of the Other entering from Russia.

Tarasti (1990, 207), a semioticist, provides one explanation for Finnish exclusiveness, arguing that the Finnish world of signs is poor and the signs tend to be interpretable through the same overlapping codes — the key for understanding the character of the Finnish national culture — while the penetration of different signs into the static semiosphere is felt to constitute a threat. The change in culture is slow, since all cultures consist of elements that are simultaneous but can be comprehended as residual, dominant and emerging (see Williams 1988). All cultures consist of elements that are expressions of tradition and constituents of the present, expressions of the present and constituents of the future, expressions of the future and constituents of the present. Similarly, the local is a constituent of the global, and the global a constituent of the local.

The fundamental point of departure in the present book has been to connect the general structural analysis of social consciousness with the construction of local everyday life, and accordingly to analyse the changing roles of Finland's eastern boundary in the dialectic frame of these two viewpoints: theory and practice. The aim has been to pass behind the 'official narratives' that manifest themselves in the socio-spatial consciousness of Finnish society and to reveal what the boundary means on the local scale and in the everyday lives of the people living in its immediate vicinity. This helps to show how localities are constructed in regional transformation as a manifestation of socio-spatial processes taking place on various spatial and historical scales, how local worlds may differ from those contextualizing themselves on larger scales, how territories and boundaries on various scales are produced and reproduced through societal activity and political struggle, etc. Smith (1993, 99) is doubtless right when arguing that the continual production and reproduction of geographical scale expresses the social just as much as geographical competition to establish boundaries between different places, locations and sites of experience. In addition, he maintains that geographical scale defines the boundaries and delimits the identities around which control is exerted and contested (1993, 101).

It was found in the analysis of the institutionalization of Värtsilä as a locality that the operations and institutional practices taking place on various spatial scales are in practice inseparable, even though our abstract thinking usually aims at making these distinctions more real than they actually are. This basic opposition also guides our thinking in the form of other oppositions: distinctions between global–local, macro–micro, abstract–concrete, political economy–ethnographic accounts etc. have dominated much of our geographical discourse, even though suggestions have been put forward more recently on the need to overcome these dichotomies.

An attempt has been made in the present book to move both in history and space with the aim of scrutinizing these dichotomies as a means of finding out the changing nature of geographies. I hope that in this way we have been able to reach something of what Massey (1993) labels the 'progressive sense of place', an understanding of the character of place (and region) which can only be constructed by linking that place to places beyond:

A progressive sense of place would recognize that [link] without being threatened by it: it would be precisely about the *relationship* between place and space. What we need, it seems to me, is a global sense of the local, a global sense of place.

APPENDIX

IN THE FIELD

The pre-understanding for analysing the local level, and particularly the commune of Värtsilä, is based on theoretical and empirical research carried out since the mid-1980s (see Paasi, 1988, 1989). The choice of Värtsilä as an object of research took place on both theoretical and practical grounds. My previous work on the institutionalization of regions had been concentrated on the provinces of Finland, but on its completion some dissatisfaction had been felt with the concrete analysis of the relations between institutional practices and individual actors. The level of *cultural practice*, the key to the distinction between region and place in the institutionalization theory, could not be shaped thoroughly on the basis of the technical survey performed as part of the project (Paasi 1986b). The simultaneous debate on locality studies and the nature of various territorial units — and the rather banal analogy of localities as 'laboratories' that was put forward by Newby (1986) — led to the thought that the best possible subject for analysing the connections between structural and interpretative perspectives, and hence the 'cultural', would be one small locality where the historical sedimentation of various state-engendered larger-scale social practices in everyday life and the manifestations and forms of territorial transformation would be extensive.

Värtsilä appeared to be an ideal locality for these purposes, inasmuch as the social processes taking place on various scales seemed to manifest themselves in its development. The industrial community emerged along with the Wärtsilä company on the rise of capitalism. After World War II the administrative commune was divided by the new state boundary and the company that had once created the whole industrial locality withdrew its capital from it. Also, in view of its small size and population (about 900 in 1987), it was felt that it would be possible to 'capture' it sensitively by divergent interpretative methods.

A backgound to the fieldwork was provided by various sets of statistical and historical material used to outline the past and present of the locality. The basic idea of the fieldwork and interviews was to construct a corpus of the life-experiences recounted by inhabitants of different ages, with the expectation of obtaining a great deal of information from the first interview, and filling this out with a few further interviews (cf. Bertaux 1981). Since the aim was to study the role of actors in regional transformation, it was logical to employ oral history (Thompson 1988, Henige 1982) and the methodological perspectives developed in biographical studies, as put forward mainly in anthropology and sociology.

The primary aim was not to construct and analyse the life-histories of individual actors as such, but to set their life-histories and the socio-spatial representations and experiences in the context of a regional transformation generated by diverse institutional processes (politics, economy, administration, culture etc.) originating on different spatial scales. Thus these basic methods had to be exploited in an 'applied' way, the key idea being that the information gathered by these methods was merely one among many sets of qualitative and quantitative material collected from archives, various documents, newspapers, novels, statistics and so on.

The *Leitmotiv* of the use of all these sets of material is in principle the same: to interpret the spatiality of everyday life and particularly the role of the new border in the historical sedimentation of the spatial experi-ence of the inhabitants living in Värtsilä. Furthermore, all the material was aimed at integrating the structural and interpretative approaches in a historical context. Whereas the historical approach in the case of the structural analysis of Finnish nation building meant above all outlining the large-scale transformation of the Finnish socio-spatial consciousness and territory, history makes a difference in varying, even 'interruptive' ways in the case of individual actors, owing to the fact that space makes a difference simultaneously with history.

An applied 'snowball strategy' was chosen for selecting people born before World War II, i.e. other people living in the community were asked to make recommendations (cf. Bertaux 1981). The intention was to find key persons who had — and who were known to have — memories, who had as long a horizon of experience as possible and who had been involved in the bitter historical episodes that had taken place in the locality. The selection of interviewees born after the war was based on a random sample. In addition, all 13 families living in the actual frontier zone areas were interviewed.

The aim of the interviews was to collect information about the spatial identification of the people of Värtsilä and the role of the Finnish–Soviet/Russian boundary in the social representations that people carry

with them. The material consisted altogether of more than 70 interviews, most of which were also recorded. On many occasions, however,
several members of the same family participated in the discussions, so
that the number of interviewees exceeded a hundred, i.e. more than
13% of the whole population.

After performing some 15–20 interviews in each selected 'sub-group',
it seemed that the enquiry had reached what Bertaux (1981, 37–38)
calls a saturation of knowledge of the socio-structural relations that
have prevailed in the locality. We had a picture of the essential
memories of the older inhabitants, how the old factory community and
its social relations were shaped, and how the establishment of the new
border after World War II was experienced. The simultaneous collection of other material (statistics, police records etc.) to illustrate the
historical construction of the community and the episodes involved in
this process was critical in the interpretation of the results.

During the first three years of the interviews, i.e. 1987–9, the
Finnish–Soviet boundary was virtually closed and the only connection
from Värtsilä to the Soviet Union was the railway which was used for
transporting goods across the boundary. It was exciting for a researcher
to enter this area. I had visited it twice before with colleagues from
other countries who wanted to experience the 'exotic' atmosphere of the
'border between East and West', but the restrictions on movement and
photography in the frontier zone were made evident to us.

Nevertheless, coming to Värtsilä was characterized by a curious, even
naïve feeling of excitement — a feeling that prevailed for as long as the
Soviet Union existed. In trying retrospectively to identify the grounds
for this feeling, the following contexts are probably of importance. The
drive through dark forests of tall spruces to the open fields which ended
with the forests that concealed the border itself created a mysterious
atmosphere and a feeling of its *continuous control* over the area. Both
the Finnish and Russian watchtowers continually suggested the idea of
a *panopticon*: movement in the border area was under the perpetual
surveillance of border guards. All in all, this was the context for a non-
reflective 'sense of place'. The village of Värtsilä itself always seemed as
if it was enveloped in a slow, somnolent afternoon.

When working in the area and conducting the interviews in 1987–9, it
was impossible to have any idea of the events that would occur in
Eastern Europe and the Soviet Union in the next few years and how
these changes would manifest themselves in the social consciousness of
the Finns in general and in the local structures of expectations in Värtsilä
in particular. The atmosphere of coming to the area always seemed to be
characterized by the same excited feeling. This was partly owing to the
fact that there were also significant restrictions on how the research was
to be done: it was necessary to think where to go and what to do.

The situation changed dramatically after the dispersal of the Soviet Union. As has already been seen in earlier chapters, the change in the social consciousness of the Finns was a fairly radical one. Many taboos regarding Finnish–Soviet relations since the war were removed, and it became fashionable for historians to study the nature of 'Finlandization', a matter that Finns rarely accepted as existing previously. Discussions of possible EU membership and potential military cooperation with the Western world and even NATO became an essential part of the daily discourse of politicians, researchers and journalists. Discussion of the destiny of the ceded Karelian areas became a part of public debate, and the threat posed by millions of potential Russian refugees entering the West and the possible concentration of Russian troops being moved to Karelia from the Baltic States and Germany became important topics. The ceded Karelian areas were opened up to tourists, and thousands of former refugees and their children visited the fields and forests of ceded Karelia to search for signs of their lost roots and old homes — usually in vain, because what had not been destroyed during the years of World War II had usually been broken up by Soviet authorities.

Hence the interviews conducted in 1991 and 1992 were intended to provide a complementary perspective with regard to the previous ones and to illustrate the experience of the opening of the border as a consequence of Perestroika. 10 people were interviewed in 1991, 7 of whom had visited the Russian part of the Värtsilä area for the first time since the end of the war, and a personal one-day visit was made to the ceded area of Värtsilä in summer 1992 with a small group of old inhabitants, some of them visiting their home area for the first time since the war. An exploration was made into both the results of the first interviews and those of the participant observations and interviews conducted during this intensive visit.

Besides the material described above, it was also possible to make use of more than 40 autobiographies or topical autobiographies concentrating explicitly on certain events or episodes that the person concerned had experienced in Värtsilä. These written documents vary in length from a single page to almost 100 pages, and most were written by former inhabitants of the Värtsilä area in the early 1970s, when it was intended to compile a history of Värtsilä. This project was never successfully concluded, however, and only a brief history book was written on the basis of this autobiographical material (Kasanen 1981). This material also contains some accounts of earlier interviews, the oldest of them illustrating personal experiences from the 1870s. All in all, the texts in this corpus vary from personal life-histories through depictions of various events and objects in Old Värtsilä to poems and the words to songs.

The aim of employing the interviews and other personal documents was to make use of a mix of various interview techniques, something which as a whole can perhaps be labelled as a contextualized life-history/oral history approach (see Thompson 1982, 1988, Bertaux 1981). This method was employed in an attempt to contextualize the experiences of the people interviewed within a broader frame provided by a number of sources that are internally linked, depicting trends in Finnish society and the daily life in Värtsilä. The most important material is undoubtedly that connected with the history of the old industrial community and its factory environment and with the present administrative commune, images that are sedimented in a complicated way in the collective memory of various generations and classes.

Why is it important to enter into the local lives of ordinary people and their understanding of their personal histories when analysing the historical transformation of territorial structure and localities? As Thompson (1988, 2) writes, ordinary people seek through history to understand the upheavals and changes which they experience in their own lives: war, social transformations such as the changing position of young people, technological changes in society or adaptation to new communities. In the case of Värtsilä, too, the memories and interpretations of the collapse of the industrial community, compulsory migration from homes in the ceded area and adaptation to new environments would be very superficial without understanding the ways in which people have experienced these processes in their daily lives and the meanings that they have attached to the various events and phenomena. Thompson (1988) remarks that oral history also brings the voice of ordinary people into written history, which usually tends to be written at the level of 'public' experience.

The local milieu includes in the case of Värtsilä the landscape of the border area, which mediates history, the vernacular ways of life and the influences of the Finnish state. Finally, the local context reflects strongly the framework of international politics, i.e. the boundary between the Eastern and Western worlds. It is a landscape that constitutes (and reconstitutes) social order, and the meanings connected with it also shape and structure everyday experience (cf. Eyles 1988, 113).

Hence, by exploiting various sets of material, an attempt is made to trace the institutionalization of the Värtsilä commune and factory community of Värtsilä, i.e. its historical construction, and how the features of nature and social processes (economics, politics and administration) have influenced this. An attempt is also made to trace how the histories of individual actors and generations are related to the institutional practices that manifest themselves in the institutionalization of regions — briefly, how people produce, reproduce and destroy regions in conscious and routinized action. The concept of community is employed here heuristically; I do not begin from a naïve assumption that the workers or the industrial community as a whole automatically constitute a community in the traditional sense of the word. The community is thus an analytical construction which aims at conceptualizing those dimensions which are relevant from the perspective of the research problem and subject (cf. Paasi 1991c).

REFERENCES

Abrams, P. 1982, *Historical Sociology*, Cornell University Press, Ithaca, N.Y.

Agnew, J. A. 1987, *Place and Politics: The Geographical Mediation of State and Society*, Allen and Unwin, Boston.

Aho, J. A. 1990, 'Heroism, the construction of evil, and violence', in Harle, V. (ed.), *European Values in International Relations*, Pinter Publishers, London.

Alapuro, R. 1973, *Akateeminen Karjala-Seura: Ylioppilasliike ja kansa 1920- ja 1930-luvulla*, WSOY, Porvoo.

Alapuro, R. 1988, *State and Revolution in Finland*, University of California Press, Berkeley.

Alapuro, R. 1993, 'Mihin kansa katosi?', *Kalevalaseuran vuosikirja* 72, pp. 7–12.

Alapuro, R. and Stenius H. 1987, 'Kansanliikkeet loivat kansakunnan', in Alapuro, R., Liikanen, I., Smeds, K. and Stenius, H. (eds), *Kansa liikkeessä*, Kirjayhtymä, Helsinki.

Alasuutari, P. 1989a, *Erinomaista rakas Watson*, Vastapaino, Jyväskylä.

Alasuutari, P. 1989b, 'Menneisyyden jäsentäminen identiteetin ilmauksena', in Heiskanen, P. (ed.), *Aika ja sen ankaruus*, Gaudeamus, Helsinki.

Alexander, L. M. 1957, *World Political Patterns*, Rand McNally and Co, Chicago.

Allardt, E. and Littunen, Y. 1975, *Sosiologia*, WSOY, Porvoo.

Allardt, E. and Starck, C. 1981, *Vähemmistö, kieli ja yhteiskunta*, WSOY, Juva.

Allison, R. 1985, *Finland's Relations with the Soviet Union 1944–84*, Macmillan, London.

Alsmark, G. 1982, 'Folktraditionens roll vid utformandet av nationell och regional identitet', in Nenola-Kallio, A. (ed.), *Folktradition och regional identitet i Norden*, NIF, Åbo.

Alter, P. 1989, *Nationalism*, Edward Arnold, London.

Aminoff, T. G. 1943, *Karjala lännen etuvartiona: 700-vuotinen taistelu Karjalasta*, Otava, Helsinki.

Anderson, B. 1991, *Imagined Communities: Reflections on the Origin and Spread of Nationalism*, Verso, London.

Anderson, J. 1986, 'On theories of nationalism and the size of states', *Antipode*, 18, pp. 218–32.

Anderson, J. 1988, 'Nationalist ideology and territory', in Johnston, R. J., Knight, D. B. and Kofman, E. (eds), *Nationalism, Self-Determination and Political Geography*, Croom Helm, London.

Anderson, P. 1991, 'Nation-states and national identity', *London Review of Books*, 13(9), pp. 4–8.

Andersson, R. 1980, 'Plats, identitet och förändring: Om att sätta identiteten på plats', *Institutionen för Nordiska språk vid Uppsala Universitet, FUMS rapport Nr.82*, Uppsala.

Apunen, O. 1977, 'Maantieteelliset ja poliittiset tekijät Suomen idänsuhteissa', *Ulkopolitiikka*, 14, pp. 33–47.

Archer, M. S. 1990, 'Theory, culture and post-industrial society', *Theory, Culture and Society*, 7, pp. 97–119.

Arnkil, E. 1984, 'Varustelun ja ajattelun ristiriita', *Tutkijaliiton julkaisusarja*, 28, Jyväskylä.

Aro, J. E. 1913, *Kansakoulun Venäjän maantieto*, Weilin & Göös, Helsinki.

Atlas Istorii SSSR, 1990, Glavnoe Upravlenie Geodezü i Kartografi i pri Sovete Ministrov SSSR, Moskva.

Atlas of Finland, 1899, Tilgmann, Helsinki.

Auer, V. 1941, 'Tuleva Suomi talousmaantieteellisenä kokonaisuutena', *Terra*, 53, pp. 206-17.

Auer, V. and Jutikkala, E. 1941, *Finnlands Lebensraum: Das Geographische und Geschichtliche Finnland*, Alfred Metzner Verlag, Berlin.

Auer, V. and Merikoski, K. 1951, *Maantieto kansakouluja varten*, Otava, Helsinki.

Auer, V. and Poijärvi, A. 1937, *Suomen maantieto*, Otava, Helsinki.

Augelli, J. P. 1980, 'Nationalization of Dominican borderlands', *Geographical Review*, 70, pp. 19–35.

Austin-Broos, D. 1987, 'Clifford Geertz: culture, sociology and historicism', in Austin-Broos, D. (ed.), *Creating Culture: Profiles in the Study of Culture*, Allen and Unwin, London.

Bachelard, G. 1969, *The Poetics of Space*, Beacon Press, Boston.

Barnes, T. J. and Duncan, J.S. 1992, 'Introduction: Writing Worlds', in Barnes, T. J. and Duncan J. S. (eds), *Writing Worlds: Discourse, Text and Metaphor in the Representation of Landscape*, Routledge, London.

Barth, F. 1969, 'Introduction', in Barth, F. (ed.), *Ethnic Groups and Boundaries: The Social Organization of Culture Difference*, Allen and Unwin, London.

Bauman, Z. 1990, 'Modernity and ambivalence', *Theory, Culture and Society*, 7, pp. 143–69.

Beattie, J. 1964, *Other Cultures: Aims, Methods and Achievements in Social Anthropology*, Routledge and Kegan Paul, London.

Becher, T. 1989, *Academic Tribes and Territories*, Open University Press, Milton Keynes.

Bender, T. 1978, *Community and Social Change in America*, Johns Hopkins, Baltimore.

Bennington, G. 1990, 'Postal politics and the institution of the nation', in Bhabha, H. K. (ed.), *Nation and Narration*, Routledge, London.

Berdoulay, V. 1989, 'Place, meaning, and discourse in French language geography', in Agnew, J. A. and Duncan, J. N. (eds), *The Power of Place: Bringing Together Geographical and Sociological Imaginations*, Unwin Hyman, London.

Berger, P. L. and Luckmann, T. 1976, *The Social Construction of Reality: A Treatise in the Sociology of Knowledge*, Penguin Books, Harmondsworth.

Bergman, E. F. 1975, *Modern Political Geography*, WCB, Dubuque, Iowa.

Bertaux, D. 1981, 'From the life-history approach to the transformation of sociological practice', in Bertaux, D. (ed.), *Biography and Society: The Life History Approach in the Social Sciences*, Sage Publications, London.

Best, S. and Kellner, D. 1991, *Postmodern Theory: Critical Interrogations*, Macmillan, London.

Bhabha, H. K. 1990, 'DissemiNation: time, narrative, and the margins of the modern nation', in Bhabha, H. K. (ed.), *Nation and Narration*, Routledge, London.

Bhaskar, R. 1983, 'Beef, structure and place: notes from a critical naturalist perspective', *Journal for the Theory of Social Behaviour*, 13, pp. 81–95.

Billig, M. 1991, *Ideology and Opinions: Studies in Rhetorical Psychology*, Sage Publications, London.

Birch, A. H. 1989, *Nationalism and National Integration*, Unwin Hyman, London.

Bird, J. 1989, *The Changing Worlds of Geography: A Critical Guide to Concepts and Methods*, Clarendon Press, Oxford.

Bird, J., Curtis, B., Putnam, T. et al. 1993, *Mapping the Futures: Local Culture, Global Change*, Routledge, London.

Bleicher, J. 1989, 'The cultural construction of social identity: the case of Scotland', in Haferkamp, H. (ed.), *Social Structure and Culture*, Walter de Gruyter, Berlin and New York.

Bloom, W. 1990, *Personal Identity, National Identity and International Relations*, Cambridge University Press, Cambridge.

Bocock, R. 1986, *Hegemony*, Ellis Horwood, Chichester.

Borg, O. (ed.) 1970, *Mitä puoluetta äänestäisin: Tietoja ja tutkimuksia puolueista, politiikasta ja vaaleista*, Otava, Helsinki.

Bourdieu, P. 1977, *Outline of a Theory of Practice*, Cambridge University Press, Cambridge.

Bourdieu, P. 1991, *Language and Symbolic Order*, Polity Press, Cambridge.

Bourdieu, P. and J-C. Passeron 1990, *Reproduction in Education, Society and Culture*, Sage Publications, London.

Bowman, I. 1928, *The New World: Problems in Political Geography*, Yonker-on-Hudson, New York.

Brander, U. B. 1932, 'Yhteiskunnallinen huoltotoiminta ja rajaseutujen elämäntason kohottaminen', *Rajaseutu*, 9, pp. 79–83.

Braudel, F. 1980, *On History*, Weidenfeld and Nicolson, London.

Brennan, T. 1990, 'The national longing for form', in Bhabha, H. K. (ed.), *Nation and Narration*, Routledge, London.

Briggs, C. L. 1986, *Learning How to Ask: A Sociolinguistic Appraisal of the Role of the Interview in Social Science Research*, Cambridge University Press, Cambridge.

Brown, R. H. 1987, *Society as Text: Essays on Rhetoric, Reason, and Reality*, University of Chicago Press, Chicago.

Brunn, S. 1984, 'Future of the nation-state system', in Taylor, P. and House, J. (eds), *Political Geography: Recent Advances and Future Directions*, Croom Helm, London.

Brunn, S. 1991, 'Peacekeeping missions and landscapes', in Rumley, D. and Minghi, J. V. (eds), *The Geography of Border Landscapes*, Routledge, London.

Burgess, J. and Wood, P. 1988, 'Decoding Docklands: Place advertising and the decision-making strategies of the small firm', in Eyles, J. and Smith, D. M. (eds), *Qualitative Methods in Human Geography*, Polity Press, Cambridge.

Burke, P. 1992, 'We, the people: popular culture and popular identity in modern Europe', in Lash, S. and Friedman, J. (eds), *Modernity and Identity*, Blackwell, Oxford.

Buttimer, A. 1978, 'Home, reach and the sense of place', in Regional identitet och förändring i den regionala samverkans samhälle, *Acta Universitatis Upsaliensis, Annum Quingentesimum Celebrantis* 11, Almqvist and Wiksell International, Stockholm.

Calhoun, C. 1991, 'Indirect relationships and imagined communities: large-scale social integration and the transformation of everyday life', in Bourdieu, P. and Coleman, J. (eds), *Social Theory for a Changing Society*, Westview Press, Boulder.

Capel, H. 1981, 'Institutionalization of geography', in Stoddart, D. (ed.), *Geography, Ideology and Social Concern*, Blackwell, Oxford.

Carlson, L. 1958, *Geography and World Politics*, Prentice-Hall, Englewood Cliffs, N.J.

Carr, D. 1986, *Time, Narrative, and History*, Indiana University Press, Bloomington, Indianapolis.

Castells, M. 1989, *The Informational City: Information Technology, Economic Restructuring, and the Urban-Regional Process*, Basil Blackwell, Oxford.

de Certeau, M. 1984, *The Practice of Everyday Life*, University of California Press, Berkeley.

Clark, G. L. and Dear, M. 1984, *State Apparatus: Structures and Language of Legitimacy*, Allen & Unwin, Boston.

Clifford, J. 1988, *The Predicament of Culture: Twentieth Century Ethnography, Literature and Art*, Harvard University Press, Cambridge, Mass.

Cohen, A. 1982, 'Belonging: the experience of culture', in Cohen, A. (ed.), *Belonging: Identity and Social Organization in British Rural Cultures*, Manchester University Press, Manchester.

Cohen, A. 1985, *The Symbolic Construction of Community*, Ellis Horwood, Chichester.

Cohen, A. (ed.) 1986a, *Symbolising Boundaries: Identity and Diversity in British Cultures*, Manchester University Press, Manchester.

Cohen, A. 1986b, 'Of symbols and boundaries, or, does Ertie's greatcoat hold the key?', in Cohen, A. (ed.), *Symbolising Boundaries: Identity and Diversity in British Cultures*, Manchester University Press, Manchester.

Cohen, S. B. 1964, *Geography and Politics in a Divided World*, Methuen, London.

Cohen, S. B. 1982, 'A new map of global geopolitical equilibrium: a developmental approach', *Political Geography Quarterly*, 1, pp. 223–41.

Cohen, S. B. 1991, 'Global geopolitical change in the post-Cold War era', *Annals of the Association of American Geographers*, 81, pp. 551–80.

Cohen, S. B. and Kliot, N. 1992, 'Place-names in Israel's ideological struggle over the administered territories', *Annals of the Association of American Geographers*, 82, pp. 653–80.

Comaroff, J. and Comaroff, J. 1992, *Ethnography and the Historical Imagination*, Westview Press, Boulder, Colorado.

Cooke, P. 1987, 'Individuals, localities and postmodernism', *Environment and Planning D: Society and Space*, 5, pp. 408–12.

Cooke, P. (ed.) 1989a, *Localities: The Changing Face of Urban Britain*, Unwin Hyman, London.

Cooke, P. 1989b, 'Nation, space, modernity', in Peet, N. and Thrift, N. (eds), *New Models in Geography: The Political-Economy Perspective (Volume 1)*, Unwin Hyman, London.

Coser, L. A. 1992, 'Introduction', in Coser, L. A. (ed.), *Halbwachs, Maurice: On Collective Memory*, University of Chicago Press, Chicago.

Cosgrove, D. 1989, 'Geography is everywhere: culture and symbolism in human landscapes', in Gregory, D. and Walford, R. (eds), *Horizons in Human Geography*, Macmillan, London.

Crang, P. 1992, 'The politics of polyphony: reconfigurations in geographical authority', *Environment and Planning D: Society and Space*, 10, pp. 527–49.

Dalby, S. 1988, 'Geopolitical discourse: The Soviet Union as Other', *Alternatives*, XIII, pp. 415–42.

Dalby, S. 1990, *Creating the Second Cold War: The Discourse of Politics*, Pinter Publishers, London.

Daniels, S. 1988, 'The political iconography of woodland in later Georgian England', in Cosgrove, D. and Daniels, S. (eds), *The Iconography of Landscape: Essays on the Symbolic Representation, Design and Use of Past Environments*, Cambridge University Press, Cambridge.

Daniels, S. and Cosgrove, D. 1988, 'Introduction: the iconography of landscape', in Daniels, S. and Cosgrove, D. (eds), *The Iconography of Landscape: Essays on the Symbolic Representation, Design and Use of Past Environments*, Cambridge University Press, Cambridge.

Dannholm, O. 1889, *Maantiede kansakouluja varten*, K. E. Holm, Helsinki.

Dannholm, O. 1894, *Maantiede kansakouluja ja muita oppilaitoksia sekä itsekseen lukemista varten*, K. E. Holm, Helsinki.

Dayan, D. and Katz, E. 1988, 'Articulating consensus: the ritual and rhetoric of media events', in Alexander, J. C. (ed.), *Durkheimian Sociology: Cultural Studies*, Cambridge University Press, Cambridge.

Dear, M. 1986, 'State'/'State apparatus', in Johnston, R. J., Gregory, D. and Smith, D. M. (eds), *The Dictionary of Human Geography*, Blackwell, Oxford.

Dear, M. and Wolch, J. 1989, 'How territory shapes social life', in Wolch, J. and Dear, M. (eds), *The Power of Geography: How Territory Shapes Social Life*, Unwin Hyman, London.

Deutsch, K. W. 1963, 'Some problems in the study of nation-building', in Deutsch, K. W. and Foltz, W. J. (eds), *Nation-Building*, Atherton Press, New York.

Doherty, J., Graham, E. and Malek, M. (eds) 1992, *Postmodernism and the Social Sciences*, Macmillan, London.

Douglas, M. 1987, *How Institutions Think*, Routledge and Kegan Paul, London.

Driver, F. 1988, 'The historicity of human geography', *Progress in Human Geography*, 12, pp. 497–506.

Driver, F. 1992, 'Geography's empire: histories of geographical knowledge', *Environment and Planning D: Society and Space*, 10, pp. 23–40.

Duchacek, I. D. 1975, *Nations and Men: An Introduction to International Politics*, Dryden Press, Hinsdale, Ill.

Duijker, H. and Frijda, N. 1960, *National Character and National Stereotypes*, North Holland Publishing Company, Amsterdam.

Dumont, L. 1986, 'Collective identities and universalist ideology: the actual interplay', *Theory, Culture and Society*, 3, pp. 25–33.

Duncan, J. 1993, 'Sites of representation: Place, time and the discourse of the Other', in Duncan, J. and Ley, D. (eds), *Place/Culture/Representation*, Routledge, London and New York.

Duncan, J. and Ley, D. (eds) 1993. *Place/Culture/Representation*, Routledge, London and New York.

Duncan, S. and Savage, M. 1991, 'New perspectives on the locality debate', *Environment and Planning A*, 23, pp. 155–64.

Durkheim, É. 1964, *The Division of Labor*, Free Press, New York.

Durkheim, É. 1966, *The Rules of Sociological Method*, Free Press, New York.

Eagleton, T. 1991, *Ideology: An Introduction*, Verso, London.

Ehn, B. 1989, 'National feeling in sport: The case of Sweden', *Ethnologia Europaea*, XIX, pp. 57–66.

Elias, N. 1993, *Kuolevien yksinäisyys* (orig. Über die Einsamkeit der Sterbenden in unseren Tagen, 1982), Gaudeamus, Helsinki.

Enckell, O. 1939, *Rajan vartio: Jalkamatkoja Raja-Karjalassa*, Otava, Helsinki.

Entrikin, J. N. 1991, *The Betweenness of Place: Towards a Geography of Modernity*, Johns Hopkins University Press, Baltimore.

Ervasti, O. 1993, 'Kelloselkä — Kuolan portti: Odotuksia ja asenteita Sallassa ja Kemijärvellä 1992', unpubl. MSc thesis, Dept. of Geography, Univ. of Oulu.

Eskola, A. 1963, *Maalaiset ja kaupunkilaiset*, Kirjayhtymä, Helsinki.

Essén, R. 1941, *Venäjän arvoitus*, WSOY, Porvoo.

Eyles, J. 1985, *Senses of Place*, Silverbrook Press, Cheshire.

Eyles, J. 1988, 'Interpreting the geographical world: qualitative approaches in geographical research', in Eyles, J. and Smith, D. M. (eds), *Qualitative Methods in Human Geography*, Polity Press, Cambridge.

Eyles, J. 1989, 'The geography of everyday life', in Gregory, D. and Walford, R. (eds.), *Horizons in Human Geography*, Macmillan, London.

Eyles, J. and Smith, D. M. (eds.) 1988, *Qualitative Methods in Human Geography*, Polity Press, Cambridge.

Featherstone, M. 1991, *Consumer Culture and Postmodernism*, Sage Publications, London.

Ferrarotti, F. 1990, *Time, Memory and Society*, Greenwood Press, Westport, Connecticut.

von Fieandt, K. 1946, *Lauma ja yhteisö*, Otava, Helsinki.

Forsman, V. I. 1941, 'Väärät ja oikeat rajat', *Rajaseutu*, 18, pp. 177–8.

Foucault, M. 1980a, 'Questions on geography', in Gordon, C. (ed.), *Michel Foucault: Power and Knowledge, Selected Writings 1972–1977*, Pantheon Books, New York.

Foucault, M. 1980b, 'Truth and power', in Gordon, C. (ed.), *Michel Foucault: Power and Knowledge, Selected Writings 1972–1977*, Pantheon Books, New York.

Frampton, K. 1985, 'Towards a critical regionalism: six points for an architecture of resistance', in Foster, H. (ed.), *Postmodern Culture*, Pluto Press, London.

Fredrikson, E. 1993, *Suomi 500 vuotta Euroopan kartalla: Kuvaus Suomen karttakuvan vaiheista keskiajalta nykypäivään*, Gummerus, Jyväskylä.

Friedrich, C. J. 1963, 'Nation-building?', in Deutsch, K. and Foltz, W. J. (eds), *Nation-Building*, Atherton Press, New York.

Fromm, E. 1962, *Vaarallinen vapaus* (orig. Escape from Freedom, 1941), Kirjayhtymä, Helsinki.

de Gadolin, A. 1952, 'The solution of the Karelian refugee problem in Finland'. *Publications of the Research Group for European Migration Problems V*, Martinus Nijhof, the Hague.

Geertz, C. 1973, *The Interpretation of Cultures: Selected Essays*, Basic Books, New York.

Gellner, E. 1983, *Nations and Nationalism*, Basil Blackwell, Oxford.

Giddens, A. 1984, *The Constitution of Society: Outline of the Theory of Structuration*, Polity Press, Cambridge.

Giddens, A. 1987a, *The Nation-State and Violence: Volume Two of a Contemporary Critique of Historical Materialism*, University of California Press, Berkeley and Los Angeles.

Giddens, A. 1987b, *Social Theory and Modern Sociology*, Polity Press, Cambridge.

Giddens, A. 1991, *Modernity and Self-Identity: Self and Society in the Late Modern Age*, Polity Press, Cambridge.

Gilbert, M. 1989, *On Social Facts*, Princeton University Press, Princeton N.J.

Glassner, M. I. and de Blij, H. J. 1980, *Systematic Political Geography*, John Wiley and Sons, New York.

Gottmann, J. 1973, *The Significance of Territory*, University Press of Virginia, Charlottesville.

Granö, J.G. 1941, 'Sotamarsalkka Mannerheimin matkateos', *Terra*, 53, pp. 70–78.

Gregory, D. 1989, 'Areal differentation and post-modern human geography', in Gregory, D. and Walford, R. (eds), *Horizons in Human Geography*, Macmillan, London.

Gregory, D. and Urry, J. (eds) 1985, *Social Relations and Spatial Structure*, Macmillan, London.

Haavikko, P. 1984, *Wärtsilä 1834–1984: Wärtsilä-yhtiön ja siihen liitettyjen yritysten kehitysvaiheita kansainvälistyväksi monialayritykseksi*, Oy Wärtsilä AB, Helsinki.

Haavio-Mannila, E. 1973, 'Etnologia ja sosiaaliantropologia', in Alapuro, R., Alestalo, M. and Haavio-Mannila, E. (eds), *Suomalaisen sosiologian juuret*, WSOY, Porvoo.

Habermas, J. 1988, 'Historical consciousness and post-traditional identity: remarks on the Federal Republic's orientation to the West', *Acta Sociologica*, 31, pp. 3–13.

Hakalehto, A. and Salmela, A. 1936, *Isänmaa ja maailma*, WSOY, Porvoo.

Hakalehto, A. and Salmela, A. 1951, *Isänmaan ja maailman maantieto*, WSOY, Porvoo.

Hako, M., Huhtanen, H. and Nieminen, M. (eds) 1975, *Jako kahteen: Työmiehen 20-luku*, Tammi, Helsinki.

Hakovirta, H. 1975, *Suomettuminen: Kaukokontrollia vai rauhanomaista rinnakkaiseloa?*, Gummerus, Jyväskylä.

Hakulinen, K. 1984, 'Maat ja kansat: Maantieteilijä J. E. Rosbergin maailmankuva', in Melasuo, T. (ed.), *Wallinista Wideriin: suomalaisen kolmannen maailman tutkimuksen perinteitä*, Suomen rauhantutkimusyhdistys, Tampere.

Halén, H. 1989, 'Altain huiput siintelee', in Löytönen, M. (ed.), *Matka-arkku: suomalaisia tutkimusmatkailijoita*, SKS, Helsinki.

Halfacree, K. H. 1993, 'Locality and social representation: space, discourse and alternative definitions of the rural', *Journal of Rural Studies*, 9, pp. 23–37.

Halila, A. 1980, 'Kansanopetus', in Tommila, P., Reitala, A. and Kallio, V. (eds), *Suomen kulttuurihistoria*, II, WSOY, Porvoo.

Hämynen, T. 1993, 'Liikkeellä leivän tähden: Raja-Karjalan väestö ja sen toimeentulo 1880–1940', *Suomen Historiallinen Seura, Historiallisia tutkimuksia*, 170, Tampere.

Hamilton, N. 1990, 'The price of independence: Finland–USSR', in Eyre, R., Gordimer, N., Hamilton, N. et al. (eds), *Frontiers*, BBC Books, London.

Hareven, T. and Langenbach, R. 1981, 'Living places, work places and historical identity', in Lowenthal, D. and Binney, M. (eds), *Our past before us: Why do we save it?*, Temple Smith, London.

Harle, V. 1990, 'European roots of dualism and its alternatives in international relations', in Harle, V. (ed.), *European Values in International Relations*, Pinter Publishers, London.

Harley, J. B. 1989, 'Deconstructing the map', *Cartographica*, 26, pp. 1–20.

Härö, E. 1991, 'Högfors teollisuushistoriallisena muistomerkkinä', in Schulman, H. (ed.), *Teollisen yhdyskunnan murros. Tapaus Högfors*, Otava, Helsinki.

Harré, R. 1978, 'Architectonic man: on the structuring of lived experience', in Brown, R. H. and Lyman, S. M. (eds), *Structure, Consciousness and History*, Cambridge University Press, Cambridge.

Harvey, D. 1973, *Social Justice and the City*, Edward Arnold, London.

Harvey, D. 1987, 'Three myths in search of a reality in urban studies', *Environment and Planning D: Society and Space 5*, pp. 367–76.

Harvey, D. 1989, *The Condition of Postmodernity: An Enquiry into the Origins of Cultural Change*, Basil Blackwell, Oxford.

Hassner, P. 1993, 'Beyond nationalism and internationalism: ethnicity and world order', *Survival*, 35, pp. 49–65.

Hegel, G. W. F. 1978, *Järjen ääni: Hegelin historianfilosofian luentojen johdanto* (orig. Vorlesungen über die Philosophie der Geschichte), Gaudeamus, Helsinki.

Heikinheimo, A. L. 1953, 'Toimittajan esipuhe', *Rajaseutu*, 30, pp. 83–4.

Heikkilä, H. 1986, 'Neuvostoliiton ja Suomen väliset taloussuhteet 1945–55', *Suomen Historiallinen Seura, Historiallinen Arkisto*, 88, pp. 109–44.

Heikkinen, R. 1987, 'Maantiedon asema ja maantiedon oppikirjojen sisältö yleissivistävässä koulutuksessa Neuvostoliitossa', unpubl. MSc thesis, Dept. of Geography, Univ. of Joensuu.

Heikkinen, S. and Hjerppe, R. 1986, 'Suomen teollisuus ja teollinen käsityö 1860–1913', *Bank of Finland Publications, Studies on Finland's Economic Growth*, XII, Helsinki.

Heller, A. 1984, *Everyday Life*, Routledge and Kegan Paul, London.

Henige, D. 1982, *Oral Historiography*, Longman, London.

von Hertzen, G. 1921, *Karjalan retkikunta*, Gummerus, Jyväskylä.

Herzfeld, M. 1986, 'On definitions and boundaries: the status of culture in the culture of the state', in Chock, P. P. and Wyman, J. R. (eds), *Discourse and the Social Life of Meaning*, Smithsonian Institution Press, Washington, D.C. and London.

History Workshop 1990, Issue 29.

Hobsbawm, E. J. 1983, 'Introduction: inventing traditions', in Hobsbawm, E. J. and Ranger, T. (eds), *The Invention of Tradition*, Cambridge University Press, Cambridge.

Hobsbawm, E. J. 1990, *Nations and Nationalism since 1780: Programme, Myth, Reality*, Cambridge University Press, Cambridge.

Hoffman, K. 1982, 'Sawmills — Finland's proto-industry', *Scandinavian Economic History Review*, 30, pp. 35–43.

Homen, T. 1921, 'Final survey of economic, religious and political life', in Homen, T. (ed.), *East Karelia and Kola Lapmark*, *Fennia*, 42, pp. 253–64.

House, J. W. 1968, 'A local perspective on boundaries and the frontier zone: two examples from the European Economic Community', in Fisher, C. A. (ed.), *Essays in Political Geography*, Methuen, London.

Hudson, R. 1988, 'Changing spatial divisions of labour and their impacts on localities', *Nordisk Samhällsgeografisk Tidskrift*, 7, pp. 3–14.

Hummon, D. M. 1990, *Common Places: Community, Ideology and Identity in American Culture*, State University of New York Press, Albany.

Hustich, I. 1972, *Suomi tänään: Kuinka maamme on muuttunut*, Tammi, Helsinki.

Hustich, I. 1974, *Yksi maailma, monta maailmaa*, Tammi, Helsinki.

Hustich, I. 1983, 'An autobiographical sketch of the "life-path" of a geographer', in Buttimer, A. (ed.), *The Practice of Geography*, Longman, London.

Ignatius, K. E. F. 1890, *Suomen maantiede kansalaisille*, G. W. Edlund, Helsinki.

Immonen, K. 1987, *Ryssästä saa puhua. . .Neuvostoliitto suomalaisessa julkisuudessa ja kirjat julkisuuden muotona 1918-39*, Otava, Helsinki.

Jääskeläinen, M. 1961, *Itä-Karjalan kysymys*, WSOY, Porvoo.

Jackson, P. 1989, *Maps of Meaning: An Introduction to Cultural Geography*, Unwin Hyman, London.

Jakobson, M. 1984, *Finland Survived: An Account of the Finnish–Soviet Winter War 1939-1940*, Otava Publishing, Helsinki.

James, P. and Martin, G. 1981, *All Possible Worlds: A History of Geographical Ideas*, John Wiley and Sons, New York.

Jameson, F. 1984, 'Postmodernism, or the cultural logic of late capitalism', *New Left Review*, 146, pp. 53–92.

Jansson, J-M. and Tuomioja, E. 1981, *Tulevaisuuden varjossa*, Tammi, Helsinki.

Jeans, D. N. 1988, 'The First World War memorials in New South Wales: centres of meaning in the landscape', *Australian Geographer*, 19, pp. 259–67.

Joenniemi, P. 1993, 'Euro-Suomi: rajalla, rajojen välissä vai rajaton?', in Joenniemi, P., Alapuro, R. and Pekonen, K. (eds), Suomesta Euro-Suomeen: Keitä me olemme ja mihin matkalla, *Rauhan- ja konfliktintutkimuslaitos, tutkimustiedote No. 53*, Tampere.

Johnson, N. C. 1992, 'Nation-building, language and education: The geography of teacher recruitment in Ireland, 1925-1955', *Political Geography*, 11, pp. 170–89.

Johnston, R. J. 1986, 'Power', in Johnston, R. J., Gregory, D. and Smith, D. M. (eds), *The Dictionary of Human Geography*, Blackwell, Oxford.

Johnston, R. J. 1989, 'The state, political geography, and geography', in Peet, R. and Thrift, N. (eds), *New Models in Geography (Volume I)*, Unwin Hyman, London.

Johnston, R. J. 1991, *A Question of Place: Exploring the Practice of Human Geography*, Blackwell, Oxford.

Johnston, R. J., Knight, D. and Kofman, E. 1988, 'Nationalism, self-determination and the world political map: an introduction', in Johnston, R. J., Knight, D. and Kofman, E. (eds), *Nationalism, Self-determination and Political Geography*, Croom Helm, London.

Joll, J. 1990, *Europe Since 1870: An International History*, Penguin Books, London.

Jones, S. B. 1959, 'Boundary concepts in setting time and space', *Annals of the Association of American Geographers*, 49, pp. 241–55.

Joyce, P. 1980, *Work, Society and Politics: The Culture of the Factory in Later Victorian England*, Rutgers University Press, New Brunswick, N.J.

Julku, K. 1987, 'Suomen itärajan synty', *Studia historica septentrionalia 10*, Oulu.

Jussila, O. 1979, 'Nationalismi ja vallankumous venäläis-suomalaisissa suhteissa 1899–1914', *Suomen Historiallinen Seura, Historiallisia tutkimuksia 110*, Helsinki.

Jutikkala, E. 1976, 'Myytti entisajan ihmisen liikkumattomuudesta', *Kanava* 4, pp. 88–91.

Juvonen, J. 1991, *Vanhan Tohmajärven historia*, Pieksämäki.

Kailas, U. 1941, *Isien tie: Kokoelma isänmaallisia runoja*, WSOY, Porvoo.

Kalliola, R. 1969, *Kotimaa: Suomen maantieteen oppikirja*, WSOY, Porvoo.

Kalpio, O. 1988, 'Ruukista naulatehtaaksi', *University of Jyväskylä, Institute of Ethnology, Research Report*, 22, Jyväskylä.

Kanter, R. M. 1972, *Commitment and Community: Communes and Utopias in Sociological Perspective*, Harvard University Press, Cambridge, Mass.

Karjalainen, P. T. 1988, 'Lähtökohtia siirtokarjalaisten paikkakokemusten tutkimiseksi', in Jokipii, M. (ed.), Tutkielmia sodanjälkeisestä asutustoiminnasta Suomessa, *Jyväskylän yliopiston historian laitos, Suomen historian julkaisuja No. 13*, Jyväskylä.

Karjalainen, P. T. and Paasi, A. (1994). 'Contrasting the images of a written city: Helsinki in regionalist thought and as a dwelling-place', in Preston, P. and Simpson-Housley, P. (eds), *Writing the City: Eden, Babylon and the New Jerusalem*, Routledge, London and New York.

Kärnä, O. A. 1932, 'Vaikeneeko virtemme?', *Rajaseutu*, 9, pp. 101–3.

Kasanen, T. I. 1981, *Värtsilä. Muistelmia kylän ja kunnan vaiheista vuoteen 1945*, Värtsilä-seura, Helsinki.

Katajamäki, H. 1988, 'Alueellisen työnjaon muotoutuminen Suomessa', *Turun yliopiston maantieteen laitoksen julkaisuja*, 121, Turku.

Kaukoranta, T. 1935, *Piirteitä Värtsilän seudun asutuksesta ja teollisuudesta*, Otava, Helsinki.

Kaukoranta, T. 1944, *Itä-Karjalan vapaudentie: Poliittis-kronologinen yleiskatsaus*, Karjalan ulkomaisen valtuuskunnan julkaisuja, Helsinki.

Keith, M. 1988, 'Racial conflict and the "no-go areas" of London', in Eyles, J. and Smith. D. M. (eds), *Qualitative Methods in Human Geography*, Polity Press, Cambridge.

Keith, M. and Pile, S. (eds) 1993, *Place and the Politics of Identity*, Routledge, London and New York.

Kekkonen, U. 1943, 'Good neighbourliness with the "hereditary enemy"', in Vilkuna, T. (ed.), 1973, *Neutrality: The Finnish Position — Speeches by Dr Urho Kekkonen*, Heinemann, London.

Kemiläinen, A. 1964, 'Nationalism: Problems concerning the word, the concept and classification', *Studia Historica Jyväskyläensia III*, Jyväskylä.

Kemiläinen, A. 1984, 'Initiation of the people into nationalist thinking', in Nationality and nationalism in Italy and Finland from the mid-19th century to 1918, *Societas Historica Finlandiae, Studia Historica*, 16, pp. 105–20.

Kemiläinen, A. 1986, 'Onko Suomen kansa "epähistoriallinen"? — Hegelin ja Engelsin jäljillä', in Kemiläinen, A. (ed.), Kansallisuuskysymyksiä ja rotuasenteita, *Jyväskylän yliopiston historian laitos, yleisen historian tutkimuksia*, 6, Jyväskylä.

Kemiläinen, A. 1989, 'Nationalism in nineteenth century Finland', in *Societas Historica Finlandiae, Studia Historica*, 33, pp. 93–127.

Kirjavainen, I. 1969, 'Rajavartiolaitos 1944–1969', in *Rajavartiolaitos 1919–1969*, Rajamme vartijat ry, Mikkeli.

Kivi, A. 1991, *Seven Brothers: A Novel*, Braun-Brumfield, Ann Arbor, Michigan.

Klementjev, J. E. 1991, *Karely: Etnograficeskii ocerk*, Karelija, Petrozavodsk.

Klineberg, O. 1950, *Tensions Affecting International Understanding: A Survey of Research*, Social Science Research Council, New York.

Klineberg, O. 1951, 'The scientific study of national stereotypes', *International Social Science Bulletin*, 3, pp. 505–515.

Klinge, M. 1972, *Vihan veljistä valtiososialismiin: Yhteiskunnallisia ja kansallisia näkemyksiä 1910- ja 1920-luvuilta*, WSOY, Porvoo.

Klinge, M. 1975, *Bernadotten ja Leninin välissä: Tutkielmia kansallisista aiheista*, WSOY, Porvoo.

Klinge, M. 1980, 'Poliittisen ja kulttuurisen Suomen muotoutuminen', in Tommila, P., Reitala, A. and Kallio, V. (eds), *Suomen kulttuurihistoria II*, WSOY, Porvoo.

Klinge, M. 1981, *Suomen sinivalkoiset värit: Kansallisten ja muidenkin symbolien vaiheista ja merkityksestä*, Otava, Keuruu.

Klinge, M. 1982, *Kaksi Suomea*, Otava, Keuruu.

Klinge, M. 1984, 'Let us be Finns!', in Nationality and nationalism in Italy and Finland from the mid-19th century to 1918, *Societas Historica Finlandiae, Studia Historica*, 16, pp. 121–33.

Kliot, N. and Watermann, S. 1983, 'Introduction', in Kliot, N. and Watermann, S. (eds), *Pluralism and Political Geography: People, Territory and State*, Croom Helm, London.

Knight, D. 1982, 'Identity and territory: geographical perspectives on nationalism and regionalism', *Annals of the Association of American Geographers*, 72, pp. 514–31.

Knight, D. 1984, 'Geographical perspectives on self-determination', in Taylor, P. J. and House, J. (eds), *Political Geography: Recent Advances and Future Prospects*, Croom Helm, London.

Knuuttila, S. 1989, 'Periferia + menneisyys = satumaa?', in Saarelainen, P. (ed.), *Aluesuunnittelupäivät 1988: Yhdyskuntarakenne-rakennemuutos, Joensuun yliopisto, Kulttuuri-ja suunnittelumaantiede, tiedonantoja*, 8, Joensuu.

Knuuttila, S. 1992, *Kansanhuumorin mieli. Kaskut maailmankuvan aineksena*, SKS, Helsinki.

Kobayashi, A. and Mackenzie, S. (eds) 1989, *Remaking Human Geography*, Unwin Hyman, London.

Koljonen, A. 1985, 'Rajaseututyö: 60 vuotta kehitysaluepolitiikkaa', *Nordia tiedonantoja Sarja B*, 2, Oulu.

Komiteanmietintö 1932:6, *Oppikoulukomitean mietintö*, Helsinki.

Komiteanmietintö 1952:3, *Kansakoulun opetusuunnitelmankomitean mietintö*, Helsinki.

Konttinen, E. 1991, *Perinteisesti moderniin: Professioiden yhteiskunnallinen synty Suomessa*, Vastapaino, Tampere.

Korhonen, A. 1938, *Suomen itärajan syntyhistoriaa*, WSOY, Porvoo.

Korhonen, K. 1966, *Naapurit vastoin tahtoaan: Suomi neuvostodiplomatiassa Tartosta talvisotaan I 1920–1932*, Tammi, Helsinki.

Korhonen, K. 1973, 'Kannattaa yrittää: Puhe rajaseututoiminnan 50-vuotisjuhlassa 28.2.1973', *Rajaseutu*, 50, pp. 165–8.

Koskinen, T. 1987, 'Tehdasyhteisö', *Vaasan korkeakoulun julkaisuja, tutkimuksia, 123*, Vaasa.

Koskinen, Y. 1876. 'Onko Suomen kansalla historiaa?', *Historiallinen arkisto* V, pp. 1–9.

Kosonen, M. and Pohjonen, J. 1994, *Isänmaan portinvartijat: Suomen rajojen vartiointi 1918–1994*, Otava, Helsinki.

Kristeva, J. 1992, *Muukalaisia itsellemme*, Gaudeamus, Helsinki.

Kukkonen, A. 1933, 'Joulu: rajaseudunkin riemujuhla', *Rajaseutu*, 10, pp. 125–6.

Laine, A. 1982, *Suur-Suomen kahdet kasvot: Itä-Karjalan siviiliväestön asema suomalaisessa miehityshallinnossa 1941–1944*, Otava, Keuruu.

Laine, A. 1992, 'Karjala takaisin-keskustelun historia: argumenttianalyysi Karjala takaisin-keskustelun historiasta', Studia Generalia Lecture, University of Joensuu, 1 April.

Laine, A. 1993, 'Tiedemiesten Suur-Suomi — Itä-Karjalan tutkimus jatkosodan vuosina', *Historiallinen Arkisto*, 102, pp. 91–202.

Laine, A. 1994, 'Die Deutschlandbeziehungen der finnischen Wissenschaftler während des zweiten Weltkriegs', mimeo, Karelian Institute, Univ. of Joensuu.

Laine, E. 1907, *Piirteitä Suomen vuoritoimen historiasta 19-vuosisadan ensipuoliskolla: oas I rautateollisuus 1808–1831*, Helsinki.

Laine, E. 1948, 'Suomen vuoritoimi 1809–1884, II ruukit', *Historiallisia tutkimuksia*, XXXI:2.

Laine, E. 1950, 'Suomen vuoritoimi 1809–1884, I yleisesitys', *Historiallisia tutkimuksia*, XXXI:1.

Lakio, M. 1975, 'Teollisuuden kehittyminen Itä-Suomessa 1830–1940', *Itä-Suomen Instituutti, sarja A*, 5, Mikkeli.

Lakoff, G. and Johnson, M. 1980, *Metaphors We Live By*, University of Chicago Press, Chicago.

Lambert, T. A. 1972, 'Generations and change: toward a theory of generations as a force in historical process', *Youth and Society*, 4, pp. 21–45.

Lash, S. and Friedman J. (eds) 1992, *Modernity and Identity*, Blackwell, Oxford.

Lee, R. 1985, 'The future of the region: regional geography as education for transformation', in King, R. (ed.), *Geographical Futures*, Geographical Association, Sheffield.

Lefebvre, H. 1991, *The Production of Space* (orig. *La production de l'espace*, 1974), Basil Blackwell, Oxford.

Lehtinen, A. 1991, 'Northern natures: A study of the forest question emerging within the timber-line conflict in Finland', *Fennia*, 169, pp. 57–169.

Lehto, L. and Timonen, S. 1993, 'Kertomus matkasta kotiin: Karjalaiset vieraina omilla maillaan', *Kalevalaseuran vuosikirja*, 72, pp. 88–105.

Lehtonen, K. R. 1983, 'Valtiovalta ja oppikirjat: Senaatti ja kouluhallitus oppi- ja kansakoulun oppikirjojen valvojina Suomessa 1870–1884', *Helsingin yliopiston opettajankoulutuslaitos, tutkimuksia*, 9, Helsinki.

Leikola, A. 1980, 'Luonnontieteet', in Tommila, P., Reitala, A. and Kallio, V. (eds), *Suomen kulttuurihistoria II*, WSOY, Helsinki.

Leimgruber, W. 1991, 'Boundary, values and identity: The Swiss–Italian transborder region', in Rumley, D. and Minghi, J. V. (eds), *The Geography of Border Landscapes*, Routledge, London.

Leiviskä, I. 1937, *Nykypäivien maat ja kansat*, WSOY, Porvoo.

Leiviskä, I. 1938a, *Poliittinen maantiede (Geopolitiikka)*, WSOY, Porvoo.

Leiviskä, I. 1938b, 'Maantieteellinen katsaus', in Leiviskä, I. (ed.), *Suomen kirja*, Oy Suomen Kirja, Helsinki.

Leiviskä, I. 1941, 'Itäkarjala', *Rajaseutu*, 18, pp. 124–7.

Leiviskä, I. 1949, *Maantiede, sen historia, olemus ja tehtävät*, WSOY, Porvoo.

Lilja, F. E. 1942, 'Valoon ja voittoon', *Rajaseutu*, 19, pp. 135–6.

Linna, V. 1957, *The Unknown Soldier*, WSOY, Porvoo.

Löfgren, O. 1989, 'The nationalization of culture', *Ethnologia Europaea*, XIX, pp. 5–23.

Lowenthal, D. 1961, 'Geography, experience, and imagination: towards a geographical epistemology', *Annals of the Association of American Geographers*, 51, pp. 241–60.

Lowenthal, D. 1985, *The Past is a Foreign Country*, Cambridge University Press, Cambridge.

Luckmann, B. 1978, 'The small life-worlds of modern man', in Luckmann, T. (ed.), *Phenomenology and Sociology: Selected Readings*, Penguin Books, Harmondsworth.

Luho, V. 1964, 'Suomen synty', in Korhonen, A. (ed.), *Suomen historian käsikirja*, WSOY, Porvoo.

Luostarinen, H. 1986, *Perivihollinen: Suomen oikeistolehdistön Neuvostoliittoa koskeva vihholliskuva sodassa 1941–44, tausta ja sisältö*, Vastapaino, Tampere.

Luostarinen, H. 1989, 'Finnish Russophobia: the story of an enemy image', *Journal of Peace Research*, 26, pp. 123–37.

Lutheran Hymnal (Suomen Evankelis-luterilaisen kirkon virsikirja), 1981, Pieksämäki.

Mach, Z. 1993, *Symbols, Conflict and Identity: Essays in Political Anthropology*, State University of New York Press, Albany.

Mac Laughlin, J. 1986, 'The political geography of "nation-building" and nationalism in social sciences: structural vs. dialectical accounts', *Political Geography Quarterly*, 5, pp. 299–329.

Mäkelä, J. (ed.) 1947, *Luokkasodan muisto*, Kansankulttuuri, Helsinki.

Makkonen, M. 1984, 'Kansakoulun ja peruskoulun maantiedon oppikirjojen asenteellinen suuntaavuus kansainvälisyyskasvatuksen kannalta 1882–1982', unpubl. MSc thesis, Dept. of Geography, Univ. of Joensuu.

Mann, M. 1984, 'The autonomous power of the state: its origins, mechanisms and results', *European Journal of Sociology*, 25, pp. 185–213.

Mannerheim, G. 1951, *Muistelmat: ensimmäinen osa*, Otava, Helsinki.

Mannheim, K. 1952, 'The problem of generations', in Kecskemeti, P. (ed.), *Essays on the Sociology of Knowledge by Karl Mannheim*, Routledge & Kegan Paul, London.

Manninen, J. 1977, 'Maailmankuvat maailman ja sen muutoksen heijastajina', in Kuusi, M., Alapuro, R. and Klinge, M (eds), *Maailmankuvan muutos tutkimuskohteena*, Otava, Keuruu.

Manninen, O. 1980, *Suur-Suomen ääriviivat*, Kirjayhtymä, Helsinki.

Manninen, O. 1987, 'Turvallisuuspoliittiset ratkaisut "välirauhan" aikana ja jatkosodan alkaessa (1940–41)', in *Suomen turvallisuuspolitiikka*, Otava, Helsinki.

Manninen, O. 1993, 'Suomi, keisarinna ja itsenäisyys', *Helsingin Sanomat*, 5 December.

Marcus, G. 1992, 'Past, present and emergent identities: requirements for ethnographies of late twentieth-century modernity worldwide', in Lash, S. and Friedman, J. (eds), *Modernity and Identity*, Blackwell, Oxford.

Markusen, A. 1987, *Regions: The Economics and Politics of Territory*, Rowman & Littlefield, Totowa, N.J.

Marx, K. 1963, *The Eighteenth Brumaire of Louis Bonaparte*, International Publishers, New York.

Mason, P. 1990, *Deconstructing America: Representations of the Other*, Routledge, London and New York.

Massey, D. 1984, *Spatial Divisions of Labour: Social Structures and the Geography of Production*, Macmillan Education, London.

Massey, D. 1993, 'Power-geometry and a progressive sense of place', in Bird, J., Curtis, B., Putnam, T. et al. (eds), *Mapping the Futures: Local Cultures, Global Change*, Routledge, London and New York.

Mead, W. R. 1991, 'Finland in a changing Europe', *Geographical Journal*, 157, pp. 307–15.

Meinig, D. W. 1979a, 'Symbolic landscapes: some idealizations of American communities', in Meinig, D. W. (ed.), *The Interpretation of Ordinary Landscapes: Geographical Essays*, Oxford University Press, New York and Oxford.

Meinig, D. W. 1979b, 'The beholding eye: ten versions of the same scene', in Meinig, D. W. (ed.), *The Interpretation of Ordinary Landscapes: Geographical Essays*, Oxford University Press, New York and Oxford.

Meyrowitz, J. 1985, *No Sense of Place: The Impact of Electronic Media on Social Behaviour*, Oxford University Press, Oxford.

Mihailov, N. N. 1950, *Kotimaamme kartan ääressä*, Karjalais-suomalaisen SNT:n Valtion kustannusliike, Petroskoi.

Mikesell, M. W. 1983, 'The myth of the nation state', *Journal of Geography*, 82, pp. 257–60.

Mikkola, J. J. 1942, *Lännen ja idän rajoilta*, WSOY, Porvoo.

Milgram, S. 1984, 'Cities as social representations', in Farr, R. M. and Moscovici, S. (eds), *Social Representations*, Cambridge University Press, Cambridge.

Miller, D. and Branson, J. 1987, 'Pierre Bourdieu: culture and praxis', in Austin-Broos, D. (ed.), *Creating Culture: Profiles in the Study of Culture*, Allen and Unwin, London.

Minghi, J. V. 1991, 'From conflict to harmony in border landscapes', in Rumley, D. and Minghi, J. V. (eds), *The Geography of Border Landscapes*, Routledge, London.

Mitchell, D. 1992, 'Iconography and locational conflict from the underside. Free speech, People's Park, and the politics of homelessness in Berkeley, California', *Political Geography*, 11, pp. 152–69.

Moberg, K. A. 1899, 'Metalliteollisuus ja kivilouhokset', *Suomen Kartasto, teksti*, F. Tilgmann, Helsinki.

Morley, D. and Robins, K. 1989, 'Spaces of identity: communications technologies and the reconfiguration of Europe', *Screen*, 30, pp. 10–34.

Morley, D. and Robins, K. 1990, 'No place like *Heimat*: images of home(land) in European culture', *New Formations*, 12, pp. 1–24.

Moscovici, S. 1981, 'On social representation', in Forgas, J. P. (ed.), *Social Cognition: Perspectives on Everyday Understanding*, Academic Press, London.

Mosse, G. L. 1982, 'Nationalism, Fascism and the radical right', in Kamenka, E. (ed.), *Community as a Social Ideal*, Edward Arnold, London.

Muir, R. 1975, *Modern Political Geography*, Macmillan, London.

Murphy, A. B. forthcoming, 'The sovereign state system as political-territorial ideal: historical and contemporary considerations', in Biersteker, T. and Weber, C. (eds), *State Sovereignty as Social Construct*, Cambridge University Press, Cambridge.

Mustelin, O. 1973, *Nils Ludvig Arppe: Karjalan teollisuuden perustaja*, WSOY, Porvoo.

Myrsky, V. and Ulvinen, A. 1957, *Oma maamme*, WSOY, Porvoo.

Nevakivi, J. 1984, 'Miten Suomen ja Neuvostoliiton välinen YYA-sopimus otettiin vastaan Englannin ulkoministeriössä helmi-maaliskuussa 1948', *Suomen Historiallinen Seura, Historiallinen Arkisto*, 84, pp. 137–53.

Nevakivi, J. 1987, 'Euroopan suurvallat ja Suomen talvisota', in *Suomen turvallisuuspolitiikka*, Otava, Helsinki.

Newby, H. 1986, 'Locality and rurality: the restructuring of rural social relations', *Regional Studies*, 20, pp. 209–15.

Niemi, J. 1980, *Kullervosta rauhan erakkoon: Sota ja rauha suomalaisessa kirjallisuudessa kansanrunoudesta realismin sukupolveen*, SKS, Helsinki.

Nietzsche, F. 1989, *Iloinen tiede* (orig. Die fröhliche Wissenschaft), Otava, Helsinki.

Niiniluoto, Y. 1957, *Mitä on olla suomalainen*, Otava, Helsinki.

Nisbet, R. 1990, *The Quest for Community: A Study in the Ethics of Order and Freedom*, ICS Press, San Francisco, CA.

Norberg-Schulz, C. 1971, *Existence, Space and Architecture*, Praeger, New York.

Noro, M. 1968, *Kaitselmusaate Topeliuksen historianfilosofiassa*, WSOY, Porvoo.

Numelin, R. 1929, *Poliittinen maantiede*, Kustannusosakeyhtiö Kirja, Helsinki.

Numminen, J., Impola, H., Linna, M., et al. (eds) 1983, *Suomen ja Neuvostoliiton väliset suhteet 1948–1983: Vuoden 1948 ystävyys-, yhteistoiminta ja avunantosopimus käytännössä. Asiakirjoja ja aineistoa*, Valtion painatuskeskus, Helsinki.

Nuolijärvi, P. 1986, *Kieliyhteisön vaihto ja muuttajan identiteetti*, SKS, Helsinki.

Nygård, T. 1978, *Suur-Suomi vai lähiheimolaisten auttaminen: aatteellinen heimotyö itsenäisessä Suomessa*, Otava, Helsinki.

Official Statistics of Finland (SVT) 1979, 'Population by industry and commune in 1880–1975', Statistical Surveys No 63, Helsinki.

Oksanen, A. J. 1982, 'Paikallishistoria koulun historian opetuksessa', *Historiallinen Aikakauskirja*, 80, pp. 340–42.

Olsson, G. 1994, 'Heretic cartography', *Fennia*, 172, pp. 115–30.

ÓTuathail, G. and Agnew, J. 1992, 'Geopolitics and discourse. Practical geopolitical reasoning in American foreign policy', *Political Geography*, 11, pp. 190–204.

Paasi, A. 1984a, 'Kansanluonnekäsitteestä ja sen käytöstä suomalaisissa maantiedon kouluoppikirjoissa: Tutkimus alueellisista stereotypioista', *University of Joensuu, Research Reports of the Faculty of Education, Sociology of Education*, No. 2, Joensuu.

Paasi, A. 1984b, 'Opiskelijoiden tilapreferenssit aluetietoisuuden heijastajana', *Suunnittelumaantieteen yhdistyksen julkaisuja no. 15*, Helsinki.

Paasi, A. 1986a, 'The institutionalization of regions: a theoretical framework for understanding the emergence of regions and the constitution of regional identity', *Fennia* 164, pp. 105–46.

Paasi, A. 1986b, 'Neljä maakuntaa. Maantieteellinen tutkimus aluetietoisuuden kehittymisestä', *University of Joensuu, Publications in Social Sciences No 8*, Joensuu.

Paasi, A. 1988, 'On the border of the Western and Eastern Worlds: The emergence of the Utopia of the Värtsilä community in Eastern Finland', *University of Joensuu, Human Geography and Planning, Occasional Papers No. 6*, Joensuu.

Paasi, A. 1989, 'On the constitution of Värtsilä locality, with special emphasis on its institutionalization', in Sihvo, H. and Pulliainen, K. (eds), *West of East*, Joensuu.

Paasi, A. 1990, 'The rise and fall of Finnish geopolitics', *Political Geography Quarterly*, 9, pp. 53–65.

Paasi, A. 1991a, 'Deconstructing regions: notes on the scales of spatial life', *Environment and Planning A*, 23, pp. 239–56.

Paasi, A. 1991b, 'Muuttuvat aluekäsitykset maantieteen kehityksen heijastajana', *Terra*, 103, pp. 293–308.

Paasi, A. 1991c, 'Yhteisötutkimuksen kielestä ja sen metodologisista implikaatioista: teoriaa ja empiirisiä esimerkkejä', *Terra*, 103, pp. 226–41.

Paasi, A. 1992, 'The construction of socio-spatial consciousness: geographical perspectives on the history and contexts of Finnish nationalism', *Nordisk Samhällsgeografisk Tidskrift*, 15, pp. 79–100.

Paasi, A. 1994, 'The challenge of the rapidly changing world political map: a contextualized comment for R. J. Johnston', *Fennia*, 172, 97–104.

Paasi, A. 1995, 'Constructing territories, boundaries and regional identities', in Forsberg, T. (ed.), *Contested Territory: Border Disputes at the Edge of the Former Soviet Empire*, Edward Elgar, Cheltenham.

Paasi, A. forthcoming, 'Deconstructing the idea of geography: dimensions of professional practice', in Berdoulay, V. and van Ginkel, H. (eds), *Geography and Professional Practice*, Nederlandsche Geografische Studies.

Paasikivi, J. K. 1986, *Toimintani Moskovassa ja Suomessa 1939–41: I Talvisota*, WSOY, Porvoo.

Paasivirta, J. 1962, *Suomen kuva Yhdysvalloissa 1800-luvun lopulta 1960-luvulle: ääriviivoja*, WSOY, Porvoo.

Paasivirta, J. 1984, *Suomi ja Eurooppa 1914–1939*, Kirjayhtymä, Helsinki.

Paasivirta, J. 1991, 'Suomi jäsentyy kansakuntana Eurooppaan', in Jokipii, M. (ed.), *Suomi Euroopassa: Talous- ja kulttuurisuhteiden historiaa*, Atena, Jyväskylä.

Paasivirta, J. 1992, *Suomi ja Eurooppa 1939–1956*, Kirjayhtymä, Helsinki.

van Paassen, C. 1957, *The Classical Tradition of Geography*, J. B. Wolters, Groningen.

Paavolainen, E. 1925, 'Karjalan kannaksen suomalaistuttaminen', *Rajaseutu*, 5, pp. 67–8.

Paavolainen, E. 1930, 'Matkustakaa kannakselle', *Rajaseutu*, 7, pp. 36–53.

Paavolainen, E. 1958, 'Luovutettu Karjala', in Linkomies, E. (ed.), *Oma maa*, osa 3, WSOY, Porvoo.

Paavolainen, E. 1960, *Sellainen oli Karjala: Luovutetun alueen vaiheita*, Otava, Helsinki.

Paavolainen, J. 1967, *Poliittiset väkivaltaisuudet Suomessa 1918, II: Valkoinen terrori*, Tammi, Helsinki.

Pailhous, J. 1984, 'The representation of urban space: its development and it role in the organization of journeys', in Farr, R. M. and Moscovici, S. (eds *Social Representations*, Cambridge University Press, Cambridge.

Palosuo, V. J. 1983, 'Rajaseututyön lähtökohdat vuosisadan vaihteessa ja 1920-luvun alussa', *Rajaseutu*, 60, pp. 20–24.

Pälsi, S. 1922, *Karjalan talviteillä*, Otava, Helsinki.

Parker, G. 1988, *The Geopolitics of Domination*, Routledge, London.

Peet, R. and Thrift, N. (eds) 1989, *New Models in Geography (Volume 1)*, Unwin Hyman, London.

Peltonen, M. 1992, *Matala katse: Kirjoituksia mentaliteettien historiasta*, Hanki & Jää, Tampere.

Peltoniemi, U. 1969, 'Suomen maarajojen vartiointi vuosina 1918–1930', in *Rajavartiolaitos 1919–1969*, Rajamme vartijat ry, Mikkeli.

Perec, G. 1992, *Tiloja ja avaruuksia* (orig. Espèces d'espaces, 1974), Loki-kirjat, Helsinki.

Petrisalo, K. 1989, 'Tietoteknisen yhteiskunnan folklorismia', *Kalevalaseuran vuosikirja*, 68, pp. 268–73.

Pickles, J. 1986, 'Geographic theory and educating for democracy', *Antipode*, 18, pp. 136–54.

Pickles, J. 1992, 'Texts, hermeneutics and propaganda maps', in Barnes, T. J. and Duncan, J. S. (eds), *Writing Worlds: Discourse, Text and Metaphor in the Representation of Landscape*, Routledge, London.

Plummer, K. 1983, *Documents of Life: An Introduction to the Problems and Literature of a Humanistic Method*, George Allen and Unwin, London.

Polvinen, T. 1964, *Suomi suurvaltojen politiikassa 1941–1944: Jatkosodan tausta*. WSOY, Porvoo.

Polvinen, T. 1981, *Jaltasta Pariisin rauhaan: Suomi kansainvälisessä politiikassa III: 1945–1947*, WSOY, Porvoo.

Polvinen, T. 1984, *Valtakunta ja rajamaa: N.I. Bobrikov Suomen kenraalikuvernöörinä 1898–1904*, WSOY, Porvoo.

Polvinen, T. 1986, *Between East and West: Finland in International Politics, 1944–1947*, WSOY, Porvoo.

Polvinen, T. 1987, 'Suurvallat ja Suomi 1941–44', in *Suomen turvallisuuspolitiikka*, Otava, Helsinki.

Poroila, H. 1975, 'Suomen ja Neuvostoliiton välisen valtiollisen rajan kehitys vuodesta 1917 nykypäiviin', unpubl. MSc thesis, Dept. of Geography, Univ. of Helsinki.

Porter, P. W. and Lukermann, F. E. 1976, 'The geography of Utopia', in Lowenthal, D. and Bowden, M. J. (eds), *Geographies of the Mind: Essays in Historical Geosophy*, Oxford University Press, New York.

Poulantzas, N. 1978, *State, Power, Socialism*, Verso, London.

Pounds, N. J. G. 1951, 'The origin of the idea of natural frontiers in France', *Annals of the Association of American Geographers*, 41, pp. 146–57.

Pounds, N. J. G. 1954, 'France and "les limites naturelles" from the seventeenth to the twentieth centuries', *Annals of the Association of American Geographers*, 44, pp. 51–62.

Pounds, N. J. G. 1972, *Political Geography*, McGraw-Hill, New York.

Pratt, A. C. 1991, 'Discourses of locality', *Environment and Planning A*, 23, pp. 257–66.

Pred, A. 1984, 'Place as historically contingent process: structuration and the time-geography of becoming places', *Annals of the Association of American Geographers*, 74, pp. 279–97.

Pred, A. 1989, 'Survey 14: The locally spoken word and local struggles', *Environment and Planning D: Society and Space*, 7, pp. 211–33.

Pred, A. 1990, *Making Histories and Constructing Human Geographies: The*

Local Transformation of Practice, Power Relations, and Consciousness, Westview Press, Boulder.

Prescott, J. R. V. 1965, *The Geography of Frontiers and Boundaries*, Aldine Publishing Company, Chicago.

Prescott, J. R. V. 1987, *Political Frontiers and Boundaries*, Unwin Hyman, London.

Pseudonym A. E. K. 1939, 'Rajat on turvattava', *Rajaseutu*, 16, pp. 86–7.

Pseudonym A. E. K. 1940, 'Rajaseutupolitiikka ja maanpuolustus', *Rajaseutu*, 17, pp. 4–5.

Pseudonym H. L. 1927, 'Maassa rauha ihmisten kesken', *Rajaseutu*, 7, pp. 209–11.

Pseudonym Tyyne P. 1925, 'Maalivahti', *Rajaseutu*, 5, p. 6.

Pulkkinen, T. 1987, 'Kansalaisyhteiskunta ja valtio', in Alapuro, R., Liikanen, I., Smeds, K. and Stenius, H. (eds), *Kansa liikkeessä*, Kirjayhtymä, Helsinki.

Pulkkinen, T. 1989, *Valtio ja vapaus*, Tutkijaliitto, Jyväskylä.

Puntila, L. A. 1971, *Suomen poliittinen historia 1809–1966*, Otava, Helsinki.

Pyhä raamattu (Holy Bible), 1976, Pieksämäki.

Rahkonen, K. 1985, 'Ernst Bloch — toivon filosofi', in Rahkonen, K. and Sironen, E. (eds), *Ernst Bloch: utopia, luonto, uskonto*, Gummerus, Jyväskylä.

Räikkönen, E. 1924, *Heimokirja*, Otava, Helsinki.

Raittila, P. 1988, *Suomalaisnuoret ja idän ihmemaa*, Valtion painatuskeskus, Helsinki.

Rancken, A. W. and Pirinen, K. 1949, *Suomen vaakunat ja kaupunginsinetit*, WSOY, Porvoo.

Rannikko, P. 1980, 'Suomen suurimpien kaupunkien väestönkasvun hidastuminen', *Joensuun korkeakoulu, Karjalan tutkimuslaitoksen julkaisuja*, 43, Joensuu.

Rantanen, I. 1950, 'Katsaus 30-vuotisen Värtsilän kunnan toimintaan', mimeo, Värtsilän kunnan arkisto.

Rasila, V. 1982, 'Liberalismin aika', in Ahvenainen, J., Pihkala, E. and Rasila, V. (eds), *Suomen taloushistoria*, 2, Tammi, Helsinki.

Rée, J. 1992, 'Internationality', *Radical Philosophy*, 60 Spring, pp. 3–11.

Reinikainen, K. 1991, 'Isänmaan sankarit — huippu-urheilijan vapaudet ja velvollisuudet', *Oulun yliopiston sosiologian laitoksen tutkimuksia No 17*, Oulu.

Reitala, A. 1983, *Suomi-neito: Suomen kuvallisen henkilöitymän vaiheet*, Otava, Helsinki.

Relph, E. 1976, *Place and Placelessness*, Pion, London.

Relph, E. 1981, *Rational Landscapes and Humanistic Geography*, Croom Helm, London.

Ricoeur, P. 1965, 'State and violence', in Ricoeur, P., *History and Truth*, Northwestern University Press, Evanston.

Rikkinen, H. 1978a, 'Maantieteen asema valtion oppikoulujen opetussuunitelmissa vuodesta 1888 vuoteen 1977', unpubl. Phil.lic. thesis, Dept. of Geography, Univ. of Helsinki.

Rikkinen, H. 1978b, 'Aluemaantieteen kouluopetuksen kehitys', *Terra*, 90, pp. 28–34.

Rikkinen, H. 1989, *Maantiede kouluissa*, Yliopistopaino, Helsinki.

Robins, K. 1989, 'Global times', *Marxism Today*, December 1989, pp. 20–7.

Robins, K. and Morley, D. 1993, 'Euroculture: communications, community, and identity in Europe', *Cardozo Arts and Entertainment Law Journal*, 11, pp. 387–410.

Rokkan, S. and Urwin, D. 1983, *Economy, Territory, Identity: Politics of West European Peripheries*, Sage, London.

Rommi, P. and Pohls, M. 1989, 'Poliittisen fennomanian synty ja nousu', in Tommila, P. and Pohls, M. (eds), *Herää Suomi: suomalaisuusliikkeen historia*, Kustannuskiila, Kuopio.

Roos, J. P. 1987, *Suomalainen elämä: Tutkimus tavallisten suomalaisten elämäkerroista*, SKS, Helsinki.

Roseberry, W. 1989, *Anthropologies and Histories: Essays in Culture, History and Political Economy*, Rutgers University Press, New Brunswick, N.J.

Rosenau, P. M. 1992, *Post-Modernism and the Social Sciences: Insights, Inroads, and Intrusions*, Princeton University Press, Princeton, N.J.

Rowles, G. D. 1978, *Prisoners of Space: Exploring the Geographical Experience of Older People*, Westview, Boulder.

Rumley, D. and Minghi, J. V. 1991a, 'Introduction: the border landscape concept', in Rumley, D. and Minghi, J. V. (eds), *The Geography of Border Landscapes*, Routledge, London.

Rumley, D. and Minghi, J. V. 1991b, *The Geography of Border Landscapes*, Routledge, London.

Runon ja Rajan tie 1992, *Palveluopas 1992*.

Ruotsi, P. 1974, *Rajan kahden puolen*, Pohjoinen, Oulu.

Rüsen, J. 1992, 'Historiatiede modernin ja postmodernin välissä', *Tiede & Edistys*, 17, pp. 270–82.

Ryden, K. C. 1993, *Mapping the Invisible Landscape: Folklore, Writing, and the Sense of Place*, University of Iowa Press, Iowa City.

Saarela, P. 1986, 'Klassismi ja isänmaa: antiikin esikuvallisuus ja sivistyksen sankaruus taiteissa Porthanin ajasta Porthanin patsaaseen', in Manninen, J. and Patoluoto, I. (eds), *Hyöty, sivistys, kansakunta: suomalaista aatehistoriaa*, Pohjoinen, Oulu.

Sack, R. D. 1986, *Human Territoriality: Its Theory and History*, Cambridge University Press, Cambridge.

Sahama, Y. 1985, 'Tohmajärven vaiheita vuosisatojen saatossa', in *Tohmajärvi tuttavaksi*, Tohmajärven kunta, Joensuu.

Said, E. 1978, *Orientalism*, Routledge and Kegan Paul, London.

Salminen, V. 1941, *Viena-Aunus. Itä-Karjala sanoin ja kuvin*, Otava, Helsinki.

Saloheimo, V. 1963, *Pälkjärven historia: Karjalaisen pitäjän 500-vuotiset vaiheet*, Pälkjärven pitäjäseura, Joensuu.

Salonen, A. M. 1977, *Paasikiven linjalla: kirjoitelmia hyllystäni*, Alea-kirja, Jyväskylä.

Salonen, E. 1970, *Suomalaisen kulttuurin säätelyn järjestelmä: Sosiologinen tutkimus*, Otava, Helsinki.

Saraviita, I. 1989, *YYA-sopimus*, Lakimiesliiton kustannus, Helsinki.

Sayer, A. 1989, 'The "new" regional geography and problems of narrative', *Environment and Planning D: Society and Space*, 7, pp. 253–76.

Sayer, A. 1992, *Method in Social Science*, Routledge, London.

Schlesinger, P. 1991, *Media, State and Nation: Political Violence and Collective Identities*, Sage Publications, London.

Schuman, H. and Scott, J. 1989, 'Generations and collective memories', *American Sociological Review*, 54, pp. 359–81.

Schutz, A. 1970, 'Some structures of the Life-World', in *Alfred Schutz, Collected*

Papers III, Studies in Phenomenological Philosophy, Martinus Nijhoff, The Hague.

Schutz, A. 1971, *On Phenomenology and Social Relations: Selected Writings*, edited by Helmut R. Wagner, University of Chicago Press, Chicago.

Schutz, A. 1978, 'Phenomenology and the social sciences', in Luckmann, T. (ed.), *Phenomenology and Sociology: Selected Readings*, Penguin Books, Harmondsworth.

Sederholm, J. J. 1923, 'Asema, pinta-ala, väestö ja jako', in Donner, A., Grotenfelt, A., Hendell, L. et al. (eds), *Suomi: maa, kansa, valtakunta*, Otava, Helsinki.

Selén, K. 1987, 'Itsenäistymisestä talvisotaan', in *Suomen turvallisuuspolitiikka*, Otava, Helsinki.

Seppälä, M. 1993, 'Aluemaantieteestä rajamaantieteeseen: ongelman asettelua', *Terra*, 105, pp. 109–16.

Shields, R. 1991, *Places on the Margin: Alternative Geographies of Modernity*, Routledge, London and New York.

Shils, E. 1981, *Tradition*, University of Chicago Press, Chicago.

Short, J. R. 1993, *An Introduction to Political Geography*, Routledge, London and New York.

Short, R. 1990, *Imagined Country: Society, Culture and Environment*, Routledge, London.

Sihvo, H. 1973, 'Karjalan kuva: Karelianismin taustaa ja vaiheita autonomian aikana', *Joensuun korkeakoulun julkaisuja A*, 4, Joensuu.

Sihvo, H. 1981, 'Karjala kirjallisuudessa', in *Karjala 1: Portti itään ja länteen*, Karisto, Hämeenlinna.

Sihvo, H. 1989, 'Karjala Suomen historiassa', in Sihvo, H. and Turunen, R. (eds), Rajamailta, *Studia Carelica Humanistica*, 1, Joensuu.

Sihvo, H. 1992, 'Karjala rajamaana', *Karjalainen viesti*, 1, pp. 1–11.

Siisiäinen, M. 1985, 'Interests, voluntary associations and the stability of the political system', *Acta Sociologica*, 28, pp. 293–315.

Simons, H. W. 1989, 'Introduction', in Simons, H. W. (ed.), *Rhetoric in the Human Sciences*, Sage Publications, London.

Smart, N. 1983, *Worldviews: Crosscultural Explorations of Human Beliefs*, Scribner's, New York.

Smeds, K. 1987, 'Joukkotapahtumat ja Suomi-identiteetti', in Alapuro, R., Liikanen, I., Smeds, K. and Stenius, H. (eds), *Kansa liikkeessä*, Kirjayhtymä, Helsinki.

Smith, A. 1978, 'The diffusion of nationalism: some historical and sociological perspectives', *British Journal of Sociology*, 29, pp. 234–48.

Smith, A. 1979, *Nationalism in the Twentieth Century*, Martin Robertson, Oxford.

Smith, A. 1983, 'Nationalism and classical social theory', *British Journal of Sociology*, 34, pp. 19–38.

Smith, A. 1990, 'The supersession of nationalism?', *International Journal of Comparative Sociology*, 31, pp. 1–31.

Smith, A. 1991, *National Identity*, University of Nevada Press, Reno.

Smith, N. 1990, *Uneven Development: Nature, Capital and the Production of Space*, Basil Blackwell, Oxford.

Smith, N. 1992, 'Geography, difference and the politics of scale', in Doherty, J., Graham, E. and Malek, M. (eds), *Postmodernism and the Social Sciences*, Macmillan, London.

Smith, N. 1993, 'Homeless/global: scaling places', in Bird, J., Curtis, B.,

Putnam, T. et al. (eds), *Mapping the Futures: Local Cultures, Global Change*, Routledge, London and New York.

Smith, S. J. 1988, 'Constructing local knowledge: The analysis of self in everyday life', in Eyles, J. and Smith, D. M. (eds), *Qualitative Methods in Human Geography*, Polity Press, Cambridge.

Soikkanen, H. 1970, *Luovutetun Karjalan työväenliikkeen historia*, Tammi, Helsinki.

Soikkanen, T. 1984, *Kansallinen eheytyminen — myytti vai todellisuus? Ulko-ja sisäpolitiikan linjat ja vuorovaikutus Suomessa 1933–1939*, WSOY, Porvoo.

Soja, E. 1971, 'The political organization of space', *Association of American Geographers, Commission on College Geography, Resource paper No. 8*, Washington.

Soja, E. 1989, *Postmodern Geographies: the Reassertion of Space in Critical Social Theory*, Verso, London.

Solitander, C. P. 1911, 'Metalliteollisuus', in *Suomen Kartasto 1910, teksti II, Väestö ja kulttuuri*, J. Simelius, Helsinki.

Sormunen, E. 1936, 'Ja maassa rauha . . .', *Rajaseutu*, 13, pp. 124–5.

Sormunen, E. 1940. *Rajalla: Rajahiippakunnan piispan kokemuksia tapahtumista rikkailta kuukausilta*, WSOY, Porvoo.

Sormunen, E. 1942, *Suomalaisen kulttuurin lähteille*, WSOY, Porvoo.

Spykman, N. J. 1942, 'Frontiers, security, and international organization', *Geographical Review*, 32, pp. 436–47.

Stenij, S. 1937, 'Zach. Topeliuksen ajatuksia maantieteestä tieteenä ja sen tehtävistä', *Terra*, 49, pp. 1–7.

Stewart, P. 1990, 'Regional consciousness as a shaper of local history: Examples from the Eastern Shore', in Allen, B. and Schlereth, T. J. (eds), *Sense of Place: American Regional Cultures*, University Press of Kentucky, Lexington, Kentucky.

Sugar, P. F. 1969, 'External and domestic roots of Eastern European Nationalism', in Sugar, P. F. and Lederer, I. J. (eds), *Nationalism in Eastern Europe*, University of Washington Press, Seattle.

Suomen Gallup 1992, *Suomalaisten suhtautuminen venäläisiin 1992: taulukkoraportti*, Helsinki.

Suutala, M. 1986, 'Luonto ja kansallinen itsekäsitys: Runeberg, Topelius, Lönnrot ja Snellman suomalaisten luontosuhteen kuvaajina', in Manninen, J. and Patoluoto, I. (eds), *Hyöty, sivitys, kansakunta: Suomalaista aatehistoriaa*, Pohjoinen, Oulu.

Sykiäinen, R. 1993, 'Itä-Karjalassa tehdään Suomesta hirviötä', *Kaleva*, 10 June.

Talve, I. 1983, 'Pieniä teollisuusyhdyskuntia I', *Turun yliopisto, kulttuurien tutkimuksen laitos, kansatiede, monisteita*, 20, Turku.

Tanner, V. 1936, 'Suomen karttakuvan kehitys', in Hilden, K., Auer, V., Eskola, P. et al. (eds), *Suomen maantieteen käsikirja*, Otava, Helsinki.

Tarasti, E. 1990, *Johdatusta semiotiikkaan: Esseitä taiteen ja kulttuurin merkkijärjestelmistä*, Gaudeamus, Helsinki.

Tarkka, J. 1987, 'Sodasta sopimukseen', in *Suomen turvallisuuspolitiikka*, Otava, Helsinki.

Tarkka, J. 1992, *Suomen kylmä sota: Miten viattomuudesta tuli voima*, Otava, Helsinki.

Taylor, P. J. 1982, 'A materialist framework for political geography', *Transactions of the Institute of British Geographers, New Series*, 7, pp. 15–34.

Taylor, P. J. 1983, 'The question of theory in political geography', in Kliot, N. and Waterman, S. (eds), *Pluralism and Political Geography: People, Territory and State*, Croom Helm, London.

Taylor, P. J. 1985, 'The value of a geographical perspective', in Johnston, R. J. (ed.), *The Future of Geography*, Methuen, London.

Taylor, P. J. 1988a, 'World-systems analysis and regional geography', *Professional Geographer*, 40, pp. 259–65.

Taylor, P. J. 1988b, 'Geopolitics revived', *University of Newcastle upon Tyne, Department of Geography, Seminar Papers, Number 53*, Newcastle.

Taylor, P. J. 1990, *Britain and the Cold War: 1945 as Geopolitical Transition*, Pinter Publishers, London.

Taylor, P. J. 1991a, 'A theory and practice of regions: the case of Europe', *Environment and Planning D: Society and Space*, 9, pp. 183–95.

Taylor, P. J. 1991b, 'The English and their Englishness: "a curiously mysterious, elusive and little understood people"', *Scottish Geographical Magazine*, 107, pp. 146–61.

Taylor, P. J. 1993a, '*Contra* political geography', *Tijdschrift voor Economische en Sociale Geografie*, 84, pp. 82–90.

Taylor, P. J. 1993b, *Political Geography: World-Economy, Nation-State and Locality*, Longman Scientific and Technical, London.

Taylor, P. J. 1994, 'The state as container: territoriality in the modern world-system', *Progress in Human Geography*, 18, pp. 151–62.

Tester, K. 1993, *The Life and Times of Post-Modernity*, Routledge, London.

Thompson, J. B. 1990, *Ideology and Modern Culture: Critical Social Theory in the Era of Mass Communication*, Stanford University Press, Stanford, CA.

Thompson, P. (ed.) 1982, *Our Common History: The Transformation of Europe*, Pluto Press, London.

Thompson, P. 1988, *The Voice of the Past: Oral History*, Oxford University Press, Oxford.

Thrift, N. 1983, 'On the determination of social action in space and time', *Environment and Planning D: Society and Space*, 1, pp. 23–57.

Thrift, N. 1986, 'Little games and big stories: accounting for the practice of personality and politics in the 1945 general election', in Hoggart, K. and Kofman, E. (eds), *Political Geography and Social Stratification*, Croom Helm, London.

Thrift, N. 1987, 'No perfect symmetry', *Environment and Planning D: Society and Space*, 5, pp. 400–7.

Thrift, N. 1990, 'For a new regional geography 1', *Progress in Human Geography*, 14, pp. 272–9.

Thrift, N. 1991, 'For a new regional geography 2', *Progress in Human Geography*, 15, pp. 456–65.

Tiihonen, S. and Tiihonen, P. 1983, 'Suomen hallintohistoria', *Valtion koulutuskeskus, julkaisusarja A*, no 3, Helsinki.

Tiitta, A. 1982, 'Suomalaisen maiseman hahmottuminen kirjallisuudessa ja kuvataiteessa', *Terra*, 94, pp. 13–26.

Tiitta, A. 1994, 'Harmaakiven maa: Zacharias Topelius ja Suomen maantiede', *Societas Scientiarum Fennica: Bidrag till kännedom av Finlands natur och folk*, 147, Helsinki.

Tillich, P. 1973, *Rajalla* (orig. Auf der Grenze, 1962), WSOY, Porvoo.

Tommila, P. 1989, 'Mitä oli olla suomalainen 1800-luvun alkupuolella', in Tommila, P. and Pohls, M. (eds), *Herää Suomi: suomalaisuusliikkeen historia*, Kustannuskiila, Kuopio.

Topelius, Z. 1981, *Maamme kirja*, WSOY, Porvoo.

Tournier, P. 1971, *Ihmisen paikka* (orig. L'homme et son lieu, 1966), WSOY, Porvoo.

Tsernous, V. N. and Rautkallio, H. 1992, *NKP ja Suomi: Keskuskomitean salaisia dokumentteja 1955–1968*, Tammi, Helsinki.

Tuan, Yi-Fu 1974, *Topophilia: A Study of Environmental Perception, Attitudes and Values*, Prentice Hall, Englewood Cliffs, N.J.

Tuan, Yi-Fu 1975, 'Place: an experiental perspective', *Geographical Review*, 65, pp. 151–65.

Tuominen, M. 1991, *'Me kaikki ollaan sotilaitten lapsia': Sukupolvihegemonian kriisi 1960-luvun suomalaisessa kulttuurissa*, Otava, Helsinki.

Upton, A. F. 1965, *Välirauha* (orig. Finland in Crisis 1940–1941, 1964), Kirjayhtymä, Helsinki.

Upton, A. F. 1981, *Vallankumous Suomessa 1917–1918, II osa* (orig. The Finnish Revolution 1917–18, 1980), Kirjayhtymä, Helsinki.

Urry, J. 1987, 'Survey 12: Society, space, locality', *Environment and Planning D: Society and Space*, 5, pp. 435–44.

Urry, J. 1990, *The Tourist Gaze: Leisure and Travel in Contemporary Societies*, Sage Publications, London.

Vahtola, J. 1988, 'Suomi suureksi — Viena vapaaksi', Valkoisen Suomen pyrkimykset Itä — Karjalan valtaamiseksi vuonna 1918, *Studia Historica Septentrionalia*, 17, Rovaniemi.

Vainio, V. 1958, *Neljännesvuosisata vapaaehtoista rajaseututyötä*, Hämeen kirjapaino, Tampere.

Väistö, P. 1993, 'Uudesta Värtsilästä katsottiin vuosikymmenten ajan kiikarilla vanhaan', *Karjalainen* 25 April.

Vanhoja Suomen karttoja 1967, Suomalaisen Kirjallisuuden Kirjapaino, Helsinki.

Varis, E. 1993, *Karjalan tasavalta tänään*, Pohjois-Karjalan lääninhallitus/ Joensuun Yliopisto, Joensuu.

Vartiainen, P. 1984, 'Maantieteen konstituoitumisesta ihmistieteenä', *University of Joensuun Publications in Social Sciences No 3*, Joensuu.

Väyrynen, R. 1987, 'Finlandization — from deterrence to confidence', in Kiljunen, K., Sundman, F. and Taipale, I. (eds), *Finnish Peace Making*, Finnish Peace Union, Helsinki.

Väyrynen, R. 1990, 'Puolueettomuuden uudet poliittiset ulottuvuudet', in Väyrynen, R. (ed.), *Suomen puolueettomuuden tulevaisuus*, WSOY, Porvoo.

Väyrynen, R. 1992, 'Regional systems and international relations', in Lindholm, H. (ed.), *Approaches to the Study of International Political Economy*, Padrigu Papers, University of Gothenburg.

Vehvilä, S. and Castren, M. J. 1972, *Suomen historia*, WSOY, Porvoo.

Vennamo, V. 1970, 'SMP:n ulkopoliittisista perusnäkemyksistä', in Vennamo, V., Enävaara, R., Lemström, R. et al., *Asialinjalla*, Weilin & Göös, Helsinki.

Vihavainen, T. 1988, 'Suomi Neuvostolehdistössä 1918–1920', *Suomen Historiallinen Seura, Historiallisia tutkimuksia*, 147, Helsinki.

Vihavainen, T. 1991, *Kansakunta rähmällään: suomettumisen lyhyt historia*, Otava, Helsinki.

Virolainen, J. 1971, *Ainoa vaihtoehto — poliittinen keskusta*, Kirjayhtymä, Helsinki.

Virrankoski, P. 1975, *Suomen taloushistoria: kaskikaudesta atomiaikaan*, Otava, Keuruu.

Visuri, P. 1989, *Totaalisesta sodasta kriisinhallintaan: Puolustusperiaatteiden kehitys läntisessä Keski-Euroopassa ja Suomessa vuosina 1945–1985*, Otava, Helsinki.

Vloyantes, J. P. 1975, *Silk Glove Hegemony: Finnish–Soviet Relations, 1944–74. A Case Study of the Theory of the Soft Sphere Influence*, Kent State University Press, Ohio.

Voionmaa, V. 1919, *Suomen uusi asema: maantieteellisiä ja historiallisia peruspiirteitä*, WSOY, Porvoo.

Voionmaa, V. 1922, *Suomen talousmaantieto*, WSOY, Porvoo.

Voionmaa, V. 1933, 'Suomen historian maantieteellinen pohja', in Suolahti, G., Aaltonen, E., Renvall, P. et al. (eds), *Suomen kulttuurihistoria*, K. J. Gummerus, Jyväskylä-Helsinki.

Vuoristo, K. V. 1979, *Poliittiset ja taloudelliset alueet*, Gaudeamus, Helsinki.

Walker, R. B. J. 1993, *Inside/Outside: International Relations as Political Theory*, Cambridge University Press, Cambridge.

Wallerstein, I. 1990, 'Culture is the World-System: a reply to Boyne', *Theory, Culture and Society*, 7, pp. 63–5.

Wallerstein, I. 1991, *Geopolitics and Geoculture: Essays on the Changing World-System*, Cambridge University Press, Cambridge.

Warf, B. 1988, 'Regional transformation, everyday life, and Pacific Northwest lumber production', *Annals of the Association of American Geographers*, 78, pp. 326–46.

Waris, H. 1952, *Siirtoväen sopeutuminen: tutkimus karjalaisen siirtoväen sosiaalisesta sopeutumisesta*, Otava, Helsinki.

Waris, H. 1968, *Muuttuva suomalainen yhteiskunta*, WSOY, Porvoo.

Webb, J. W. 1976, 'Geographers and scales', in Kosinski, L. A. and Webb, J. W. (eds), *Population at Microscale*, International Geographical Union Commission on Population Geography and New Zealand Geographical Society, Hamilton.

Weigert, H. W., Brodie, H., Doherty, E. W. et al. 1957, *Principles of Political Geography*, Appleton-Century-Crofts, New York.

Weilenmann, H. 1963, 'The interlocking of nation and personality structure', in Deutsch, K. and Foltz, W. J. (eds), *Nation-Building*, Atherton Press, New York.

Westerholm, R. 1978, *Kristilliseen yhteiskuntaan*, Kirjayhtymä, Helsinki.

White, H. 1987, *The Content of the Form: Narrative Discourse and Historical Representation*, Johns Hopkins University Press, Baltimore.

Williams, C. and Smith, A. D. 1983, 'The national construction of social space', *Progress in Human Geography*, 7, pp. 502–18.

Williams, R. 1961, *The Long Revolution*, Chatto and Windus, London.

Williams, R. 1976, *Keywords: A Vocabulary of Culture and Society*, Fontana Press, London.

Williams, R. 1979, *Politics and Letters*, Verso, London.

Williams, R. 1988, *Marxismi, kulttuuri ja kirjallisuus* (orig. Marxism and Literature, 1977), Vastapaino, Jyväskylä.

Wilson, W. A. 1976, *Folklore and Nationalism in Modern Finland*, Indiana University Press, Bloomington and London.

Wilson, W. A. 1993, 'Suomalaiset uudessa maailmassa — folkloristinen näkökulma', *Kalevalaseuran vuosikirja*, 72, pp. 182–94.

Wood, D. 1992, *The Power of Maps*, Routledge, London.

von Wright, G. H. 1989, 'Venäjä ja Eurooppa', in Wright, G. H., *Ajatus ja julistus*, WSOY, Porvoo.

Wuorinen, J. 1935, *Suomalaisuuden historia* (orig. Nationalism in Modern Finland, 1931), WSOY, Porvoo.

Ylikangas, H. 1993a, *Tie Tampereelle: Dokumentoitu kuvaus Tampereen antautumiseen johtaneista sotatapahtumista Suomen sisällissodassa*, WSOY, Porvoo.

Ylikangas, H. 1993b, 'Vuoden 1918 vaikutus historiatieteessä', in Ylikangas, H. (ed.), *Vaikea totuus: Vuosi 1918 ja kansallinen tiede*, SKS, Helsinki.

Zilliacus, B. 1984, *Wilhelm Wahlfors*, Oy Wärtsilä AB, Porvoo.

Zukin, S. 1992, 'Postmodern urban landscapes: mapping culture and power', in Lash, S. and Friedman, J. (eds), *Modernity and Identity*, Blackwell, Oxford.

NAME INDEX

SUBJECT INDEX

Printed and bound by CPI Group (UK) Ltd, Croydon, CR0 4YY

27/10/2024

14580206-0003